Cool North Wind
Morley Nelson's life
with birds of prey

Cool North Wind

Morley Nelson's life with birds of prey

Stephen Stuebner

Caxton Press
Caldwell, Idaho
2002

Library of Congress Cataloging-in-Publication Data

Stuebner, Stephen.
Cool North wind : Morley Nelson's life with birds of prey / Stephen Stuebner.
p. cm.
ISBN 0-87004-426-5 (alk. paper)
1. Nelson, Morley. 2. Falconers--United States--Biography. 3. Birds,
Protection of--United States. I. Title.
SK17.N45 S74 2002
799.2'32'092--dc21

2002000997

Lithographed and bound in the United States of America
CAXTON PRESS
Caldwell, Idaho
168355

Bill Meiners

Morley Nelson prepares to release the golden eagle "Clyde" for a flight over Idaho's Snake River Canyon.

TABLE OF CONTENTS

Illustrations

Color Section

Introduction

By Jim Fowler

When you visit the Nelson home near Boise, and sit down in the living room filled with built-in memorabilia, the house seems to be just another part of nature. When I was there last, I remembered the hot dry air moving through windows open to the wooden porch, the view of dry yellow grass, gray sagebrush and a few scattered trees outside—just as it had been years before. Morley Nelson and his always present birds of prey were still very much a part of the home's "ecosystem." He and they are part of the scenery. I was glad he had not moved away from there.

The first time I met Morley he was already a legend to me. Even though I had grown up in the East, and developed a passion for birds of prey on my own, I had already heard about his reputation as an eagle expert. When Marlin Perkins and the crew of *Wild Kingdom* and I stopped by in 1964 to film a segment on birds of prey, he and his sons flew his golden eagle free right in front of the house from the top of the ridge down across the bright yellow grass to his fist. Not only had jack rabbits vacated the area, there weren't many small dogs or cats around either. His golden was a big female with a 7-foot wingspread that could tackle almost anything. We filmed her in slow motion coming right into the camera. By that time, the Nelsons had worked with many film crews and to some extent Morley had already turned his passion into a profession.

When visiting with him, though, it's sometimes difficult to leave the house. The temptation, whether Morley is carrying a 14-pound golden eagle on his fist or enthusiastically showing you his prized falcon perched nearby, is to talk. There couldn't be a house with a setting such as that without a man in it who is passionate about life and always willing to converse. Not only does he know almost everything there is to know about his avian raptor friends (I say that because he talks to them the same way he talks to people), Morley also is an educator who communicates effectively with people of all ages and persuasions. Above all, he is a rare "spokesperson" for the natural world who speaks from experience, not hearsay. Sometimes he's talking birds, explaining how to protect eagles from being killed on powerlines. He might be promoting the Snake River Canyon Birds of Prey Sanctuary, talking politics or describing the uncontrolled invasion of inedible "cheet grass" into the high and dry grasslands of the West. It quickly becomes obvious to his listeners that Morley is concerned and involved. Although he has added a few years to his life since the '60s, when you're around him you still get the feeling that he's just getting started!

I know how it is to be interested in hawks, owls, eagles and vultures. I wrote a paper on them when I was in the fourth grade. But it's still hard to say how and why some people become "hooked" on predatory birds, usually for life. Most passions probably develop as a result of instinct, experience and examples set by people around you. In the case of birds of prey and falconry, I'm convinced people such as Morley are programmed genetically at birth to be passionate about them. Whether it's because these birds are independent, functional and perfectly designed for survival, is not certain. In my case, I found out that my name "Fowler" comes from the old German-English word that actually means "someone who trains birds of prey." Morley's ancestors were probably royal falconers who lived during the Middle Ages.

If you're fortunate enough to spend time with Morley, you can feel his knowledge and interest in everything on our planet, its plants, animals and people. The problem is, we humans are apparently the only species on Earth that can contemplate birth and death and the time period in between, which means that we are both "blessed" and "burdened" with this awareness. Morley loves life, but he also knows that we must correct our course and act now if we are going to provide a "quality of life" for our descendants and all other people on Earth. Covering his beloved wilderness with urban sprawl, eliminating other life forms and severing our basic connection with nature will not provide future generations with the freedom to pursue knowledge and happiness the way Morley has been able to do.

There was only one time when filming a *Wild Kingdom* sequence that I questioned Morley's advice. When preparing to rappel part way down a 400-foot cliff on the Snake River Canyon to a golden eagle's nest, I notice there were no trees or big rock outcrops to tie my rope to. In typical Morley fashion, showing moderate disdain for my quandary, he explained that considering the physics involved and the knowledge he had gained while serving as a commander in an elite mountain division during the invasion of Italy in World War II, I could safely tie off to a flat rock, lying not more than ten feet from the cliff edge. The rock weighed less than I did. He said as long as the flat part of the rock was on the ground when pulled against, it would provide enough friction to hold me. At that point I looked down into the canyon and decided to bring over a Jeep to tie to with a longer rope after tying to his rock. To my amazement, he was right. The flat rock never moved, but my confidence in his trust of flat rocks remained shaky.

Not only did Morley spark my interest in rappelling down cliffs, but he also helped many people form deep connections with his birds of prey. People who probably would not have cared are now supporting efforts to guarantee that generations to come will be able to gaze in wonder at an eagle soaring and know what our national symbol stands for. They will have a chance to experience the always rare spectacle of a wild falcon stooping at its prey and realize that, above all, saving birds of prey means saving ourselves from the unsustainable future, without natural resources, without trees, without animals and without scenic beauty.

Morley's extraordinary story shows us that once you have the love of nature in your heart, you can always have a "quality of life" that can be experienced no matter what happens to Wall Street or our material world. His wealth comes from within, not from without. Knowing Morley, I'm sure he would agree that, "how we treat the Earth" is the essence of an important new message for humankind that says that the future of our social welfare depends upon our understanding and respect for the natural world.

Morley's message comes from deeds, not words, as you'll see when you read his story. No one could have lived such a life without first being blessed with common sense, adaptability, intense curiosity, a love of people and a passion for his birds. Morley Nelson is real to everyone—from presidents, Saudi kings, children and politicians to Hollywood celebrities and TV stars. Probably because the falcon is his queen, the eagle his king and Mother Nature, in all her glory, his mentor.

Jim Fowler
April, 9, 2002

Idaho Public Television

Morley Nelson "chats" with a prairie falcon.

Chapter One

Centurion

Kuna, Idaho

Nell Newman, a precocious and slender thirteen-year-old girl, peers over a cliff's edge into the yawning Snake River Canyon. It's an awesome, intimidating sight—a 500-foot sheer drop into house-sized rocks and a rushing river below. At this moment of truth, Nell feels nervous and excited—all at once. It's the first time she's rappelled down a cliff, and she's got to do it with the cameras rolling.

The daughter of Hollywood legend Paul Newman and actress Joanne Woodward, Nell is the co-star of the *The Eagle and the Hawk*, a General Electric-sponsored Monogram Special for NBC-TV. It's an adventure tale featuring a city girl's discovery of the beauty and inspirational qualities of birds of prey in this wild canyon in Idaho.

Dressed in a buckskin coat, jeans and tennis shoes, Nell's long blond hair flies in the wind as she bear-hugs a slab of brown basalt rock on the precipice of the cliff. "Should I go now?" she asks, her voice audibly shaking.

"You bet Nell, go real slow and easy," Morley Nelson says in his calm, baritone voice. A master falconer, expert rock climber and birds of prey conservationist, Morley co-stars in the film as

Newman's mentor. He's a ruggedly handsome guy with a flat-top crewcut who looks very comfortable with a climbing rope around his waist. Morley belays Newman from above, letting out an arm's length of rope as she begins to descend toward a golden eagle nest, some fifty feet below.

"Lean back into the rope and away from the cliff," Morley says. "That's a girl. That's beautiful."

"My hands are going to freeze," she mutters quietly, continuing to drop down the vertical wall of rock. Her tennis shoes find good footholds as she descends. Morley continues to coach.

"Lean way back now. You're doing great."

Cameras cut to the next scene: Morley hangs from a rope on the opposite side of the eagle nest from Newman. Together, they marvel at a solitary golden eagle chick staring up at them. "Go ahead and pick it up, Nell," he says.

The tiny white furry bird—only a couple days old—looks so helpless and lonely in the stick nest. It hunkers down and makes little high-pitched chirps as Newman looks down with fascination with her big pale blue eyes. She picks up the chick and cradles it in the palm of her hand. It fits perfectly.

"OK Nell, we better get up now and let mother eagle come back and keep her warm," Morley says. "If we stay here too long, she might get killed."

Joanne Woodward, the film's special guest narrator, notes that her daughter's quest to visit a cliff-side eagle nest and hold a baby eagle gives her an extra measure of bravery, more than she thought possible. Meanwhile, Morley and Nell Newman return to the top of the cliff. He has taught yet another youngster how to rappel down a cliff to experience an unforgettable moment with a bird of prey.

Placing his arm around Newman, he says, "I'll tell you girl, you are really something else. Wow!"

It was springtime in southern Idaho. May 1972. Morley was excited about the prospects for *The Eagle and the Hawk*. With solid financial backing through the sponsorship of General Electric and the National Audubon Society, the film was certain to reach millions of Americans on prime-time network television on a Sunday night. Scenes of a cute young girl rock-climbing, horseback riding, trout fishing and exploring, combined with excellent close-up footage of eagles and hawks, would capture the collective heart of America.

Later on, the film was released to public schools across America. Recommended by the National Education Association, the film taught youngsters about predator-prey relationships in the wild, and instilled an ethic about the need to protect birds of prey from wanton illegal shooting. During her month-long stay at Morley's home, Nell quickly learned that when people found wounded hawks and eagles on the roadside, they invariably were told to call Morley, the "birdman" in Boise. He would know what to do.

In the movie, Newman visited veterinarian Doctor John Lee with Morley to learn about what happens to birds whose wings are shot off, or whose tail feathers are ripped away. Lee finds a way to give birds a second chance with high-tech surgery.

The film must end on a hopeful note, of course. Newman and Morley ride horseback to the top of a desert butte, with a rehabilitated red-tailed hawk on Morley's fist. Looking every bit like a true mountain girl, Nell lets go of the reins at a high rocky point, takes the hawk and throws it headlong into the wind. She smiles with glee as the hawk takes flight, its wings flapping rapidly and working perfectly. She looks over at a grinning Nelson—a touching moment, indeed.

Cameras pan from a broad view of the rocky Snake River Canyon to Nell and Morley riding horseback along the river, as John Denver strums his folk guitar, building to a crescendo. He breaks into song at the top of his lungs:

"Oh, I am the eagle
I live in High Country
In rocky cathedrals
That reach for the sky

I am the hawk
And there's blood on my feathers
But the tide is still turning
They, too, will be dry

All those who see me

And all those who believe in me
Carrying the freedom
I feel when I fly

Come dance with the west wind
Perch on the mountain top
Sail over the canyon
and up to the stars

And reach for the heavens
And hope for the future
And all that we can be
And not just who we are.

This was the theme song for the movie, one of several that Denver wrote for the film. The popular singer-songwriter gained inspiration for his songs by rappelling into golden eagle aeries with Morley in the Snake River Canyon. The song, *The Eagle and the Hawk,* became a top hit in 1972.

Springtime 2000

It's a typical weekday afternoon at Morley Nelson's ranch-style home in the foothills of Boise. It's time to fly Zackar, a prized gyrfalcon, and feed Morley's stable of birds. As always, Morley pulls on his falconers coat, grabs a pair of heavy leather gloves, and strolls out on a well-worn path to the mews, an elaborate hawk house on the back slope of his yard.

Even at 82, Morley cuts a handsome athletic figure as he flashes a smile and strides along a well-worn narrow dirt pathway to the mews. His blue eyes twinkle at the thought of flying Zackar. A thinning crop of brown hair on top, no longer kept in a crewcut, is brushed straight back like the hair on a golden eagle's head. Long, white wishbone-style sideburns grow out sideways, hinting at an independent streak. Morley is still remarkably agile. He's stayed in good physical condition by tromping around the hills of Idaho, flying his birds, and following the physical discipline of a soldier—he does twenty-five sit ups and pushups, and twenty-five sets of stairs every morning, without fail.

Zackar hears Morley coming. The slate-gray bird suddenly appears from behind a wool curtain and perches behind the horizontal bars of his birdhouse window, sizing up the scene with his pure black eyes. Zackar stands erect, puffs out his white-speckled breast and holds his head high—looking noble and proud for his master.

"Heh, heh. He's saying I want to go," Morley says, a big grin creasing his face. "He thinks I'm his mate."

Morley has special admiration for gyrfalcons, rulers of the arctic, because they are the most spectacular hunters and fliers among all birds of prey. "They can outfly anything in the world," he says. "And they're the fastest in the world, too."

Even faster than a peregrine?

"The terminal velocity is about the same," he says. "Once they go more than 800 feet in a vertical dive they're going 180 mph. I've seen them dive from 10,000 feet, and they probably go 200 mph or more, but no one has ever measured that."

Morley goes into Zackar's private room, and the bird eagerly jumps on his gloved fist. The leather glove is a hoot—the top half of a cowboy boot sewn to a leather glove to provide protection for his forearm. We walk through a grove of fruit trees to an open ridge. Morley takes a moment to chat with the bird, chuckling and purring as he lifts the bird to his face. The bird nuzzles with Morley, chirps and tilts its head sideways. It's the closest thing possible to an embrace.

"He thinks I'm a falcon," Morley says. "He thinks I can do no wrong."

Now it's time for Zackar to fly. "Go anytime you want big hawk. Heh. Heh. Heh." It's Morley's patented falcon talk. He's in a different zone when he talks to his birds—he speaks quietly and gently when he's hooding a bird, but when it's time for the birds to fly, Morley gets excited and shouts a few choice words as he lifts a bird into the air. This time, Zackar is a bit pensive. He looks around the open grass and sagebrush draw for a moment and takes off. His dark blue wings are so powerful you can hear a "whoosh-whoosh-whoosh" as he takes flight. In seconds, the bird is flying 1,500 feet above the draw, racing over the top of homes on the rim above. Pigeons and songbirds scatter in all directions.

"Gyrfalcons are very territorial," Morley says. "If there's an eagle, or a red-tail, or a Merlin out there, boy, they'd better go land somewhere and get out of its way or there's going to be trouble."

Morley provides a running commentary as Zackar cruises in gray skies over Boise's North End neighborhood for a fresh meal. The private bowl-shaped mountain draw is a perfect spot for flying birds. The adjacent slopes are too steep for any other homes to encroach on the natural open space, and the area is large enough to support small game and wild birds. Morley picked the isolated nook in the 1950s, when Boise was a sleepy little town of 20,000 people. Now, it's a bustling city, in Rocky Mountain terms, of about 200,000 residents.

To watch Morley fly his birds is to see him totally in his element. Here's a man who has caught, trained and hunted with countless birds of prey since he was a teen—a span of time that extends for sixty-five years—and he gets just as excited to watch Zackar take a pigeon in a vertical dive as he was the first time he saw a peregrine falcon take a teal duck as a young lad in North Dakota. Everything about his coveted birds is so utterly *profound*.

Morley's youngest son, Tyler, forty-three, marvels at how the birds inspire his dad. "He's so passionate it's unbelievable. He has this totally undeniable love and dedication to the birds of prey. I mean, he can watch a gyrfalcon go 800 feet into the sky, roll over and make this incredible stoop to nail a duck or a pheasant, and man, every time, Morley says directly from his heart, "Gawd, that is the most magnificent thing I've ever seen in my life." Every time, it's the most magnificent thing he's ever seen in his life. It's that special to him, every time.

"No one has ever had more dedication to his subject matter than Morley. I mean, this guy knows what he's talking about, and he believes it to the core."

Morley's devotion to birds of prey, and the many strides he's made over the years to protect and conserve them, stem from a personal relationship he's developed with raptors as a falconer. He quickly discerned that if he could teach his birds to fly and hunt like the premier predators do in the wild, they would return the gift by opening a window to the wild.

"The birds can make you feel the same way your children do when you teach them to do things," Morley says. "The wonderful part about it is, they come to understand you, and what you're trying to do for them. They are your partner, and they work for you as a partner in the deepest kind of consideration for each other. And when you do that, you find out things so deep it makes you realize how close we are to all life on this planet."

Morley the philosopher frequently reminds us about the chilling natural law that birds of prey observe, nature's imperative: To kill or be killed. It's a concept that Morley personally understands all too well from his near-death combat experience in World War II, and it's something he instantly feels with a twelve-pound golden eagle perched on his gloved fist. "Their talons have tremendous strength—1,700 pounds per square inch of raw power. They can grab a wolf by the head and kill them right now," he says, gritting his teeth at the thought.

Pour Morley a glass of whiskey, and he might just spill a tale or two about a frightening close encounter with an eagle. One evening in Boise in the 1960s, Morley went out to fetch "Clyde," his first golden eagle, in the mews. He placed the big brown bird on his gloved fist and walked down the path by the fruit trees toward the open ridge. Suddenly, Morley went ass-over-tea-kettle on the icy path, and unwittingly stuck out his tongue while trying to hold onto the eagle. Clyde pounced on Morley's face with its powerful feet, mistaking Morley's tongue for a piece of fresh red meat.

"He took the bottom of my chin and the talon went right through my tongue to the roof of my mouth, and the other one grazed the side of my eye and tore a line through my face," he says.

"That talon was so sharp, it was like a doctor's needle. It was a great big cut. I'm lucky Clyde didn't take a second whack at me."

Morley's boys, Norman and Tyler, ran to fetch their mom, while Morley got back on his feet and placed the eagle on the gloved fist. Blood poured from his face and jaw as the eagle looked at him with those piercing yellow and black eyes, "What's the big deal there, Nelson. What are you getting so worked up about?"

The cut was so clean it barely left a scar. Today, it folds into a deep vertical wrinkle on the right side of Morley's leathered face. Behind every wrinkle is a story, and a storehouse of knowledge. The wrinkles run deep on this man because he's got a lot of stories to tell, and he tells them with a great sense of animation and humor. Even in his eighties, Morley is prone to leap out of his chair and act out the parts—of both the animals and the people.

Morley begins to swing the light lure to bring Zackar back out of the sky. "Hi-yo! Hi-yo!" Planting his feet wide, Morley swings the lure of duck wings in a counter-clockwise direction in the style of a calf roper. Zackar zooms in an angle toward the lure at high speed. "Hi-ya!" Morley pulls the lure away from Zackar at the last second, causing the bird to speed by his left side, and fly up into the sky for a second try. Morley swings the lure again, and slows its rotation as Zackar comes in flying fast, and snags the wings with her feet.

"Isn't that the most beautiful thing," Morley says, crouching down to offer his bird a piece of fresh meat. "Step up, big hawk."

This is a story about Morley Nelson, a passionate man who was a major force in convincing America that birds of prey occupy a special place in the universe, and in our hearts. To understand Morley as a person, we must visit a number of key watershed events that shaped his life. We must get to know the favorite birds in his personal clutch—birds that he trained to hunt in the wild and perform for cameras in movies and TV commercials. We must delve into his many conservation struggles and achievements to understand the outcome. We must follow him on a number of exciting bird-chasing adventures in the outdoors, the place where he felt most at home. We must discover how he developed such a strong conservation ethic, a sixth sense for working with animals and an uncommon zeal for action.

By taking this journey, we can peer into the inner soul of

Morley Nelson, a flamboyant salt-of-the-earth character, a father of four, a husband, a widower, a pioneer, a visionary.

Like many Americans who came of age in the first half of the 20th century, lived through the Depression and served in World War II, Morley possessed a strong work ethic, and a deep sense of duty to his country. The personal relationship he developed with birds of prey, as an expert falconer and raptor-rehabilitator, fueled a fire in his belly to protect birds of prey from many threats: wanton shooting, electrocution, DDT, loss of habitat. It became a singular cause that superceded everything else.

As Tom Brokaw writes in *The Greatest Generation,* World War II veterans returned to America with uncommon zeal. "They were a new kind of army now, moving onto landscapes of industry, science, art, public policy, all the fields of American life, bringing to them the same passions and discipline that had served them so well during the war."

It became Morley's duty to convince Americans through films, speeches and politicking that eagles, hawks, falcons and owls deserved better than being routinely shot off the top of fence posts for mere kicks. As a war veteran, Morley understood what it meant to get shot. A company commander in the elite 10th Mountain Division, he suffered multiple wounds in April 1945 as American troops seized the Italian Alps. He received a Silver Star, a Purple Heart and a Bronze Star for his leadership and bravery. So to Morley, crippled birds were akin to soldiers injured in battle—they deserved quick medical care and a chance to recover to full health. At many times, Morley's hawk house resembled a hospital, providing shelter for dozens of sick and injured birds. But the birds always paid him back—they gave him strength and joy and inspiration every day. Occasionally, they gave him grief, too, painful lessons that caused a deeper understanding of nature and life.

Morley Nelson is not a household name as a leading American conservation figure in the twentieth century, but many prominent people think he should be. His achievements in falconry, nature films, raptor-rehabilitation, conservation laws, raptor-electrocution issues and the recovery of endangered peregrine falcons have made a major impact on the welfare of birds of prey in North America and around the world.

Beyond his conservation achievements, Morley has inspired literally hundreds of people to get involved in various aspects of birds of prey conservation. He is widely considered to be the dean of raptor conservation.

"There are few people who have entered my life who have left a more indelible impression than Morley Nelson," says Roy Disney, vice chairman of the board for the Walt Disney Company. "His lifelong work for, and as an advocate of, the birds of prey could, in my opinion, be favorably likened to the work of Rachel Carson or Jacques Cousteau."

"Morley is the leading figure in North America with respect to birds of prey. He's known to everybody in this country and the world for what he's done," adds longtime friend and falconer Tom Cade, founder of The Peregrine Fund.

Maurice Hornocker, a prominent wildlife biologist who recently completed a study of Siberian tigers in Russia, says Morley motivated him in many ways during their work together in the Snake River Canyon. "He had this uncanny ability to instill enthusiasm in everyone," Hornocker says. "It was a treat just to go out with him just for that. I'd find myself refreshed and energized by every trip, in my own work."

Hornocker, who went on to write for *National Geographic* and gain media coverage of his wildlife research around the world, says he learned the importance of broadcasting wildlife work to the mainstream from Morley. "He'd always tell you, you have to tell people, you gotta show people the beauty of these birds. You gotta show them how regal, and how truly beautiful these birds are, and how beautiful the whole world is. He always had that infectious enthusiasm and optimism. It's a gift. It really is. An absolute gift. It's obvious from the people he was able to influence—politicians and business people with no connection to wildlands or wildlife, previous to their association with Morley—and they became absolute crusaders for the cause. He's really an amazing guy."

When Roger Caras visited Boise to produce a TV segment on Morley in the 1970s, he referred to him as a "centurion," someone who comes along once every 100 years.

Of course, Morley didn't save birds of prey all by himself.

Other pioneers who made key contributions to protecting raptors include:

- John and Frank Craighead, famous Yellowstone National Park grizzly bear ecologists, invented raptor research and photography techniques as young men in the late 1930s. Their first book, recently reprinted, titled *Hawks in the Hand* (Lyons and Burford 1998), is a falconry classic.
- Rosalie Edge and Richard Pough, working with the Emergency Conservation Committee in New York, purchased a key tract of private land on the summit of Hawk Mountain in the 1930s to halt the slaughter of raptors as they migrated by the rocky peak in eastern Pennsylvania. In the fall, biologists take daily counts on the mountaintop to keep track of raptor migrations, a practice that has spread to many other key migration points from Canada to Mexico through groups like the Hawk Migration Association of North America..
- Charles Broley, a retired banker from Winnepeg, Manitoba, banded more than 1,000 bald eagles in treetop nests from 1939-1958 along the Florida coast. Broley's banding work showed that over a period of time, bald eagle populations were plummeting at the same time that ornithologists observed a widespread problem with the effects of DDT on song bird and raptor reproduction.
- Joseph Hickey, a professor of wildlife at the University of Wisconsin, brought attention to the deleterious effects of DDT on peregrine falcons and other raptors by holding a world conference on the issue at the University of Wisconsin in 1965. The outgrowth of the conference, combined with environmental lobbying efforts spurred by Rachel Carson's *Silent Spring,* led to a nationwide ban on DDT in 1972.
- In the early 1970s, Tom Cade and several key associates developed a pioneering large-scale captive-breeding program for restoring peregrine falcons throughout North America, one of the nation's greatest environ-

mental success stories. Morley has served on The Peregrine Fund's board of directors for twenty years and convinced the non-profit organization to relocate to Boise in 1983. Under the leadership of President Bill Burnham, Cade and others, The Peregrine Fund expanded its operations in the mid-1980s to restore many raptor species around the globe.

- Beginning in 1974, Doctor Patrick Redig, director of The Raptor Center at the University of Minnesota, developed the world's premier raptor-rehabilitation center, where more than 800 wounded birds of prey are treated each year. The Center provides specialized training in raptor medicine and surgery for veterinarians from around the world, and it reaches 300,000 people nationwide each year through public education programs and events. The center's success has led to the outgrowth of many other raptor-rehabilitation clinics in North America, including California, Florida, Alaska, New York and Oregon.

In a marvelous tribute to Morley on his seventieth birthday at the elegant Morrison Center concert hall in Boise, a long line of influential people took turns at the microphone, telling stories about their experiences with their friend. People like Roy Disney, Jim Fowler of Mutual of Omaha's *Wild Kingdom*, four-term Idaho Governor Cecil Andrus, Nell Newman and Tom Cade saluted Morley. And he received well wishes from former President Ronald Reagan and Vice President George Bush. Equal praise from Democrats and Republicans, scientists, citizens and film producers.

"When we were working on films about birds of prey, and someone had a question, the first thing we'd do is call Morley," Disney told the packed house of 2,000 attendees. "He'd always give you a straight answer, and then he'd give you about ten answers to questions you hadn't asked yet."

"Idaho and the nation are a better place because of your tenacity," said Logan Lanham, former vice president of Idaho Power Company. "He showed us how to walk a tightrope between people who were pro-development and environmentalists. I think he's a real hero."

Looking back at Morley's life, it's remarkable how much he accomplished, especially in view of the fact that his crusade to protect raptors was a "hobby." It was a mission that he wedged into a busy family life and his profession as Snow Survey Supervisor of the Columbia River Basin for the U.S. Department of Agriculture. Like a golden eagle spiraling ever higher into an azure sky, Morley's efforts to protect birds of prey took on an ever-widening radius, increasing scope, influence and success as time went on. When it was time to act, he pounced.

Experts say Morley's most important achievements include:

- His work on more than thirty movies and TV nature specials starring one or more of Morley's eagles, hawks and falcons. "When we did those films, we got every principle across that I'd ever dreamt of—bam!—in one hour," he says. "You impact more people in one hour than you could writing about it for twenty years!"
- His personal discovery of the Snake River Birds of Prey National Conservation Area, the premier nesting site for more than 500 pairs of hawks, falcons and golden eagles in North America. Morley convinced Andrus to set aside the birds of prey area by executive order in 1980 when he was Interior secretary under President Carter. The area received permanent protection from Congress in 1994.
- Pioneering work with the Idaho Power Company, Edison Electric Institute (EEI) and the National Audubon Society to modify powerline designs to reduce a high incidence of large raptor electrocutions. Solutions developed by Morley and utility company engineers were quickly embraced by other utilities in the United States and in Western Europe. The environmental success story was one of the first national examples, if not *the* first, of the "green business" movement.
- Playing a key role in educating Americans about the value of birds of prey through countless lectures to

schools, service clubs, ranchers, hunters and community groups throughout the United States to halt the relentless shooting of birds of prey.

- Advancing the cause of raptor rehabilitation by calling to national attention the need for places to treat injured birds of prey, a call that was answered by the outgrowth of numerous raptor-rehabilitation clinics, including The Raptor Center at the University of Minnesota.

It's important for the public to know Morley's story because today birds of prey are safe from a multitude of threats they faced half a century ago. Clearly, Morley's efforts had a major impact. At the dawn of the twenty-first century, the thought of bounty hunters shooting golden eagles from the air seems unthinkable. Efforts to restore bald eagles and peregrine falcons from near extinction stand as our nation's greatest conservation achievements. Raptor-powerline solutions are now well-known around the world. Public attitudes toward birds of prey have changed 180 degrees to the point where they are widely revered and treasured. Most people have no idea how the seeds of these success stories were sown by Morley Nelson.

On a personal level, Morley's story is intriguing. How could one man achieve so much? What drove him to act every day like there's no tomorrow? How did he find the strength to push on despite the fact that he suffered more than his share of personal tragedies—debilitating injuries in the war, the death of a brother, two children and a painful divorce. Just like anyone else, Morley had to work through many painful moments in life, episodes that would cause weaker people to give up.

In the end, Morley's visionary leadership stands as a shining example for tomorrow's conservation leaders.

Morley is back on the south rim of the Snake River Canyon, making yet another film. This time, he is making a special appearance by invitation. The late John Denver starred in a PBS film about his favorite conservationists and natural areas. Released in January 1999, the film was called *Let This Be A Voice*.

The hour-long picture opens with Denver revisiting Jacques Cousteau, whose legendary conservation work on behalf of the oceans and sea life inspired Denver's song, *Calypso*. Next, Denver visits Morley Nelson and rappels into the Snake River Canyon twenty-seven years after he visited Idaho for the film, *The Eagle and the Hawk*. Morley rigs up the climbing ropes, and they descend a sheer basalt cliff to a golden eagle aerie containing one juvenile, just days before it's ready to fly the coop.

"Hello there big one," Morley says as they arrive overhead. He whistles a series of quick notes at the bird as if he's mother eagle. Morley is calm and casual on the side of a cliff, focusing on the bird, and his demeanor wears off on the bird, putting it at ease.

Denver tells Nelson that eagles inspired him to learn how to fly gliders and small airplanes. "If I come back, I'll be an eagle," Denver says. "You can count on it."

The film fades to a shot of Morley's bald eagle, Pearl, soaring high in a blue sky. Hidden from view, Morley's son, Norm, of Echo Films, trains his camera on the bird. Falconer and friend Lucy Nickerson, who was inspired by Morley's talk to her college class, is Pearl's personal trainer. She visits Nelson's home several times a week to feed and exercise the stately bird.

Maybe John Denver will be reincarnated as an eagle. But many people think Morley *is* a living reincarnation of a golden eagle. His good friend, Dick Thorsell, retired vice president of public affairs for the Edison Electric Institute, said as much when they held a reception in Washington D.C. after the release of *Silver Wires and Golden Wings*, an Emmy-winning film about raptors and powerlines. "He saw the Eagle God in his kingdom, and the Eagle God said it was OK for him to come back to earth as a man, as long as he had the heart and soul of an eagle," Thorsell says.

That's Morley Nelson, all right, a man who's got the heart and soul of an eagle, a man who will defend the glorious birds to the end, and do it in a way that wins over his worst enemies. Now that's a trait a golden eagle could never learn. Political finesse. The eagle would say, as Nelson has so many times, "It's either you or me, boy. What's it going to be?"

Chapter Two

Boy on a farm

1917-1938

A flash of golden light swept over the Steen Nelson farm at dawn. Young Morley Nelson, one of Steen's favorite grandsons, gazed at the morning sky from a fine perch atop his tan thoroughbred horse named "Slim." Morley watched the light fill a wedge-shaped draw that falls away from flat wheat fields above and flows in a downhill slant toward the Sheyenne River. A few Hereford cows grazed on grassy slopes above a small pond in the bottom of the draw. Morley could see the early-morning glow in their great soft eyes. The closest steer fed next to a tiger lily, whose reddish-orange black-spotted leaves lit up like a candle. The tips of green sedges bordering the pond glowed like diamonds, a reflection of the morning dew.

Fantastic beauty held Morley's attention as he marveled at the good fortune to see it at all.

Slim started for the water. "All right Slim drink all you can," Morley said, stroking behind his ears. Slim's nose came up from the water and pushed Morley backward as he stood up. This was his way of saying, "Let's go."

Morley mounted him. Slim's hooves made a powerful sucking sound as he stepped away from the pond, causing four teal ducks to flush. They flew down the draw and banked toward the northern sky. Morley knew first-hand about these colorful fine fliers. Smaller than mallards, green-winged teal were lightning-fast in the air. He marveled at the speed and agility of the teal as they turned into the cool north wind in symmetrical formation.

Then, Morley saw a vertical flash of cerulean blue slice open the sky like a lightning bolt, directly over the ducks. Feathers exploded in the air as a peregrine falcon struck a teal. As the bird fell toward the ground, the falcon made an inverted loop above the sinking bird, swooped in and caught its quarry in mid-air and flew off.

Morley stood there with his mouth and eyes wide open, completely awe-struck. He was so impressed—and inspired—by what he had just seen that he had to ponder the wonder of it all.

"Jesus, I was so shook, I couldn't believe it," he says.

He had just witnessed the fastest single action in nature, a peregrine falcon diving with authority to kill a duck instantly. Would his father and grandfather ever believe it? He had seen hawks kill gophers and snakes, but never a duck. The image of that rocket-like hawk made an indelible impression in his mind. It was an epiphany.

"I've got to have one of them," he whispered in the wind.

Brimming with excitement, Morley leaped on top of Slim, his right leg sailing over the saddle. He slapped the reins, kicked Slim in the ribs and raced home in a gallop.

Morley spent the first half of his childhood on Steen Nelson's 1,280-acre farm, located about 120 miles northwest of Fargo, North Dakota, between the small towns of Finley and Cooperstown. A tall, sturdy and proud man of Norwegian descent, Steen and his wife, Bertha, bought the place in 1896. The Midwest proved to be popular with settlers from Norway and Sweden due to its productive, deep black soil and similar northern climate (brutally cold winters; hot humid summers).

Over the years, Steen built quite an empire. It was a self-sufficient, diversified operation, with dairy cows, chickens, goats, hogs, a herd of beef cattle, and big crops of wheat, oats, flax, barley and hay. The farm animals provided eggs, milk, butter, pork and meat to feed the family. Oats and hay provided feed for livestock, while the grain crops, flax and beef cattle provided income and—if prices were right—good profits.

All the work in those days was done by manual labor. Teams of draft horses provided horsepower for tilling the soil, planting and harvesting.

The stock was kept in an enormous T-shaped barn, one of the largest in North Dakota. It was really two barns built into one. The main section of the barn ran east-west for about 200 feet. It contained stalls for fifty-five head of draft and riding horses. The T-section ran north-south on the east end of the main barn. It contained the dairy operation and provided space for equipment storage. In an extensive loft above both barns, the Nelsons stockpiled a small mountain of hay, enough to feed all the animals through North Dakota's long winters. A high ceiling, built with an impressive superstructure of wooden beams and lumber, provided plenty of room for the hay piles. A rope swing tied to one of the beams created a perfect launching point for the kids to perform aerial acrobatics into the hay pile.

Steen and Bertha Nelson raised four children in a spacious and elegant white three-story home, flanked on two sides by an extensive windbreak of poplar, spruce, willow, oak and maple trees. Morley's father, Norris, was the second-oldest of the four. Norris had an older brother, Steve, a younger brother, Art, and a sister, Lulu. All of them had dark hair. Norris and Steve, handsome lads, looked so much alike they could have been twins. They both had a round face, a strong jaw, bold arching dark eyebrows and close-cropped dark hair, parted on the left side.

Just two years after the Steen Nelson family settled into their farm, Bertha became ill and died, when all of her children were still young and living at home. Norris was only fourteen. Steen had to hire nannies to care for the children and cook for the workmen.

Norris and the Nelson kids went to school at a small one-room schoolhouse next to a white Lutheran Church, about a half mile from the farm. Norris was the only child of the brood who went on to get a college education. Norris chose the University of North Dakota in Grand Forks, where he became captain of the varsity basketball team. He also played third base and pitched for the UND Fighting Sioux.

"Dad was very intelligent and one hell of an athlete," Morley says.

"The fingers on his left hand—his catching hand—were all broken up and disfigured from catching lots of hard-hit balls with those old mitts," adds Morley's son, Tyler.

While earning a law degree at UND, Norris met Agnes Hoglund, a peppy Swedish gal who studied English and drama. She was seven years younger than Norris. He was a big man on campus because of his accomplishments on the UND basketball and baseball teams—plus he was a handsome fella. Agnes was spunky and attractive with dark curly hair. Both of them came from a farming background.

Agnes, the youngest of six children, was the only Hoglund child to attend college. The daughter of Swedish homesteaders Hans and Hannah Hoglund, Agnes grew up on a 560-acre farm in the town of Prosper, just on the outskirts of present-day West Fargo. Starting out with a mere $60 in 1871, Hans Hoglund built a log cabin to provide shelter for his family and began to acquire farm land along the Sheyenne River. Much like the Nelson clan, the Hoglunds were tough, hard-working folks who expanded their farm operation into a successful enterprise.

Following law school, Norris played semi-pro AAA baseball for a short time in North Dakota, and then he began to practice law full-time. He and Agnes got married in 1912. It took four years for Agnes to conceive her first child, Morlan Wendell Nelson, on October 5, 1917, in Munich, N.D., a tiny farm town near the Canadian border. It was likely a home birth. Soon afterward, Norris and Agnes moved to Steen's farm near Finley.

A typical day on the Nelson farm started at 5:30 a.m. When he got old enough, Morley woke up and did the morning chores—shoveling manure, feeding the chickens and the hogs and the horses—before breakfast. Agnes cooked for the extended family and the farm workers in a large kitchen. She was an excellent cook, having been raised in a household that thrived on good food, particularly meat, poultry, and all kinds of tasty desserts.

Young Morley developed a strong work ethic on Steen's

ranch, and a keen sense of how to work with animals of all kinds. He remembers forming an unusually close bond with a young bull. Morley had been around the bull since it was born. "I fed him the milk, and he thought I was his mother," Morley says. "He'd follow me around everywhere in the fields, and I didn't really think anything of it, except that I liked him. I called him "Bull." Of course, later, I looked back on that and realized that he had imprinted on me, but we didn't know anything about that at the time."

Nelson family collection
Morlan Nelson, age two

When winter arrived, the bull was full grown and ready for slaughter. Steen shot it, skinned it, and hung it up on the north side of the big barn, just as they always did until it was time cut it into steaks. Morley swore he'd never eat his friend Bull.

But one day, several weeks later, it was brutally cold—forty-two degrees below zero. Morley was out in the ice and snow, working all day. By the time he came in for supper, his friend Bull was served up for dinner. "I was so god damned hungry, and Jesus, I don't think there was anything else to eat. They brought out this beautiful part of this friend of mine, and I ate him. But I felt bad. I told my grandfather, I don't understand having to kill for everything we eat," Morley says.

One evening after supper, Morley sat down with Steen and Norris in the spacious wood-paneled living room. Several hardwood logs glowed orange and red. Sparks popped with a loud "crack" in the fireplace. Morley told his father and grandfather about the spectacular sight of watching a hawk take a teal duck in the back draw. To Morley's surprise, they believed him. Steen and Norris had observed hawks catching pigeons in the air and prairie chickens as well.

Norris mentioned that Shakespeare had written about falconry—the art of using birds of prey to catch quarry for food. It was a centuries-old practice dating back to King Arthur's time and Genghis Kahn, the twelfth century Mongol conqueror, Norris said. But after gunpowder was invented, rifles and shotguns replaced the art of falconry as a way to capture food or hunt for sport. Neither Norris nor Steen had even heard of anyone hunting with hawks in America.

Morley told them that he wanted to try to catch a hawk and see if he could train it. He was a strong, lean and agile kid, with blondish-brown hair. His mother cut his bangs straight across, leaving a generous mop on top and on the sides. He was 12 years old now.

Riding home towards his grandfather's homestead he remembered seeing a pair of hawks in Johnson's wind break. Every spring, the hawks screamed at him when he passed by on his horse during the nesting season. Just a month earlier, he remembered that hawks had squawked at him in the draw leading down to the Sheyenne River. This area of the range could be checked tomorrow if his father would let him leave the cattle for a while.

Steen told Morley it would be dangerous to capture a hawk, and it would be hard to feed and care for it.

"Don't you like to hunt with the old double-barrel twelve?" Steen asked.

"Sure, but that hawk was different, it flew like the fastest horse I ever dreamed of," Morley replied. "I would like to try keeping a hawk just to see if I could learn how to do it right."

The long day cut the conversation short as everyone was tired.

Norris gave him the approval he needed. "All right, go down and see what you can find, but be careful. If you don't get home by supper time we'll come down looking for you."

It was hard to sleep that night for Morley as his thoughts went over how he could catch a young hawk in the nest. He'd have to climb the tree, peer into the nest and see what he found. Time to get some sleep: 5:30 a.m. always came early.

The next day, he searched the sky for hawks all morning. Several soared high in the sky in the wind. Strange how they could go so fast when they wanted to but sailed so slow most of the time. In the afternoon, the cattle settled down. It was time to check on the hawk nest in the draw.

"Let's go Slim!" Morley said, kicking his heels into the horse's sides.

The north-facing slopes had protection from the sun and wind, resulting in a robust crop of chokecherries and June berries. Mallards flushed from the swampy areas on the way down. He searched the sky for a hawk above the ducks but never saw any. Deeper in the canyon, close to the winding Sheyenne River, a clutch of big poplar trees grew close together. That's where he had seen the hawk nest before.

A quarter of a mile away, he saw the first hawk rise up out of the largest tree. Morley's pulse quickened like it never had before. The hawk soared into the sky and began to circle. Soon, it was joined by its mate. Morley heard the hawks issue a repetitive shrieking call. Were they trying to warn Morley to stay away?

No doubt. As he rode into a clearing next to the tall poplars, one of the hawks dove at Slim and Morley. The horse shied sideways away from the diving hawk. The hawk's battle cry—CAW!—rose in crescendo as it dove closer. It was a very frightening attack but the hawk pulled out of its dive ten feet or more above the rider.

"Steady now, Slim, steady," Morley said softly.

The hawks flew directly above, screaming incessantly, far more than he had ever heard before. I must be really close to the nest, he figured.

"All right Slim, let's get you out of here." Morley rode away from the big trees and tied Slim loose so he could eat while the search for the nest continued.

The hawks became more excited as Morley walked into the center of the poplar grove. One of them swooped down with its wings folded and feet extended right at his hat. Morley ducked instinctively—oh, that was close! He looked up to watch for another dive, and he saw the big nest. At least two young hawks were inside!

Morley quickly checked on Slim. He was eating grass, completely unconcerned about all the excitement in the trees. There was only one thing to do—he had to climb the tree.

It was easy going. Thirty feet up the tree, the branches thinned out. Mother hawk came down, feet and talons extended, and cut under the branches going by at great speed only a few feet from Morley's face. He saw the red-tail's yellow feet and long black talons. He was amazed at how big the bird was up close. But there was no mistaking the message: Get out of my house.

For a moment, Morley reconsidered. Was this worth the obvious hazard either alone or with his younger brother, Norm? Yes. Time to take one of the young hawks now or forget the whole idea.

He placed one hand around the outer edge of the nest, took a firm hold of the branch, and he swung out and saw three young hawks with their beaks open, hissing at him with their wings spread. They were fierce-looking but also had a baby look, too. Suddenly, one of the adult birds ripped Morley's hat off in a violent strike, and one of the young birds jumped out of the nest. It made a clumsy effort to fly, but it was too young, and having never flown before, the bird crashed into the ground next to Slim.

Morley quickly retreated. He hit the ground and ran towards Slim. Just beyond the horse, the young hawk stood in the open with wings flared. It did not run or move, just looked straight at his opponent with a defiant attitude. Morley found his hat in good shape and put it on.

When Morley approached the hawk, it turned to face him but it didn't move. Morley grabbed a large branch in hopes the young hawk would sit on it. He knew its big talons would hurt him, and he hadn't thought of bringing a glove. Morley presented the branch under the hawk's feet multiple times. Finally the bird latched onto the limb, hissing and flapping its wings. Just like a wild horse, slow steady movements did not bother the bird, but a normal quick movement resulted in hissing. Now, Morley faced the next problem. How could he mount Slim, a tall horse, with the hawk on the limb? He decided to stick the branch in the crotch of a tree, and to his delight, the bird stayed on the perch.

This gave Morley time to find a way to mount Slim with the hawk held on the limb by hand. He needed a cut in the soil so he could step over the saddle from a position above the horse. A cut bank on the edge of the little intermittent stream was nearly ideal. With the horse in position, Morley grabbed the limb from the tree notch. The hawk flared its wings and hissed, but Morley kept the branch as steady as possible. Very slowly, he walked toward Slim. He had put the reins up on the saddle horn so he could grab them easily. He stretched his leg over the saddle very carefully, but suddenly, Slim moved backward, causing Morley's hand to jar the bird, and it hissed and flapped its wings, spooking the horse. Slim jumped around the brush and headed home in a fast trot.

Morley fell backward and dropped the branch, but fortunately, the bird remained on the perch. He slid down the face of the cut bank and watched Slim disappear toward the barn. The riderless horse would be an obvious sign to his dad that something went wrong. Morley would have to walk home with his bird. Maybe his dad would come for him.

A half-hour later, sure enough, Norris, Morley's younger brother, Norm, and several farm employees came out in a pickup and on horseback looking for the lost rider. Norm spotted Morley and the hawk walking in a very direct and slow manner. "Hey, you got a hawk!"

"Go slow, Norm, he's worse than a wild colt," Morley replied.

The young hawk flared and hissed but held its position on the perch.

"Slim all right?" Morley asked.

"Sure, he's home waiting for you but the reins were up and we were worried you might have been hurt in the fall."

"I never did get on," Morley said. "The hawk scared him off when I was trying to mount him. Norm, clean out that big box in the back of the pickup and I think we can put the limb, hawk and all in it to get home."

"Good idea." Norm took shovels, ropes, and all sorts of gear out of the big box. The box had a top and went all the way across the bed of the pickup. The young hawk stood steady as Morley lowered him very slowly into the box, limb and all.

"Well, I'll be darned," Norm said. "He's already calming down."

"Yep, maybe so."

Everyone wanted to see the "chicken hawk" but Norris suggested that they leave the bird in the box until they got home. "We'll have to look for a rabbit on the way home so we can feed the hawk," Morley said.

Rabbits were common on the farm. Morley had an easy shot along the way. The young hawk would have a fresh meal.

"Can we put the hawk in the spare stallion stall in the barn, dad?" Morley asked.

"OK, son. That should work out for now."

Morley took the red-tailed hawk into the giant barn and placed it on the edge of a wooden feeding trough in the first horse stall on the left side. The bird immediately flew up to the edge of the manger and stood facing Morley. The hawk stared at him with its gold and black eyes, while Morley quickly cleaned the rabbit, skinned it, and chopped up some meat into small pieces.

Morley offered a piece of meat to the hawk. It had a large black beak that hooked downward toward a sharp tip. It was interesting how the bird opened its beak, and once the meat touched its tongue, it immediately gulped it down. Morley's fingers got pinched a few times by the sharp beak when offering the meat. He learned after a couple attempts that if he held the meat in front of the hawk with an open hand, it would reach out and take it, swallowing it immediately, without trying to bite his hand.

My, how Morley was pleased.

Grandpa Steen and Norris were quite proud of Morley. "So how bad did those hawks beat you up while you were climbing around the nest?" Steen asked during dinner that night.

"They were dive-bombing me, and shrieking and really carrying on," Morley said. "One time, one of them dove so close it

Stephen Stuebner
Morley visits his boyhood home, the Steen Nelson farm, about 120 miles northwest of Fargo, North Dakota. The house was moved not long after this photograph was taken.

knocked my hat off. I was pretty scared, but I kept a strong grip on the tree. Luckily, I didn't fall off."

"Well, now you're going to have to learn how to train that bird," Norris said.

First, he needed to fashion a large glove for holding the bird. Morley found a large left-hand leather glove in the coat closet. He tied some big pieces of leather around the wrist area to extend the glove and protect his forearm. The hawk's talons were as sharp as a needle.

Morley's experience with other animals helped him understand the need to move slow and steady around the hawk. In the early stages, each time he approached the hawk, it would hiss. But when he offered some cut-up meat on his fist, the bird stepped up on his glove and ate. After a week or so, the bird began to anticipate Morley's afternoon feedings, and it stopped hissing at him at feeding time. It surprised Morley that the bird did not want to get out of the barn or leave the area.

Norm, who was three years younger than Morley, had a keen interest in the hawk as well. He was Morley's assistant in teaching the bird to fly to a perch with food. They closed the doors of the large barn, and got about ten to fifteen feet apart. Norm took the bird on his gloved fist, and faced Morley, who had fresh bits of meat in his glove.

"Much to our surprise the bird would fly to us immediately when it was hungry," Morley says.

Without any background in the art of falconry, Morley and Norm had taken the first step of calling a trained hawk to the fist. The next step was to try flying the bird outside the barn. They set the bird on a wooden post and Morley called her to his fist, about 20 feet away. It worked every time. "We had great reservations about going out in the open," Morley says, "but much to our surprise, the bird paid no attention to that. It could have flown anyplace, but it looked to us for food and flew immediately to the fist for something to eat as it had in the barn."

Now the boys were ready to see if they could teach the hawk to chase wild prey, such as a rabbit or a duck. Morley didn't know the difference between hawks and falcons at the time, so he hoped that the hawk could learn how to take a duck in a vertical dive.

"I think you have to have a different kind of hawk to catch teal and other birds," Norris said. "I think this bird would catch ground squirrels and rabbits easier, or at least attempt to catch them, since there are so many ground squirrels and rabbits around here."

Norm and Morley began to carry the hawk into the field, looking for a rabbit. Sure enough, the first time a rabbit jumped, the young hawk flew after it, but it was easily outmaneuvered. However, every time that the bird chased a rabbit, it seemed a little more possible that it could catch it. Due to its daily exercise routine, the hawk was becoming more skilled in flight. He improved his acrobatic flying skills, his control in the air.

One day Norm suggested, "Why don't you try cottontail rabbits? They're smaller and they don't run as fast as jack rabbits."

That proved to be an excellent idea. The hawk learned to catch cottontail rabbits after a few days of training. As time went on, Morley let the bird fly high in the air, and circle above the farm. He taught the hawk to come back to the fist when he whistled.

In the fall hunting season, the hawk took a keen interest in the boys when they hunted for pheasants. It would fly high overhead and watch the dogs flush birds and the boys gun them down. If one of the boys wounded a pheasant, the hawk pounced on the running bird. Sometimes, the hawk perched in a tree and just watched the boys hunt. It was the best bird-hunting season Morley and Norm had ever experienced.

As the days grew shorter and colder, Steen told Morley, "You can't keep this bird here in the winter, it will be too cold for it."

The boys were disappointed to hear that, but they knew it was true. Still, how would the bird migrate south on its own, when it had been trained by humans?

One day in November, a tremendous storm swooped in from the North. Morley and Norm went out to fly the hawk in the field. The bird flew up into the cool north wind, blowing ever so strong, and the hawk got blown far south out of sight. They never saw it again.

For the moment, Morley was crushed. "But the experience of watching the hawk evolve from a wild hissing juvenile bird to one that looked at me with very, very strong intelligent eyes was very interesting to me at the time," he says. "The bird still had something to say in terms of let's work together, but with a tremendous fierce determination."

It was a different feeling than he had experienced working with horses and dogs and cattle, whose eyes would tell him when they were feeling friendly and calm. Morley could not see a soft look in the hawk's eyes, yet there was a message there that led to the same understanding that a hawk certainly could be trained like a wild horse. Now Morley knew enough about the hawk to inspire a complete fascination with birds of prey. He wanted to learn so much more.

Steen taught Morley a lot about life, and particularly

strength of character, on the farm. His grandfather, being tall, smart and strong, often had to seize control of situations with ornery animals. One summer day, Morley and his brother were helping Norris and Steen herd cattle in a pasture above the banks of the Sheyenne. They had all dismounted from their horses and were working the rear of the herd on foot. Somehow, a big orange-and-white Hereford bull with an impressive set of horns got separated from the herd, and it decided it was going to charge the Nelson boys. In a matter of seconds, Norris, Morley and Norm scurried up a nearby tree to avoid the ornery bull. Not Steen. He took it personal.

Steen grabbed a six-inch fence post lying on the ground and stood his ground. He was going to fight the bull! Norris and the boys yelled at him to climb the tree.

"Grandpa! Look out!"

He didn't hear a thing. He stared at the bull, which pawed the dirt, lowered its head and charged.

"Hee-yah!" Steen yelled as the bull thundered full-bore across the pasture, coming right at him.

"Steen's eyes got real big and he stared right at that old bull with complete concentration," Morley says, eyes bulging and crouched on the floor with his feet planted wide, acting out the scene. "That bull came charging right for him, and at the last second, Steen side-stepped the bull, and swung that post at the bull's head, breaking off one of the horns. Jesus, that bull hit the ground so fast he didn't know what hit it.

"And Steen says, "Johnny be quick or dead. You've got to be quick in a life-threatening situation, or you're dead. It's that simple." From that point on, boy, I'll tell you what, I was going to be quick."

Another time, Steen was working on a thrasher and accidentally got his hand caught in a spinning belt, pinning his hand to a metal pulley and crushing his fingers. With help from his farm workers, he got his hand free, but the bones had been crushed.

"Steen didn't say a thing," Morley says. "He just wrapped his hand in a towel, went over and got on his horse, and rode ten miles to the nearest doctor. My dad wanted to drive him in the

Model T, but he wouldn't hear of it. He just got on his horse and rode into town. Gawd, he was a tough son of a bitch."

The next year, 1929, proved to be one of the driest years on record in North Dakota. All of the crops withered without rain. With an average rainfall of 16.3 inches, the farms normally fared just fine with natural rain and snow. No irrigation was needed. But in 1929, the actual beginning of the "Dustbowl" years during the Great Depression, less than ten inches of precipitation occurred. The drought would have a devastating effect on farms, the economy and family income.

It soon became obvious that even Steen's large farm holdings would not be able to support the extended Nelson family. Steen was seventy-five years old now, and his son, Art, had agreed to take over responsibility for running the farm. Since Norris was trained as a lawyer, he decided to move his family to Fargo, an attractive border town on the banks of the Red River.

Norris bought a two-story three-bedroom house on Eighth Avenue in Fargo, just a half block from the Red River. Moorhead, Minn. lay on the east side of the river, and Fargo on the west side. Both towns were urban trade centers that served the agricultural economy. Fargo also was the home of North Dakota State University, where Morley would go to college.

Morley kept several red-tailed hawks and a kestrel—all of which he caught along the Red River near Fargo—in the garage behind the house.

In Fargo, Morley and Norm got to know their Hoglund cousins much better, and they often visited the Hoglund's farm in West Fargo to play with them. The Hoglunds had lots of children in their extended family, and many of the cousins were about the same age as Morley and Norm. On weekends, they chased each other around on horseback, shot BB guns, performed flips and jumps on haystacks, swam in the Sheyenne River, directly adjacent to the Hoglund farm, and went bird-hunting in the fall.

One time Chester Hoglund and Morley were playing in the Hoglund's barn, and they decided to build a parachute for a stray cat. They found some string, and a piece of cloth, and rigged up a little parachute for the unwitting kitty. Then they tossed the cat out of the upper hay chute in the loft of the barn and watched it fall to the ground, amidst much laughter. The parachute didn't employ very effectively, but the cat survived. "We never saw that cat again," Morley says.

Ah, yes, but Morley nearly shot his eye out with a BB gun while playing with his cousins. They were running around Chester's house with their BB guns, and Morley's gun went off while he was climbing the stairs. The BB pierced the skin deep behind his right eyebrow. "I didn't dare tell my mom and dad that I'd almost shot a BB in my eye—I would have been in big trouble," Morley says.

Fortunately, the BB hole didn't bleed very much, and the eyebrow masked the wound. His parents never found out. And the BB is still there, under his eyebrow, almost seventy years later.

After school recessed in the summer, Morley and Norm went back to Steen's place to help work on the farm, catch and train hawks, and horse around. The drought continued in the early 1930s, and that made things real tough on Steen, due to poor growing conditions, pests and low wheat and cattle prices.

One of Morley's favorite chores was to run a two-bottom plow with a team of six horses. He enjoyed watching the stocky, thick-muscled draft horses till that North Dakota soil. The double plow worked two furrows at a time. "I had a beautiful feeling working with the horses because it was hard work, and yet it was so honest and strong," Morley says. "In a wonderful way, it was possible to feel a part of the environment there on the farm, and you also felt that you developed a friendly way with the horses as they worked more and more."

On Sundays, the horses had the day off. Steen always took the Nelson clan to the Lutheran church, about a half mile from the farm. They usually rode on horseback to church, and they'd hear the bell ringing as they approached. The classic one-room white church had a spacious well-kept lawn surrounding it, bordered on the edges with fir and pine trees. It had a high vaulted ceiling, and a steeple above the bell tower, with a gold cross on top. The church even had stained-glass windows, with large panes of red, orange and blue glass. Inside, the church had an elegant feel, with finely varnished wooden pews lined up in one row on either side, an altar with candles glowing, and a large cross hanging from the ceiling.

Steen Nelson was one of the first members of the Lutheran church, as evidenced by the fact that his late wife, Bertha's grave was located on the far corner of the property, near other family plots that were set up after the church had been built. "Steen was a very religious man," Morley says. "Sundays were reserved for church, worship and family get-togethers."

On Sunday afternoons, Norris liked to take his boys swimming in the Sheyenne River. "Norris was a great swimmer," Morley says. "He taught Norm and I how to swim in the river, how to work with the current, and how to dive very early in our life because of the heat of summer. We often went swimming after a long day of work rather than taking a shower because it was so refreshing."

On a Saturday night, Morley and Norm saw a movie that showed a horse jump off a relatively small cliff and dive into the water. Afterward, the boys thought, hey, let's give that a try!

After church the next day, they went down to the Sheyenne River with Morley's horse, Slim. "We arrived on a big gravely cliff directly above the water," Morley says. "We decided that we could dive off that cliff easily because it was only fifteen or twenty feet high, but of course, it would be quite something to ride a horse off that cliff. It looked so easy in the film. So we rushed to put on our swimming suits. We decided to ride bareback and head for the cliff at a full gallop and soar into the water, just as we had seen in the movie."

Morley went first. "I rode my horse to the edge of the cliff at a gallop, but the horse halted at the edge and I went flying over the cliff into the water. I landed far enough out in the river that it was fairly deep, so there didn't seem to be any particular problem with the water being too shallow. But how could we get the horse to jump? Norm got the brilliant idea to cover the horses eyes, because he was obviously afraid to make the jump.

"So he covered Slim's eyes and rode hard for the cliff. The horse was reluctant to go, but it ran right up to the edge of the cliff, where it tried to stop. But it was too late. Its front feet went over the cliff, and the horse flipped upside down. Norm had the presence of mind to bail off the horse just before it hit the water. The horse came up immediately; Norm did not.

"I thought maybe he got a foot stuck in the soft silt bottom of the river. I dove down immediately to where Norm landed in the river, and sure enough, Norm was down there with his arms outstretched straight down in the mud struggling to get his arms free. He was holding his breath, so he wasn't taking on any water, but boy was he stuck!"

Morley helped Norm get his arms out of the muck, and they surfaced amid much laughter. "We figured that our horses needed a little more training before we tried that again," Morley says.

On August 13, 1929, Steen Nelson died in his sleep. He had lived a full life, much longer than many in those days. He was seventy-five. The following passage from the Bible was inscribed on his gravestone: "The suffering of the present time is not worthy to be compared with the glory which shall be revealed in all of us." A pigeon-like bird is depicted in flight in the bottom panel of the four-foot tall gravestone, located next to his wife's plot.

When Morley turned sixteen, he entered Fargo High School. He was passing through the final stages of puberty. He was about five-foot ten inches tall, thin but muscular, and well-groomed. Due to the many hours he spent outdoors, his skin complexion was better than most.

Morley and Norm continued to train and fly hawks without any formal or technical knowledge. In hopes of finding a faster-flying hawk than a red-tail, Morley and Norm caught a female juvenile Cooper's hawk, a quick flier, along the Red River. Morley was in the process of training the new bird—called manning—in his first year of high school.

In a serendipitous coincidence, Morley's high school class received a visit that year from Captain C. W. R. Knight from England, a famous falconer. Captain Knight brought along a golden eagle named "Ramshaw" as he visited with school kids about birds of prey and the art of falconry. How it was that Knight happened to visit a high school in Fargo of all places—way off the beaten path of a typical lecture circuit—turned out to be a fortuitous situation for Morley. He thirsted for knowledge about training hawks, and suddenly, the best possible source dropped into his class.

After the guest lecture, Morley went up and introduced himself. He mentioned that he had a Cooper's hawk at his house, and Captain Knight expressed interest in seeing the bird. So he paid a visit to Morley's house, located just a few blocks from school.

"He was very much impressed with the bird's agility and the fact that it was just slightly smaller than a goshawk," Morley says. "We told him it could take pheasants in flight, even though it was just a small hawk. He was very impressed."

Captain Knight gave Morley a book on the art of falconry, published in England. The book described all of the different subspecies of falcons and the soaring hawks, the buteos and the accipiters. Now Morley finally learned why he and Norm weren't able to catch teal ducks with red-tailed hawks. It had been a falcon that took the teal duck in such a spectacular way. Although peregrine falcons and gyrfalcons didn't nest in North Dakota, they passed through in the fall on their way south from the Arctic and Canada to warmer climes.

Morley soaked up every word on every page. He realized that he needed to make a few things such as leather hoods for his birds, jesses for their feet, and a few other basic items that all falconers must have.

In the winter, Morley played hockey for the varsity high school team. He had learned to skate at some ponds near Steen's farm. But he really learned how to skate well, handle the puck and shoot with accuracy after he moved to Fargo. He played frequently with his friends in official-sized outdoor hockey rinks with exterior boards and goal nets. Morley played left wing, being fast and agile in the corners. Back in the 1930s, hockey was a tough sport: the knee pads and elbow pads weren't very thick, and players didn't wear helmets or mouth guards. Body-checking was a big part of the game. If you had the puck, you'd better be ready to get hit.

Morley remembers that his hockey coach gave him a brutal lesson about smoking. "I never did smoke, but on the way to play a game, someone brought out a cigarette, and I thought, ah, what the hell, I'll try it."

Somehow the coach found out, and he told the players that it would ruin their lungs. To prove the point, he made Morley and the other players stay on the ice, even though they got tired and needed to sit down and rest on the bench after a few minutes. "The coach didn't take me out for the whole period!" Morley says. "And I burned my lungs like you can't believe. And I'll tell you what, I've never touched a cigarette since."

Morley's mom, Agnes, encouraged him to take piano lessons even though he was more interested in hockey and birds, but he gave it a try. "She played the piano beautifully," Morley says. "She would sing with us in the church. And at home, she would play classical music, Swedish and Norwegian music. It was a beautiful thing how she filled our house with music."

When it came time for Morley to play a piano recital, his hockey buddies snuck in through the back door and filtered into the back of the audience. Morley didn't see them come in. "I wasn't any great piano player," Morley says, "but after I finished playing my recital, all of the hockey players clapped like hell. It was so embarrassing to me because I couldn't play worth a shit, and yet, those guys, they just shook the place down. I got more applause than anyone else. I had to give it up right there and then. I decided that piano was beyond me."

Classes were not difficult for Morley because he had a great interest in scientific things, such as how the soil combined with the sun and water to grow food, and the relationships he watched first-hand between predators and prey. Morley's parents had a better education than most, so they always emphasized the importance of doing well in school. They made sure that Morley got his homework done on time, and they scrutinized his report cards.

"Steen and Norris told me that I must get the best education possible in order to bring out the potential I had as a person to accomplish whatever I wanted to do in the world," Morley says.

His high school teachers helped him prepare for college, too.

"The programs at North Dakota State University were well known to the high school instructors, so when they asked me what I'd like to study, they told me to keep up with the highest levels of mathematics and science so I could start immediately in this sort of subject area at the university."

Morley began to develop an interest in girls as well. Cowboys at Steen's farm had taken him to a few barn dances on Saturday nights in the summertime, so he knew a few moves on the dance floor as well as the proper way to approach girls to ask them out on dates.

"It was wonderful to dance with all of those beautiful Norski ladies," Morley says. "Steen always said, "You'll burn in hell if you try to sleep with girls before you get married," but the cowboys told me, 'You've got to have a little lovin'. I kind of liked the way the cowboys looked at things."

Morley was just beginning to develop a sense of charm at this point in his life. He wasn't the most popular guy in school, and he didn't try to be, either. He did his school work and he was eager to come home to take his birds out flying or hunting. He also worked odd jobs in the evenings and delivered a paper route in the morning.

Dick Brown lived across the street from the Nelsons in Fargo, and became friends with both Morley and Norm. "Both of them were pretty quiet kids and they stayed around the house a lot," Brown recalls. "Morley can put on a big speech now, of course, but at that time, he wasn't really outgoing. Their family didn't have a lot of resources—no one did during the Depression—so things were pretty tough."

Brown remembers going bird-hunting with Morley quite a bit. Morley would load their shotgun shells by hand, and they'd take his father's Ford Model T to some hunting areas in South Dakota, just across the border. "We hunted pheasants, ducks and geese, and we got as many as we wanted," Brown says.

"The drought forced the farmers to leave their wheat uncut standing in the field, and that gave the pheasants all kinds of food to eat," Morley adds. "The pheasants were really thick in those fields. You'd get the birds up with our dogs, and the ground nearly shook with vibration from all of those pheasants taking off at once. It was easy shooting, that's for sure. But I remember we shot so much that my shoulder would be sore for over a week."

On nearly every Sunday evening, Morley's mom cooked up a feast for the family at the Nelson home in Fargo. In the fall, she often served roast pheasant with lots of trimmings. Usually some of the Hoglunds would join them, or she'd invite friends and neighbors over for supper. The Hoglunds would bring fresh vegetables from the farm, and the Nelsons always had an ample supply of wild game in the freezer.

"She was a marvelous mother in that she took care of everybody, and she loved to do that," Morley says. "She was very open and hospitable to everybody. She'd cook up a big dinner, and then she'd bake pies for dessert. We'd all sit around the table and have great conversations about our lives and the issues of the day."

Morley's mother and father did not drink much alcohol, for which Morley was grateful. "My mom had a little bit of wine once in a while, and my dad drank a little bourbon," he says, "but neither one of them were big drinkers. My mom saw her own uncle and brother ruin their lives drinking too much. If you hit that god damn booze like they did, you become useless to everybody."

Family gatherings were a pleasant time for Morley because he enjoyed seeing his cousins and friends. He also began to learn the art of debate by discussing political and other issues with the adults at the dinner table.

Clearly, Morley benefited from growing up in a strong family situation, in which his mother and father loved one another, and wanted everything possible for their children, despite the fact that money was tight. They instilled a strong sense of family values in their children, and commanded respect through the high personal standards they set for themselves.

"Both my mom and my dad were very expressive to me in terms of life and philosophy, and what's right and wrong," Morley says. "We learned most of our discipline and work ethic on Steen's farm because there was so much work to be done every day, and you just had to get out there and do it. That was the philosophy of the Norski's and the Swedes. You had to get the job done right now, and not half-way. Because in the cold winter, by god, you don't have time to fool around. You've got to work hard, regardless of how cold it is, or you won't survive."

The summer after Morley's junior year in high school, a

He Floats Through The Air, This Diving Major Dog

A swooshing splash into the Red river's chilly waters is just everyday fun for Major, 5-year-old police dog, owned by W. H. Wright, 335 Eighth av S. With his 'pal,' Morlan Nelson, 341 Eighth av S, Major jaunts to the river every fair day, dives and swims like an Olympic participant. Here the 80-pound animal follows Nelson off a spring board three feet above the water, but he dives just as willingly from the higher, 15-foot board.

Fargo Forum

neighbor's German shepherd began to show up at his house every day. Quite naturally, Morley petted the dog and took it swimming in the Red River. The dog's name was "Major," a five-year-old eighty-pound shepherd mix. One day Morley took Major to the Red River with him. There was a diving board on the river's edge, and a high dive. Morley peeled off his T-shirt and dove off the lower board. Major wasn't far behind. He leapt into the water in fine style, his tail and back feet extended far back behind him.

So Morley decided he'd carry the dog to the top of the high dive and see what happened. He climbed the ladder to the top of the fifteen-foot high dive, put the dog down at the back of the board, and he ran to the edge of the board and dove. When he rose up to surface, Major already had dove off the board and surfaced shortly afterward.

When Morley told his brother about the diving dog, Norm suggested they take Major to the gravel pit, where there was a thirty-foot cliff they liked to dive from into a deep, clear pool.

"Norm held Major on the top of the cliff, and I ran and made my dive," Morley says. "As soon as I was near the edge of the cliff, Norm released the dog, and sure enough, Major leapt right after me. I even felt his wet tongue on my feet. He was that close. So when I hit the water, I went deep, so the dog wouldn't land on top of me. I looked up from fifteen feet below, and the dog was already swimming for the surface, showing no sign of stress. I watched his feet swim perfectly in the beautiful clear water of the sandpit. I surfaced next to him, and his eyes seemed to say, "That was fun. Let's do it again!"

"Those great brown eyes, just like the horses and the dogs on the ranch, had a sparkle and glow that meant something to me. We went up on top of the cliff, and I dove off again, and of course, the dog sailed through the air in a beautiful arcing dive and splashed into the water. Our friends thought it was the neatest thing they'd ever seen."

Taking a cue from his buddies, Morley asked the dog's owner if he could put on a few exhibitions for the public in Fargo at area swimming pools. Mr. W. H. Wright said that would be fine. "That was a time when the city was building a number of swimming pools in town, so I was able to make some extra money for college by taking Major to the pools and taking him off the high dive. The people loved it."

So did the press. The local newspaper took a photo and put this caption next to it: "He Floats Through the Air, This Diving Major Dog."

Morley's high-flying antics with Major and his newfound celebrity status in Fargo had an unexpected benefit—a new relationship with an "older" woman, a blond-haired Norwegian girl who was a young nurse at the local hospital. She had watched one of Morley's performances with Major, and struck up a conversation with him in the park. She flirted a bit with Morley, making it obvious that she was interested in him. He could tell she was older than a high school senior. She exhibited an air of confidence and sexual zeal that he had never seen in a woman before. Morley asked her out on a date.

"I thought she was great," Morley says. "She was very tall and athletic, and she certainly knew what she wanted with me."

Morley graduated from Fargo High School in June 1934, but he couldn't get his diploma until he made up a math test in summer school. He and Dick Brown had skipped school to go pheasant hunting, and Morley missed the final exam in his algebra class. "That was the last time I took math lightly," he says.

In the fall, Morley enrolled at North Dakota State College in Fargo to begin work toward a bachelor's degree in agricultural science. "I had a great zest for knowledge in chemistry, mathematics, biology and soil science, all of which they offered at the university," Morley says. "These were all wonderful subjects that seemed to explain so many of the unknowns that we ran into on the ranch."

Even as a freshman in college, Morley was beginning to think that he'd like to get a job as a soil scientist working outdoors most of the time. The prospect of trying to run a farm, especially in the middle of the Depression, didn't seem like a wise career move at the time. He also played hockey for the NDSU varsity hockey team.

Morley had to be a serious student in light of the subjects he tackled. Many of his classes, such as chemistry and physics, required an immense amount of memorization. All of the mathematical equations had to be done by hand. "It was rugged," Morley says. "In nuclear physics, we had to know all of the weights and the valence of every atom. It helped me to write things down, and I used flash cards with my friends to study for exams. The long math was as complicated as the devil."

To pay for his college education, Morley found a well-paying job at the Armour meat-packing plant in West Fargo, yarding bulls from 6 p.m. to midnight, five days a week. He worked for thirty-three cents per hour, meaning he could make $1.80 a day, before taxes. Based on Morley's lifelong experience on the farm, he knew how to handle bulls without getting gored, and he even discovered a way to make some extra money.

"The cattle would be shipped into Fargo with anywhere from two to six bulls tied by a ring in their nose to the side of the cattle (rail) car, and the rest of the cattle, the heifers and the calves, would be loose," Morley says. "Once you unloaded the other animals, we had to cut the line that tied the bulls to the car so they could walk down the ramp without stepping on the rope and getting spooked. This became a very interesting procedure by which you climbed up the wall of the cattle car, hung on the wall next to a bull, and reached down to cut the rope."

After a few weeks on the job, Morley learned that he could make a little extra money by untying the rope from the nose ring, as opposed to cutting it. A shortage of hemp rope made the longer rope lengths quite valuable to cattle ranchers. But to untie the rope from the nose ring took more than a little more skill. Morley was the only one who dared to do it.

"I'd reach down and untie the bull, right next to his horns and right next to his head—if the bull was gentle enough to do that," Morley says. "I could almost double my pay by untying the knots and saving the entire length of rope for each bull. Ranchers paid ten cents for each hemp line that I saved. It was a very dangerous procedure for only ten cents, but in those days, ten cents was a fair chunk of change. It greatly increased my wages, too, considering I was making thirty-three cents an hour at the plant.

"You had to be careful in cutting the rope because many bulls would try to horn you and shove you against the wall of the cattle car. But I could tell from the look in a bull's eyes—just as I learned to do on the ranch—what a bull was thinking. They have soft gentle eyes when they're steady, you see, but if they're upset and on the prod, you can see it in their eyes. So I'd go to the first bull, check his eyes very carefully, and touch him on the flank. If his muscles tightened up and it became obvious he was going to try to take me, then I wouldn't dare try to untie that one. On the other hand, some bulls were very tame. When I touched them on the flank, they were relaxed and I could see in their eyes that it was safe to carefully untie the knot. Then it was a simple matter of running him out the ramp and go on to the next one. Some nights, I could make several dollars on one shift untying these bulls."

Morley finally did get fooled one time by an ornery bull in the cattle car.

"I looked him in the eye, and put my hand on the flank, and there was a slight reaction. He had a pretty good set of horns—not like a longhorn but, a very substantial set of horns. So when I reached down to untie the knot, all at once it became obvious, through the look in his eyes, that he was going to try to hook me. He just reared up and rammed that great horn against the wall of the cattle car. It missed me, fortunately, but then he turned his back end and banged my hip against the side of the cattle car. I quickly scaled the wall with my arms and hung there.

Nelson family collection

Morley Nelson, college graduate, North Dakota State University, 1938.

"By now, the bull was in a rage, and he reared back and yanked the ring from his nose. Whoa. I'd never seen that happen before. It's a very drastic thing because it's practically impossible to do. They took the mad bull out in the yard and shot it with a .30 caliber rifle. That bull was really out of its mind and dangerous. It had to be shot.

"The emotion that came out of the bull was one of those things in life where the philosophy comes out for everyone to see. If you denigrate a man or bull or anything male to the point where he is fighting for his life, then he becomes another animal altogether. And so it was here, in the frothing and heaving and the wild eyes and all the characteristics seen when man or beast is in the midst of a pitched battle. I think it's quite obvious that this is the kind of behavior we should try to avoid as we go through our lives and relationships with one-another."

The bruise on Morley's hip was bad enough that he had to leave work to see a doctor. The nasty bull didn't break his hip, but it was bruised badly enough that the doctors wouldn't allow him to return to work. Then Morley got slammed against the

Nelson family collection
ROTC officer Morlan Nelson

boards in a hockey game, reinjuring the bruise and making it much worse. So he got workman's compensation for several months. The silver lining from the injury was that Morley could enjoy a brief breather from a most rigorous work and college schedule.

In his third year in college, Morley saw two major opportunities arise at the prospect of joining the Reserve Officers Training Corps at North Dakota State. It was becoming clear, even in 1938, that Hitler was on an aggressive rampage in Europe that might draw Allied forces into the war, possibly even the United States. If that occurred, a ROTC officer would be a step ahead of a private drafted at war time. Morley also would greatly benefit from the financial assistance that ROTC officers were guaranteed for college education. That meant Morley wouldn't have to work at the Armour plant at night anymore, and ROTC would pay for the rest of his college. It was a perfect opportunity for Morley in the midst of the Depression.

"The reserve officers' work at the university was not taken seriously at the time, but obviously, there were extremely serious implications to being a reserve officer," Morley says. "My knowledge and background in the use of guns and rifles fit in very well with the military training, and the studies were a great deal easier than integral and differential calculus and other courses that I was taking at the university."

Morley had to take basic training at Fort Snelling, near St. Paul, located high on the banks of the Minnesota River. Weapons and attire seemed quite antiquated to Morley, but he went along with the training regime. "Every morning, we had a saber drill," he says. "Everyone lined up for reveille in the morning to salute the flag and to carry on our saber exercises. The sabers were large swords with a big handle, similar to those used in the Indian wars.

"Each morning we had to dress very quickly. There was one officer who was super intelligent, but very effeminate, named Bruce Darling. He always stayed ahead of the rest of us in dressing. Each day the rest of us would struggle to get these sabers and our gear on to stand for the first formation of the morning. Bruce Darling did everything so well to the point where a few of us in the platoon decided that we had to play a trick on him."

One morning Morley caught a very small garter snake, probably a foot and a half long, and he thought it'd be fun to put the snake into Darling's right-hand pocket.

"So I put the snake into his trousers, and when everyone stood up to raise his right hand for the saber salute, the snake stuck its head out of Darling's pocket and looked up. Darling looked down and saw the snake, and instead of screaming (which we thought he would do) he completely fainted and fell face-down on the ground, narrowly missing his head with the saber as he went down."

All the other officers were aghast and could not understand what had happened. Morley and his accomplice snickered as quietly as possible. Darling was sent off to the hospital for observation and sent back. He was fine. It was the first of many practical jokes that Morley would pull with snakes.

When the NDSU hockey team played the University of Minnesota in Minneapolis, it was a real highlight for the Fargo boys. They didn't get down to the city very often, so it was a rare opportunity to try to pick up a few sorority girls on campus. The University of Minnesota was a big school with lots of attractive blond-haired and blue-eyed girls. In Minnesota, college-level hockey players were considered to be damn near royalty. It wasn't hard for the guys to get a date.

By the time Morley finished his course work to graduate from NDSU with a bachelor's of science degree, he had mastered the subjects of soil science, hydrology, civil engineering and nuclear engineering. "When I came to understand the interconnections between the soil, the sun, the water and the grass, I realized that the soil is the basis for all of humanity," Morley says. "My education was the most important thing in my life. It gave me the knowledge and skills that I'd need to develop a career as a soil scientist, working outdoors most of the time, which is exactly what I wanted to do."

During his college years, Morley had taken a few trips to the Badlands, Mount Rushmore and the northeast coast of Minnesota in search of hawk and falcon nests. The small mountains and cliffs he observed there created the inspiration to search for a job in the big mountains of the West. This young man had seen enough of the flat farm country. He was ready to explore.

First, he had to wait for a job to open up with the federal Soil Erosion Service. So in the meantime, he took a job working for several cattle ranchers in the badlands of western North Dakota. His task was to study the soil and assist with designing irrigation systems for crops.

One hot summer day, Morley was out on horseback, doing some reconnaissance work in a narrow slot-canyon. He was riding along, staring at beige limestone cliffs in the midst of a day dream. When he heard the bone-chilling rattle of a large rattlesnake, it was too late to turn around.

"The canyon was only four to six feet wide, and the rattlesnake was coiled up next to the cliff wall on my right side," Morley says. "The snake didn't waste any time: it struck at my horse's foot. The horse was spooked, of course, so it reared up to the left and bucked me off. And I fell right on top of the god damned snake. I landed on my hands and feet, and tried to freeze while I hung over that damned snake. I've never been so afraid in all of my life. I thought I was a goner.

"But the snake, for some reason, turned around and slithered into the rocks next to the cliff. He didn't strike at me at all."

In the fall, Morley found a job-posting with the Soil Erosion Service in New Mexico. He called up on the phone, indicated his schooling and qualifications, and he was hired as a soil surveyor and engineer. He would be based in Albuquerque. Time to move West!

Chapter Three

Westward, ho!
1938-1943

Prior to leaving his home state of North Dakota for New Mexico, Morley bought a brand-spanking new 1939 Navy blue Chevrolet four-door sedan for $500. All of his hard work at the Armour Plant, the exhibitions with Major the flying dog and ROTC duty yielded enough cash for a new car for the big drive across the West to Albuquerque.

More than ever before, Morley was now on his own—a young man starting a new career, and a new life. Although he was only twenty-two, he was quite mature and ready for the challenge: He had a solid education, a professional job, an iron-clad work ethic, and a strong curiosity about all aspects of life. He was physically fit, savvy about women, and he had a head start on his favorite hobby, falconry, which would bloom beyond all expectations in the years ahead.

He loaded up all of his personal belongings—mainly his clothes, his six-gun, rifles and shotguns—and hit the road. Younger brother Norm would take care of the hawks in Fargo. Morley budgeted a few extra travel days so he could make the drive at a leisurely pace, check out the mountains in Colorado, Utah and Arizona and look for birds of prey.

"I had a marvelous trip," Morley says. "I saw the changes in terrain, the soils, the geology, the water and the vegetation, and of course, all the birds. So it was a unique and terrific experience for me to drive down there."

Morley followed two-lane paved U.S. highways from Fargo to Nebraska, and then he turned dead west for the Rocky Mountains in Colorado. Imagine him driving along, craning his neck to inspect every hawk, eagle and falcon sitting on fence posts and power poles along the way. He drove dirt roads to explore a few mountain canyons in the Sangre de Cristo Mountains in Colorado, and camped under the stars.

In Albuquerque, he rented a room in a house for his bachelor pad. Within weeks, he realized that his job would require him to travel to all corners of New Mexico. He was assigned to work on "straightening the Rio Grande" between the United States and Mexico on the border, near El Paso, Texas.

"Every time the river flooded, it changed course," Morley says. "And of course, that determined our border lands. If it flooded the wrong way, it could cost us a lot of land."

And he assisted the Apache, Navajo, Mescalero and Acoma Indian tribes develop irrigation systems for growing crops on their respective reservations.

Working with the Indians was fun for Morley because they were so knowledgeable about native flora and fauna. "I saw how sensitive they were toward the soil, the vegetation, the wildlife and everything else. All of those things are part of their religion. They're sacred. That's why they're so careful with things, always have been, and still are."

It impressed him that the Navajo Indians saw golden eagles as a symbolic icon of the Great Spirit. "I thought that was wonderful, and I believed it was true," he says.

In return for Morley's assistance, the tribes taught him a thing or two about surviving in the desert, namely, how to identify edible roots and plants. He learned, for example, how to cut open a barrel cactus to extract a milk-like liquid that was actually quite tasty.

In working along the Mexican border, Morley noticed a few people speaking German surveying an area next to the Rio Grande on the Mexico side of the river. Rumors were flying that the Germans were doing reconnaissance work for Hitler to scout a possible point to invade the United States. "I saw them working out there, pounding in survey stakes on the Mexican side of the river, just as I was on our side of the river," Morley says. "We knew what they were doing, but there weren't very many of them, so nobody was too alarmed at the time."

Even so, the threat seemed more genuine when Germany invaded Poland in 1939. The war had begun in Eastern Europe.

Morley didn't have a place to keep any birds of prey, nor did he have much time to train and feed them. So when he found raptor nests, he merely noted their location for future reference. "Up in the forest, I saw goshawks, which I had never seen in the Dakotas. I also found a peregrine falcon aerie, which was amazing to me, because I didn't think they lived that far south."

While in New Mexico, Morley met Luna Leopold, the second son of Aldo Leopold, the American father of ecology and author of *A Sand County Almanac*. Luna was a professor at the University of New Mexico in Albuquerque. He had developed an expertise in soil conservation and hydrology through his graduate studies at the University of Wisconsin at Madison, where his father taught wildlife conservation. He met Morley through a common interest in soil and water.

They did not get to know each other well, Morley recalls, but he remembers that Luna Leopold was quite taken with his knowledge of and interest in falconry. "He was impressed as hell that I knew so much about these falcons, and that I could fly them. Falconers were rare as hell back then."

The admiration was mutual. Morley was dazzled by Leopold's philosophy of ecology and nature—the notion that all things are interconnected, and that man's heavy-handed approach to damming and diking rivers could backfire with unintended consequences. Morley did not have the kind of naturalist training or philosophical background that Luna had learned from his father, the dean of all naturalists, but he had been raised on a farm, and he had learned many things from working with horses, cattle, hogs, chickens and hawks. He shared the same philosophy as Leopold, but he hadn't considered the deepest intricacies that Leopold spoke of with ease.

In the summer of 1940, Morley seized an opportunity to take

a new job with the Soil Conservation Service in Salt Lake City, Utah. Here, he'd have the opportunity to practice falconry—hawks and falcons were plentiful in the area—ski in the Wasatch Mountains and learn mountaineering skills.

At his first opportunity, Morley searched for a peregrine falcon nest in the steep-walled canyons of the Wasatch Mountains. He wanted to catch one of those prized birds, if he could find a nest, but to reach a high cliff, it became obvious that he'd have to learn basic rock-climbing techniques. While hiking around at Alta Ski Area in Little Cottonwood Canyon, Morley met a number of experienced rock climbers from Switzerland and Austria, who had come to the Wasatch Range in hopes of finding some interesting peaks to climb. Morley inquired if he could learn a few things about rappelling.

"This is where I learned about doing a quick rappel without a climbing harness, and I learned about using caribiners, pitons, ropes and all that," Morley says.

Morley bought some climbing rope and essential gear and kept it in his car, so if he ran across a nest site, he could go after it.

Turkey vultures created another source of curiosity for Morley. He had never visited a vulture nest, and he had heard that they had very large eggs. He wanted to see one. After pinpointing a vulture nest in a low-lying cliff in the Wasatch Front, he decided to see if he could climb into the nest and look at the eggs. "After climbing to the edge of the cliff nest, I noticed that the vultures had taken up residence in a fairly deep cave, at least six feet deep," Morley says. "There wasn't any head room. I had to lay down absolutely flat and scoot along, using my hands to go forward. As I made my way to the back of the cave, I heard the terrifying sound of a rattlesnake, which I could see was a foot or so from my face, right there in the cave!

"Obviously, I was in a jam, and my natural instinct was to stand up quick and hard and back off from the snake. I threw my head back with too much violence and I knocked myself out. I have no idea how long I was unconscious. When I came to, I put my arm in front of my face, and tried to scoot backwards with my legs and other arm. It was so hot inside the cave that sweat poured from my brow into my eyes to the point where I couldn't see. But I assumed the snake was still there and could strike anytime, although it was comforting that it was not rattling anymore. I finally reached the edge of the cliff, and when I looked inside, I didn't see the snake. Whew!"

Morley took a moment to rest in front of the cave, wiping his brow. He had a big smile on his face, realizing that he had survived a second close encounter with a deadly rattlesnake. He didn't feel the need to go back inside.

During the winter of 1940-1941, Morley took advantage of a bounty of recreation opportunities in the Salt Lake area. He played pickup hockey on weekends, and he practiced skiing steep-mountain slopes at Alta Ski Area. He even tried a little ski jumping.

A handsome and athletic fellow by the name of Alf Engen hung around Alta that winter. "He was a very strong Norwegian and a ski instructor who was also a ski jumper and a ski racer," Morley says. "Engen and his family came over from Norway, and they were some of the finest ski racers and ski jumpers in the nation at the time. I enjoyed watching them ski so gracefully in the powder snow, and it was very inspiring for me."

On typical weekends, Morley would get up early to go skiing, and then in the afternoon, he'd come back to exercise and feed his prairie falcons.

To maintain his reserve officer status, Morley was required to attend reserve officer training several times a year in Utah. It was a good idea to remain in the Army Rserve, because if a military draft occurred, it ensured that Morley would enter the service as a lieutenant. As Germany continued its aggression in Europe, it was becoming all too clear that all of Europe would become embroiled in the war, and that the United States might have to get involved as well.

On Sunday, December 7, 1941, Japan launched a sneak-attack on Pearl Harbor, drawing the United States into World War II. Morley received a letter six days after the attack, ordering him to report for duty at Fort Leavenworth, Kansas. He had to leave Utah ASAP.

Believing that it might be a long time before he could ski again, Morley went skiing at Alta for one more day on the slopes before driving to Kansas. It was a banner powder day, more than a foot of new snow fell, but that triggered a big avalanche in Little Cottonwood Canyon, smothering the road with more than 20 feet of dense, encrusted snow. Morley and many other skiers were stuck above the slide and couldn't get out. They had to stay at Alta for two days until the plows cleared the road.

By the time Morley arrived at Fort Leavenworth, he was in big trouble. He was nearly a week late. He told his superiors about the avalanche. "Of course, the commanding officer thought this was some sort of crazy thing that had no application in what possibly really happened," Morley says, "so I was confined to the post for two weeks as punishment."

As a reserve officer with experience in training men and handling weapons, Morley ran a company of privates through a series of drills every day. He led physical exercises, taught them how to disassemble, assemble and load weapons, and drilled them on presenting arms and formations.

"God damn it, lieutenant, are you deaf, dumb and blind? Can't you see the American flag and a general's flag on this car?"

All of Morley's men stopped and stood still. Morley walked over to the shiny black Packard and said, "I am Lieutenant Nelson, sir," saluting the general.

"My name is General Patton," the officer replied.

Morley had been transferred to the desert Southwest to supervise the establishment of a desert training center for the Army. His engineering skills and officer status led to the transfer. He joined eighteen men in Death Valley, near Indio, Calif., to survey the best location for the training center from a soils perspective, and lay out engineering plans for streets and buildings.

The men gathered each day by a large green wall tent with an American flag posted out front. At the end of the day, they would march by the flag, salute, and then were excused for the evening to go back to their tents to cook dinner and sleep.

On this particular evening, Morley was about to give the order to salute the flag when Patton showed up.

"After a few more cursing remarks, General Patton sat down in the back of the car and asked me how we were doing, and how my men were getting along and whether we were accomplishing what I had been sent out here to do," Morley says. "I reported that, yes, we were accomplishing our task very well. We were ready for major construction to start on the original buildings. The only problem was, it was very hot in the day time and it got very cold at night, below freezing. My men were very cold at night."

"All right, if you're cold at night, by God, tonight every man will have four blankets," Patton said.

General George S. Patton, who was nicknamed "Old Blood and Guts," was Morley's supervisor in Death Valley. Morley didn't like Patton's aggressive, paternalistic approach to supervising soldiers, but he had to bite his tongue and try to avoid any confrontations.

Once the actual training started, however, Morley enjoyed it. He taught soldiers how to identify and eat native plants and roots – all of the things he had learned from the Indians in New Mexico. He showed them how to extract liquid from barrel cactus and how to kill rattlesnakes and prepare them for a good meal, among other things.

Instead of carrying a standard-issue .45 caliber automatic pistol on his hip, Morley preferred to carry his six gun, a Colt single-action frontier .357 magnum with a hair trigger. He shot an occasional rabbit for food, and during training exercises, Morley showed the men how much easier it was to fire a pistol with a hair trigger and hit the target.

During one training exercise, the troops were practicing maneuvers and formations with Army tanks. They were trying to simulate a war exercise in the desert of North Africa, where the Germans had recently made significant advances. During the exercise, General Patton noticed Morley had an unauthorized weapon in his holster.

"God damn it, lieutenant, don't you know that only a general officer can carry an unauthorized arm? Take that gun off of him and take the bars off his shoulder and send him back to camp. This is a serious breach of military regulations," Patton said.

Morley had been humiliated in front of his own men, and appeared to be in deep trouble. They took the lieutenant bars off his shoulders, confiscated the gun and escorted him back to the desert camp. Later on, after Patton cooled off, he summoned Morley to his office. "I realize that you are training and doing an important function here in the desert infantry but it is absolutely impossible for you to carry an unauthorized arm," Patton told Morley. "You must pack that six gun away and never take it out again while you are in uniform. You may shoot it on your time off when you are either out of uniform or not on official duty, but you must never wear it again while on official duty in the Army."

Morley got his lieutenant's bars back, but he definitely did not like working under Patton. "He was a real arrogant son of a bitch. He had that high squeaky voice, and he had a totally different philosophy toward working with men than I did," Morley says. "I always felt that if I treated my men with respect, they returned the favor and would do anything I asked them to do. Patton's style was to beat people down."

Then, an opportunity came along for Morley to leave Death Valley. He discovered the Army was looking for experienced skiers and mountain climbers to serve in a new, specialized unit, the mountain infantry. The requirements fit Morley's interests perfectly, and so he put in for a transfer, detailing his experiences as a skier and mountain climber. Morley's college education as an engineer and soil scientist helped as well.

Patton recommended against the transfer, saying that Morley's skills were needed to teach basic survival techniques in the desert. But his wishes were overruled. Morley received orders to report to Fort Lewis, Washington, for special training on Mount Rainier.

Morley and Patton exchanged a few words before he left. "It was a very harsh situation, and a lot of hard feelings came out," he says. "But I sure was happy to leave that place."

Fort Lewis was a more preferable location for Morley. The post was located south of Tacoma, adjacent to the southern reach of Puget Sound. Most of the training occurred on Mount Rainier, a massive 14,411-foot volcanic peak in the North Cascades. On Mount Rainier, the troops learned to ski on glaciers, and climb ice falls and snow with ice axes and crampons. They were roped together for safety in the event that they ran across a deep crevasse.

"This was an entirely new organization being formed to perform in high mountainous and rugged terrain with or without snow," Morley says. "In a matter of less than a week, I went from 110 degree temperatures to freezing nights and very cool days on a glacier on Mount Rainier. The training assimilated all of the technical aspects of cliff climbing and ice climbing, and all of the safety procedures that one should observe in these kinds of maneuvers. These were skills that I would be able to use for the rest of my life, if I lived through the war."

Later in the year, the mountain infantry was moved to Camp Hale, near Leadville, in central Colorado. More national recruiting had occurred, thanks to the National Ski Patrol, to beef up the force. The founder of the ski patrol, Charles "Minnie" Dole, persuaded top military authorities that mountain troops would be needed to fight the Germans in the European Alps. The entire unit would be expanded to approximately 15,000 men at Camp Hale.

National recruiting efforts brought the nation's top skiers and mountain climbers to Camp Hale: Swiss-native Walter Prager, the ski coach of Dartmouth College, Torger Tokle, the nation's best ski jumper from Norway, and Austrian Friedl Pfeifer, who would later coach the U.S. Olympic Ski Team.

Few of the men were prepared for living at such an extreme altitude.

"No one in command of the unit was experienced in the problems of going from sea level at Fort Lewis to more than 10,000 feet at Camp Hale," Morley says. "This is a relatively rugged shift even for men who are in good condition. So it was really a surprise when I stood my platoon out in front of the train, and they put their heavy packs on, and I gave the order of "right face," and two men went down on their knees. They could barely breathe, much less walk with a heavy pack on."

The commanders realized that they would have to go easy on the men for a couple days. In addition, they added fifty percent more meat to the daily rations. Several weeks later, the men were ready for training again. Morley's official duty at Camp Hale was to serve as the commander of 200 men in Company I.

Morley was in heaven at Camp Hale, which was surrounded by tall mountains in all directions. Colorado's highest peak, Mt. Elbert (14,433 feet), loomed above the southern end of the post. Morley enjoyed the training—learning how to dig snow caves and build igloos, rock climbing, ice climbing, handling mules, and skiing uphill and downhill with a ninety-pound pack on your back. Most of all, he enjoyed seeing lots of raptors in the area. He found a goshawk nest in the trees, behind the hospital building at Camp Hale, and he observed golden eagles, red-tailed hawks, prairie falcons and peregrine falcons flying overhead during training exercises.

To find a goshawk nest right next to camp was a special thing for Morley. Goshawks are quite striking—they have blood-red eyes, a blue-gray body, and five dark-gray bands on a long tail. The swift, short-winged accipiters were known for being extremely fierce predators in a forested environment.

"I made frequent visits to the goshawk nest in the spring and late summer to check on the progress of the new brood," Morley says. "Mother hawk became so accustomed to seeing me climb the tree and check on the young birds that she would sit on the outer edge of the tree limb to watch for me. She'd scold me with vigor as I approached the nest, but she wouldn't strike at me like she had initially because she knew that I posed no harm to her or her brood."

One day, Morley and his company came across a falcon that was swimming in a pond. The bird was having a hard time getting out of the water, and Morley went to retrieve the bird. The men were impressed with his ability to handle a bird with sharp talons on its feet.

After that, Morley's men nicknamed him "Falcon."

Mountain infantry soldiers had different equipment and uniforms than other Army units. They dressed in white coats with fur-lined hoods and white canvas pants so they would be difficult to spot in the snow. They wore either leather ski boots or white felt-lined "Bunny Boots." No synthetic sweat-absorbing undergarments existed in those days. The men wore cotton

Denver Public Library

Tenth Mountain Division troops work on their skiing skills at Camp Hale, near Leadville, Colorado, late in 1943.

long-johns and wool shirts. They carried bitterly heavy ninety-pound rucksacks full of gear, such as down sleeping bags, heavy canvas tents, camp stoves, dehydrated food, and extra clothes. For eye protection from the sun and driving snow, they wore bug-eyed goggles. Special mittens were made with a trigger finger.

Despite the frigid cold in the winter, all units had to perform calisthenics every day at 7 a.m., stripped to the waist.

Ski equipment was rudimentary due to the state of technology for skis, bindings and poles at the time. Wide wooden skis were seven feet long, and painted white. For climbing mountains, they strapped mohair skins to the bottoms of their skis. In spring snow conditions, some men preferred to use klister wax for climbing because the skins took so long to take on and off.

The old ski bindings were unforgiving, Morley recalls. "Ski training proved to be a very difficult problem because at that time there were no safety bindings. A ski trooper had to attach his heel so it would not come up while downhill skiing, and during climbing, he had to keep the heel loose so he could climb and ski cross-country with a free heel, as the Scandinavians do. However, the bindings were not very forgiving in the event of a fall. They didn't release. The toe irons held your feet on the skis, and if your ski went sideways or backwards in a fall, your foot and leg remained attached, leading to many fractures and sprains. During the training, it became obvious that we would have to add twenty to thirty percent more men than normal to allow for injuries and attrition.

"Sleeping in the snow at 12,000 feet or higher gave many of the men colds and lung problems, which would cause them to say, I give up. I'm not cut out for this kind of rugged existence."

Indeed, a minority of the men considered Camp Hale to be a dark, hellish place. In the middle of winter, the sun went down before 5 p.m. Altitude sickness was common. Soldiers often were hospitalized for exhaustion after training exercises. Long before the days when central Colorado became home to some of the finest ski resorts in America, the high valley near Leadville was a lonely place, 120 miles from Denver.

The *Saturday Evening Post* described the camp location as "one of those whistle stops that would cause the (train) passengers from behind the car windows to remark, "Can you imagine anyone living there?" "

Robert Ellis, author of *See Naples and Die* (McFarland & Co. 1996), wrote a letter home soon after he arrived at Camp Hale in April 1943. "So far, the life here has been Hell," he said. "For a few days, I thought I would go AWOL . . . . Everyone here has asthma, or rheumatic fever, or colds, etc. Nearly everybody has a perpetual cough . . . . I may have made a serious mistake by getting into this . . . . The country is beautiful though, with snow-capped mountains all around."

All of the men had to apply to serve in the mountain troops, with three letters of recommendation, so many of them truly wanted to be there. They wanted to learn how to ski better, how to sleep in the snow, practice rock-climbing techniques, etc. The men who enjoyed skiing and rock-climbing, like Morley, actually spent their time off doing the same thing.

Francis Sargent, who would later become governor of Massachusetts, commented on the hard-cores in *Sports Illustrated*. "My God, half of the sonuvaguns in the outfit would rather go climb some rock than go down to town and look for booze and broads. I remember thinking it was the damndest thing for soldiers to act like that."

The men who volunteered for the mountain infantry also were considered to be among the best-educated in the U.S. Army. Many of them had college degrees. In I.Q. tests, ninety-two percent of the men scored higher than the average Army score of ninety-one, William Johnson wrote in *Sports Illustrated.*

Most of the men had a positive attitude and were highly motivated. They made up all kinds of songs to break up the monotony of training. That made Morley's job easier as the commander of 200 men in Company I, 87th Mountain Infantry. "We were the singiest bunch of guys you've ever seen," says Duane Lind, a member of the unit.

Lind joined Company I in November 1942. He did his basic training under Morley's supervision and served with Morley on firearms training. He shared his thoughts about Morley as a company commander.

"He had great empathy for all the men who trained with him," says Lind, a machine-gunner. "He was a man who cared for you individually as a person. When we went out on patrol, and we're going through all the training, you never felt that Morley was the commanding officer. He was just one of the guys. He knews the maps; he knew the details, the objectives, and all that. But he still was a close friend of every man in that platoon."

Lind, who had hunted pheasants as a kid in Nebraska, had a keen eye and dead aim. He shot the best score at the machine gun firing range, and the second best score in rifle training. He was the kind of guy that Morley liked to have in his company.

Lind chuckles at the memory of going out on patrol for the first time with Morley. "We were skiing up this mountain pass, and Morley wasn't even looking at the terrain. He was looking in the sky for hawks. This was Colorado. It was all an adventure to him. I hadn't ever heard of a red-tailed hawk, and neither had any of the other guys. But we learned how to identify them. Imagine a bunch of guys out on patrol, and they're all looking in the sky for birds!"

When the men got a few days of leave, the best place to go was Denver. It was a bustling city with lots of restaurants, night clubs and all kinds of young single girls.

Forbes Mack remembers the first time he had a chance to leave Camp Hale for Denver. "I heard these guys talking in the latrine, and they were talking about going to Denver on a weekend pass, I stuck my head inside the door and looked in there and I said, who are you? And the guy says, "I'm Morley Nelson." He was a first lieutenant. And I said, 'You're going to go to Denver with these guys and you're going to wear that dirty shirt?'"

Mack and Morley became extremely close friends during their time together at Camp Hale. But Mack always gave Morley a hard time about being a bit of slob. Mack, on the other hand, dressed neat, and looked sharp. His pants and sleeves were creased. But the two loved to go to Denver together and chase women.

"We went to Denver every chance we got," Mack says. "We'd go downtown and walk the streets or check out the night clubs and look for girls. Our routine on the streets worked like this—Morley would walk down one side of the street, and I'd walk down the other side, and we'd signal each other if we saw something that looked awfully good. We'd say, "Pawsay! Capisch?" And then we'd try to ask them out on a date and so forth. It worked more often than you'd think."

Mountain troops were famous for pulling rappelling stunts at the Brown Palace Hotel in downtown Denver. "When drunk, men . . . tended to do what they knew best, such as donning a pair of crampons, taking a coil of rope and—starting from one wall of a lodge, lobby or bar—crawling flylike up the wall, across the ceiling and down the other wall. Or they would climb out of a hotel window—the higher the better—and simply rappel down several stories of sheer brick to be received by gawking crowds on the street below," Johnson wrote in *Sports Illustrated.*

By all accounts, the mountain troopers appeared quite hand-

U.S. Army

White uniforms made 10th Mountain troopers difficult to spot in the snow.

some to Colorado women. They had tanned faces from training outdoors all the time, they were in tip-top physical shape, and they were snappy dressers.

For more than a year, the 87th Mountain Infantry had been training hard in Colorado, but it was repeatedly passed over for a military mission. Finally, in July 1943, the 87th was sent to Fort Ord, California, to prepare for a secret mission to Kiska, a tiny and remote volcanic island near the western tip of the Aleutian Islands. The troops engaged in amphibious training in California for several weeks. Then it was time to load up their equipment and sail for Alaska.

In the dark of night, on August 15, 1943, Companies I and K of the 87th Mountain Infantry loaded into landing craft for an attack on Lilly Beach on Kiska Island. It was a horrible night for an amphibious landing: Dense fog shrouded the dark, ghastly looking shoreline of Kiska. Black, jagged rocks pierced the heavy seas breaking on shore, and the boat coxswains had a difficult time locating the beach. The men were battle-ready, dressed in rubberized rain suits under military clothing, with bulky life jackets. They carried heavy packs with food, shelter and sleeping bags inside. Most of the men carried rifles; special firearms experts toted heavy machine guns.

Intelligence reports indicated more than 10,000 Japanese soldiers were occupying the island. The Army's mission was to seize the high ground of the island in the middle of the night.

Morley's platoon found the beach after striking a few rocks in the rough sea. The men fanned out on the beach, expecting to be fired upon as soon as they hit the sand. But instead, there was dead silence—only the continuous sound of crashing waves and howling wind. "It was obvious by the lack of fire that there were no Japanese right there down close to the ocean," Morley says. "Our intelligence from the previous day indicated that the Japs were positioned on the beach as well as all the way up the mountain. One of the first things we found was an unoccupied machine gun emplacement. That seemed strange."

Morley's men fanned out over the lower face of a round, volcanic mountain, and began to climb 2,000 vertical feet to the top of Link Hill overlooking Kiska Harbor. The fog remained thick. Fierce wind and driving rain made it difficult to get decent footing on the slippery and mushy tundra. In a rare opening in the fog, Morley spied a ridgeline that led to the top of Link Hill.

His platoon continued to climb, and still, no enemy fire. They reached the top of Link Hill and hunkered down in a little shack that had been set up by the Japanese. Here, Morley set up a command post and awaited further orders. Morley's men took shelter from the driving rain in foxholes that the Japanese had dug.

The third platoon of Company I, led by Lieutenant Wilfred Funk, summitted the mountain next to Morley's platoon. Then Morley received orders over the radio to send Lieutenant Funk's platoon back down the hill to protect battalion headquarters, adjacent to Kiska Harbor. To Morley, the order made no sense.

"This was a most difficult and unusual order under the circumstances," he says. "There had not been a single Japanese encountered in the entire trip up the mountain. Why would it be necessary, in the middle of this storm, to send the third platoon down the mountain? The headquarters already had a few men guarding it, and the whole island appeared to be vacant of any enemy troops. The bottom of the ridge was several thousand feet below in very difficult terrain. I felt the order was wrong in the first place, and in the second place, it would be dangerous."

Nelson family collection
Lieutenant Morlan Nelson, ski trooper

Nevertheless, that is what Major McClanahan ordered. So Lieutenant Funk followed the order, and the men in his platoon started down the mountain.

"The storm continued with a tremendous velocity of wind and variable fog conditions that created the effect that objects such as a man walking would suddenly jump toward you," Morley says. "It was creepy. No one could see more than a few feet in front of them."

Once Lieutenant Funk and his platoon neared battalion headquarters, they hugged a small knoll, waiting for the rest of the men to catch up. Suddenly, Funk's men heard a volley of rifle and submachine gun fire. They dove for the muddy tundra. Bullets whizzed by over their heads.

Inexplicably, Major McClanahan had ordered his men to fire on anyone who came over the knoll. The gunfire lasted only a couple minutes. When the firing ended, Lieutenant Funk and twenty-seven other men had been killed; fifty were wounded. Because of the storm, poor visibility, and a colossal mistake by McClanahan, U.S. troops had shot their own men.

Morley and his platoon didn't learn Lieutenant Funk had been killed until a radio message was delivered the next morning. Morley knew Funk personally. They had done some fencing together on the ship, on the way to Kiska, and they had talked quite a bit about their lives. Morley had been impressed with Funk. He thought Funk was smart, athletic, and was bound to succeed in life, if he had survived the war.

Morley was hopping mad about the whole incident when Major McClanahan showed up on the top of Link Hill that morning.

"When he came into my command post, I was completely beside myself to such a point that I told McClanahan that if I ever saw him again alive, I would shoot him down like a mad dog with complete disregard for all military authoritarian rules and regulations. The situation was so intense in so far as I was concerned that this had become a personal matter, which should be settled personally. I felt frustrated and trapped by an order that was foolish to say the least and by an action that was inconceivable in my mind. My men were weeping over the loss of Lieutenant Funk, and I know I had tears in my eyes because I was so mad and sad at the same time. Major McClanahan didn't say a thing. He began to cry as he turned around and went down the mountain. He was never seen again on Kiska."

A burial service was held for Lieutenant Funk and the other victims of the unfortunate episode in Lily Cove. "Everyone who heard the taps blown that morning will live their lives without ever forgetting a moment of that musical farewell to those wonderful men who fell in such a needless way," Morley says.

Eventually, there would be a military and congressional investigation of the incident. But as Johnson wrote in *Sports Illustrated*, "It would be difficult to imagine a sadder combat debut."

Fortunately for the mountain troops, the major snafu at Kiska did not undercut their future value in the minds of military leaders and strategists. The troops had made a difficult amphibious landing in the dead of night, and they had climbed to the top of Link Hill despite heavy fog, driving rain and poor footing. As time went on, it was likely that the mountain troops would be needed in the European Theater, possibly in Nazi-occupied Italy or France.

In the meantime, the mountain troops were kept on Kiska

for a short time to mop up and cache valuable weapons and equipment left behind by the Japanese (the Japanese had evacuated the island before U.S. forces landed). Morley requested to stay on Kiska because he had found a Peale's peregrine falcon, an arctic subspecies, and some bald eagles, and he wanted to study the soils and geologic formations on the island. But the request was denied because the War Department had decided to reconfigure and expand the mountain troops—now called the 10th Mountain Division. Morley, as an officer, was needed at Camp Hale.

Before Morley's platoon left the island, one of his men accidentally shot and killed a juvenile Peale's falcon. Morley preserved the bird and brought it back to Camp Hale on ice. Eventually, the bird was preserved by a taxidermist, and donated to the University of Colorado Museum of Natural History.

One of Morley's first priorities upon his return at Camp Hale was to write letters to the parents of Lieutenant Funk and Sergeant Phillip Johnson, another Kiska victim, both of whom he knew well.

"The letters that I had to write for those who were killed certainly touched the deepest feelings of humanity," Morley says. "The letter to Lieutenant Funk's family was even more difficult to write because of our close association and the things that we all believed in together."

Chapter Four

Triumph and injury in Italy
1944-1945

Training resumed at Camp Hale, Colorado, in the winter of 1944. The newly formed 10th Mountain Division combined the forces of the 85th, 86th and 87th Mountain Infantries—about 12,000 men altogether. All of the men received a new red, white and blue shoulder patch to be sewn on their uniforms. The patch depicted two bayonets crossed like Roman numeral X, over the top of a powder keg.

Morley had a new assignment as a platoon leader, Company L, 86th Mountain Infantry. Lieutenant Nelson's responsibilities were to drill his men on coping with indirect fire, mountain climbing on skis, and winter survival in the snow.

The winter of 1944 was especially brutal, even in Colorado terms. Temperatures remained below zero for weeks on end, and the snow piled higher than usual. The men had to break trail on skis or snowshoes through knee- or even hip-deep snow during training exercises.

In March and April, the 10th engaged in the "D-series" maneuvers, a month-long war-training exercise that would attempt to simulate mountain combat in the winter. Nasty weather, combined with the physical challenge of carrying heavy packs and weapons, made the D-series training exercises a most memorable experience for the men at Camp Hale.

"Military historians now say this exercise held by the 10th Division was probably the most intensive and demanding set of maneuvers in American military experience," Robert E. Ellis wrote in *See Naples and Die*.

Although the D-series began during the last week of March, when spring begins to peel away the ragged edge of winter in most parts of America, the weather turned nasty right away, with temperatures hovering at thirty degrees below zero. All 12,000 men packed their rucksacks with food, extra clothes, a tent and sleeping bag, warm boots, and a shovel. They traveled on skis or snowshoes, and carried a pistol, a rifle or machine gun. Each night, the men dug snow caves, built igloos or pitched tents for shelter.

Ellis, who served in Company F, 85th Mountain Infantry, wrote a letter home to Nebraska on April 15, 1944, describing his experience in the D-series.

"I can say with little hesitation that I have just emerged from the worst physical ordeal of my entire life," he wrote. "Easter weekend was the worst. Saturday night we started on a march on snowshoes and skis through snow that was up to our waists. We hiked until 1:30 a.m., then crawled into our bags and fell asleep exhausted in the snow. We were awoken at 4 a.m., and had to pack in a snowstorm (It snowed every single day we were out.) We then started on a brutal climb to outflank another regiment. To top it off, combined with our lack of sleep, empty stomachs, and extreme cold, the snow became a blizzard. For four hours we climbed in that blizzard. Finally soaked clear through, completely exhausted and almost frozen, I and another fellow fell out. We hiked back to the temporary base camp after rescuing a fellow who had fainted in the snow and taking him to an aid station. I never thought we could make it; you could hardly see fifteen yards ahead."

Ellis wasn't the only soldier who had to retreat, fearing that he might die of starvation or exposure. The Easter blizzard added *eight feet* of snow on top of the robust snowpack. Exactly 1,378 men got sick, nearly 200 men suffered frostbite—a remarkably low number considering the conditions—and 340 men were injured. No one died. A number of men went AWOL before and after the D-series maneuvers.

U.S. Army
Cold and short rations took a toll on 10th Mountain Division soldiers during a two-week survival exercise called the "D-series." Even water, which sometimes froze in the canteens, became a precious commodity.

Morley remembers that many soldiers got sick and complained of lung problems during the D-series. As a company commander who was in top physical condition, he didn't look back on the maneuvers as being life-threatening to him. It was more of a wakeup call to the Army officers that more men would have to be trained to keep the 10th at full strength, assuming that heavy attrition would occur in a real war situation.

"The preparation for the D-series maneuvers required excellent conditioning," Morley says. "But there wasn't enough food for the men, and a lot of them complained of having lung problems of one sort or another. It was good training for the men, however, to master the basic survival techniques in the snow and to learn how to avoid avalanches and learn to climb the mountain with skins or wax, and then put the downhill attachments on and ski down. For those who survived the whole ordeal, it proved to be very successful, and everyone agreed that the mountain infantry was going to be a part of the war effort in World War II."

Still, the 10th waited for a true combat mission nearly three years after the Japanese bombed Pearl Harbor.

At every opportunity, Morley drove to Denver with Forbes Mack and Gordie Anderson to chase girls. That spring, he met a real dandy named Betty Ann Ray, a pretty brunette. She worked for Continental Airlines at the time, and she lived with two other girls.

"Morley and the guys came down to Denver from Camp Hale wearing their fur parkas, and it was kind of a hit," says Betty Ann. "We were sitting in a café, and here comes these two guys and Morley. And he starts right off with this bird stuff."

She mimics Morley's opening line, lowering her chin and her voice, "Well, you know, they call me falcon."

"Oh, well, I know all about falconry," Betty Ann shot back. "I've read Shakespeare. My father always says, 'Anyone's an idiot if they haven't read Shakespeare.'"

Morley was impressed with Betty Ann, and they started to go out on dates. "His idea of a date was to buy some cheese and crackers and we'd go out in his Model A Ford looking for eagles," Betty Ann says. "I remember he taught me to drive a stick shift in that Ford, and I killed the motor at 12,000 feet on Loveland Pass. I thought I would die. But by the time we got back to Denver, he thought I was dandy."

Morley was twenty-eight at the time, and Betty Ann was only eighteen. "As far as I was concerned, he was older than god," she says.

Betty Ann was born in Themopolis, Wyoming, but after her parents divorced when she was five, her mom moved to Burlington, Iowa. "I was a rivertown girl," she says. She grew up and went to school in the Midwest, until she left home after high school and moved to Denver. She worked the swing shift for Continental Airlines, installing automatic detonators in the bowels of military planes.

Soon afterward, Morley introduced Forbes Mack to one of Betty Ann's roommates, named "Beachie." They hit it off immediately, and the two couples started to double-date whenever

Nelson family collection

Morley and friends join in the fun at a Denver ice skating rink.

Mack and Morley could get free time. The girls also drove up to Camp Hale for dances at the Officers Club.

"Very quickly, our relationship went into a very intimate and long-range commitment after spending many hours alone together, probing all the possibilities of the future and the changes going on around the world," Morley says.

At the same time, rumors began to swirl at Camp Hale that the 10th Mountain Division would soon be dispatched to the European Theater or the high mountains of Burma. So while Morley and Betty Ann grew closer together in their relationship, they talked about marriage. Morley wasn't sure if it was proper to get married before going off to war, but Betty Ann she was willing to take the risk. They decided to get married in June, right before the 10th was sent to Texas for more training.

Morley and Betty Ann traveled to her home town of Burlington, Iowa, to get married. They traded vows on June 24, 1944, in the backyard of her parent's home. Forbes Mack was Morley's best man. Betty Ann's sister was her matron of honor. Morley's mother, Agnes, attended the wedding, but his dad,

Norris, was working in the shipyards in San Francisco, so he did not attend. Afterwards, Morley wanted Betty Ann to meet his family, so they decided to spend a short honeymoon in North Dakota.

"It wasn't much of a honeymoon, but it was kind of fun," Betty Ann recalls. "Somebody told me once that you can stand on a beer can and see the whole state of North Dakota, you know, it's so flat. But it was pretty. You know, there's nothing prettier than a well-kept farm."

Betty Ann met the Hoglund family at their farm in West Fargo, and she visited the Nelsons at Steen's farm near Finley. Art Nelson had taken over the farm along with two of his brothers.

Being a social gal, Betty Ann particularly enjoyed the Hoglunds. "They were lots of fun," she says. "Some of them came from Minneapolis. All the Hoglunds were big drinkers. They raised a lot of hell.

"There were all these guys out working their tails off on the farm, and then they'd come in for lunch. They made as much food for lunch as they made for dinner, and they had the same thing. They'd have pork chops and mashed potatoes and corn and noodles and angel food cake for lunch, and then they'd turn around and have the same thing for dinner. I could never get over it. Everybody, everybody, knew how to bake bread and rolls. Morley's mother was a really good baker. She knew how to bake all of that stuff, and was stunned that I didn't know how to do any of it.

"The thing that I never ate more of before or since in my life was strawberry shortcake. And it was topped with fresh whipped cream from the farm."

In June, the 10th Mountain Division was transferred to Camp Swift, Texas, near the town of Bastrop, about thirty miles from Austin, the state capitol. Morley and Betty Ann returned from their wedding and honeymoon in the Midwest, and settled into a small rental house in Bastrop. Betty Ann recalls they had to buy the furnishings for their home at a second-hand store.

Why the 10th had been moved to Camp Swift was yet another source of consternation for the ski troops. Rumors were abundant about a new potential assignment in the Pacific Theater and the European Theater, but still, no real assignment to anywhere. It was a strange transition for the mountain troops to train at such a low elevation and in the heat, amid cock roaches and poison ivy.

Nelson family collection

Morley and Betty Ann exchanged vows June 24, 1944, in the backyard of her parents' Burlington, Iowa home.

On Independence Day, July 4, Morley and Betty Ann received a telegram—just ten days after the wedding. Morley's brother, Norman, had died in an airplane crash on July 2. An American C-78 was carrying Norman and three other Air Force pilots from Shoreburg, France, to England. The plane had not been shot down. The pilot became disoriented as he flew over some low-lying mountains in the clouds and crashed in a churchyard near Portsmouth, killing everyone on board. Norman, who had recently been promoted to the rank of captain, had flown more than 135 combat missions in a P-47 fighter. He was just twenty-four.

Morley immediately took leave and flew to Fargo to be with his family.

"No one is ever ready for a telegram that carries the message that your brother has been killed," Morley says. "The trip back to North Dakota to be with my father and mother was one of the saddest things that had ever occurred in my life. The finality of it was almost incomprehensible to me. The wrath that rose in my body over the war and having to pay in this most personal way was almost too much to bear."

Betty Ann remembers the situation this way: "When we got word that Norman had been killed, Morley was a basketcase. We had enough money for Morley to fly there, so he did that, and I took the train. I remember getting on the train and it was full of soldiers, and I had only one outfit with me, the one I was wearing. I mean, we just left. And I didn't have any money. I remember being on the train for two days, and I was dying of hunger, but I was too proud to ask for help. Finally, someone forced me to take a box of chicklets, and that's all I had to eat for two days."

When Betty Ann arrived at the Nelson home in Fargo, the scene was very emotional. "Morley's dad was just a wreck. I can remember Morley seeing his dad, and putting his arms around him, and I could cry just thinking about it, I felt so bad for him. It was a very hard time for the family. Morley's mom was so upset, she just stayed in bed and cried."

"Norman's death so completely demoralized my mother that she could not accept her life any longer, knowing she had just lost her son, and I was just about to go to war myself," Morley says.

The Nelsons had a memorial service for Norman, who was buried in the Riverside Cemetery in Fargo, adjacent to the Red River. Afterward, the family went to the Hoglund's farm in West Fargo. "The minute after they had the memorial service, everyone got drunk," Betty Ann says. "That was the Hoglunds for you. They went dancing. They all went out and whooped it up. I can remember going out dancing and I was still wearing that same outfit."

Several days later, Morley and Betty Ann took the train back to Bastrop, Texas. Military training continued for four months, until the 10th received word that it would be sent overseas. Major General George P. Hays had been designated as the division's new commanding general. The men liked General Hays, who had been commanding U.S. forces in Normandy. Hays told the men in the 10th that they had quite a reputation as elite mountain troops. Now they would finally have a chance to prove themselves in a combat situation. Hays promised that the men could anticipate a battle that would be "long, hard and bloody" and it would be "the greatest adventure of our lives," wrote Ellis in *See Naples and Die*.

Just before Christmas 1944, Morley left Texas for Newport News, Virginia, where the 10th would board ships and sail for Naples, Italy. Betty Ann drove back to Iowa in a second-hand Chevy to stay with her family while Morley was at war. She told Morley before he left that she had missed a period and was probably pregnant.

"I remember thinking at the time that my family was going to have a fit," Betty Ann says. "My stepfather said, 'he'll go overseas and get killed, and there you will be with a baby in your arms and no way to make a living.' He was absolutely right, but that's what happened in those days. You got married, and you got pregnant. I wondered why it took me three months to get pregnant."

When Betty Ann arrived home in Iowa, her mother took her down to a medical clinic for a "bunny test," a pregnancy test. "They drew blood from you and injected it into the bunny, and if the bunny died or whatever, you're pregnant," she says. Her test turned out positive.

Morley assumed that the baby would be born before he returned home, if he returned at all. As he sailed across the Atlantic, Morley couldn't help but think of his fallen brother, General Hays' prediction of a long and bloody battle, and how he would react in the heat of battle. "If this is the way it's going to be, maybe it's going to cost me my life, too, but if it comes down to that, I'm going to take as many of them down as possible, right up to the last second of life," he thought.

When the 10th landed in Naples, Morley (Lieutenant Nelson) was the executive officer, or second in command of Company C, 87th Mountain Infantry. The 10th's mission in Italy was to seize the high ground from the Nazis in the Apennine Mountains, starting with a raid on Riva Ridge and then Mount Belvedere, south of Bologna. If the 10th could take the high ground, it would force the Germans into a retreat, and drive them out of Italy—the last western European nation still occupied by the Nazis.

If the Allied forces recaptured Italy, the Germans were doomed.

In Naples, Morley was stunned by the look on the Italians' faces—many people were destitute without food or shelter. "I had never seen people whose eyes stared wide and looked like a scared animal in a trap," Morley says. "The desperation in their outlook was so plain to see, as if to say, will I have enough to eat today to carry on?"

A dramatic example of the plight of the Italian people occurred when one of Morley's drivers, a fellow named Redmond, disappeared while Morley went to battalion headquarters for an hour to learn more details about his company's mission. When Morley returned to the car, Redmond was no where in sight, so he honked the horn. "Redmond stuck his head out of a two-story window with a very handsome looking Pisan girl next to him," Morley says. "She waved down and Redmond waved down and he says, 'I'll be right there.' Once in the car, I couldn't help but to ask the question, how in the world in one hour's time did you meet this girl and end up in her room when you don't even speak the language?"

Redmond replied that he met a little boy on the street while he was waiting and gave the boy a couple cigarettes and some candy, and asked the boy, "Seniorita?" And the boy said, "Oh, si," and told Redmond to wave a couple cigarettes out the window when girls go by and say "Fice Fice."

So Redmond held his cigarettes out the window and said, "Fice Fice," and a girl called out from the second-story window and said, "Si si finitiqua" and motioned for me to come upstairs. The woman was so desperate for food that she performed a sexual trick in exchange for his cigarettes, which could be traded instantly for food.

While Company C was in Naples, Morley could see that the Italians would appreciate any leftover military food. He noticed that they raided the military garbage at night and swept it absolutely clean. So the next day, Morley and his soldiers set up a three garbage can system, separating liquids into one can,

Denver Public Library
Tenth Mountain Division reinforcements head for the front during the battle for Mt. Belvedere in Italy's Apennine Mountains.

sloppy food into another, and dry food such as bread and crackers into a third can.

"The next morning there was a line several hundred yards long, little boys, men of all ages, women and young girls, all carrying pails or scoops or a bottle, and they went through the garbage line to get their food and went away to eat it. A man came up to me at the end of the line and said something in Italian that I couldn't understand, but finally I realized he was saying, thank you. I'll never forget the man's desperation and the sincerity in his eyes."

In the evenings, many of the American soldiers couldn't resist the open advances of Italian women, who were eager to share their bed in return for a chocolate bar or cigarettes. Knowing there was no practical way to stop the men, Morley and other officers made sure there were plenty of condoms available to prevent sexually transmitted diseases. They needed the men to stay healthy for battle.

On February 17, the first assault in the Apennine Mountains began, on the east slope of Riva Ridge, which was very steep and rocky. Several battalions from the 86th Mountain Infantry were selected to make the assault in the middle of the night, using rock-climbing gear. The objective was to capture the top of Riva Ridge and drive the German defenders off the mountain. Strategically, Riva Ridge was an important position to hold, because the Germans could shell enemy troops in the valleys and surrounding mountains from the ridgetop. Morley's company was kept in reserve for the initial attack on Mount Belvedere, located just to the east of Riva Ridge.

A ninety percent casualty rate was expected in the first attack, if the Germans discovered the climbers as they scaled the craggy cliffs. But the Nazis never anticipated that any soldiers would attempt to climb the east face of the mountain, especially at night. The sneak attack worked far better than expected: the 86th seized the top of Riva Ridge in one night. The American casualties were far lower than anticipated, only four percent—six men killed, and twenty-seven wounded.

On the evening of February 19, the initial assault of Mount Belvedere began at midnight. Morley's Charlie Company and the 87th Mountain Infantry planned to scale the west slope of the mountain on foot, without climbing gear. The 85th would climb directly to the summit of Belvedere, also on foot, and the 86th Mountain Infantry would attack the east slope of the mountain on foot. If all went according to plan, the American troops would take the Germans by sunrise.

The sneak attack on Belvedere had been planned for a time when the moon was new and the sky black. Each company had a German shepherd dog in the lead to sniff the cool mountain air cascading down the mountain for the presence of Germans hiding in foxholes. The dogs helped the American soldiers avoid trip wires near the bottom of the mountain. The dogs would point when they smelled the Germans, hidden in foxholes, but they didn't bark.

"I remember adjusting my helmet and giving the order for fixed bayonets before the march started up the mountain," Morley says. "There was a great deal of fright in everyone eyes, and the first moments of combat turned out to be extremely fierce. There was a burst of blue machine gun fire less than fifty yards from the first escarpmental rise on Mount Belvedere. The first blast of fire sent bullets going in all directions, and it took down several casualties on each side of me. I hit the ground as everyone else did, looking around and I could see nothing except the blue spurt of the machine gun firing some fifty yards off to the right and above us."

Morley had a short conference with Captain Edwards, the commander of Company C, and they decided to go around the fringe of the machine gun, split up and try to get above it.

"Right away, a very intense firefight started, and Captain Edwards was hit in the leg and went down immediately. I was over on the edge of what was the command post of the company, attempting to move up the hill and firing back at the machine guns. I told my men, all right gentlemen, let's wait until he fires, lay low and shoot at the blaze of light. The next time the machine gun opened fire, there were a good many Browning automatic machine guns that fired back and we threw (shots from) M-1s and the officers carbines'—all fired at the flash of light and silenced it. However, Captain Edwards was already down, and the medical units picked him up and carried him out of battle."

Now Lieutenant Nelson was in charge of C Company by himself. Later, he learned that he had been promoted to captain, and officially placed in charge as company commander. The battle up the mountain continued, and Company C advanced, the men now feeling the tremendous initial shock of seeing their friends go down around them. There was only one thing to do: Continue climbing the mountain and get above the enemy fire.

"The next engagement came when there was a bigger unit of Germans that emerged from a dugout hole, and they came out into the dark to face combat with bayonets and rifle fire at extremely close range. The Germans fired pistols at us, and we fired back with rifles and fought hand-to-hand with fixed bayonets. The heat of battle geared up, and the rage of my men who were still alive began to come through. Anyone who moved in front of us and fired a gun was immediately killed by the ski troopers, who aimed at the flash of light.

"In the combat that took place, each man suddenly degenerated into a fierce animal. All of us would need instinctive fast reactions to survive."

For Morley, a flash of memory came back as he stared into the chaos of enemy fire on Mt. Belvedere. It was Steen Nelson, out in the field, facing that ornery bull.

Steen's eyes got real big and he stared right at that old bull with complete concentration. That bull came charging right for him, and at the last second, Steen side-stepped the bull, and swung that post at the bull's head, breaking off one of the horns. Jesus, that bull hit the ground so fast he didn't know what hit it.

And Steen says, "Johnny be quick or dead. You've got to be quick in a life-threatening situation, or you're dead. It's that simple."

As Charlie Company and other 10th Mountain units continued to advance up the slopes of Belvedere, it appeared that the Germans were starting to retreat. It was imperative for the American troops to hold their positions. The mountain had been taken twice before by the Americans and lost. The Germans knew it was a critical outpost for controlling the Po Valley and that region of Italy. This time around, all of the years of training were paying off for the 10th.

"The Germans were attempting to run, but they were either shot or hit with bayonets or anything we else we could throw at them," Morley says. "The horrendous use of the butt stroke—the stab of the bayonet—is one violent maneuver that no one in battle will ever forget. The butt stroke often hit a man under the chin and that would (make) his jaw to go up into his face and his eyes to pop out of his head. Once you see these things, you can not think about them again or you'd have a psychological problem forever."

In a strange paradox, Morley also felt a strong sense of elation when the tide turned on top of the mountain.

"Suddenly it was obvious that we really had the Germans on the run. We had bypassed many of the German bunkers and the foxholes in the side of the mountain. And I began to feel one of the great highs that one can feel in the world, even having seen the horrible things that I had in climbing that mountain in the

Denver Public Library
Artillery shells hammer German positions during the 10th Mountain Division advance through the Apennine Mountains of Italy during the spring of 1945.

fury of battle. I began to feel an intellectual high that comes only with such a wild outlandish opportunity of being involved with life and death to such a point where your ability to shoot and to move reaches a new height of perfection. The Germans were running in front of me and I shot my rifle at them and I felt a high overcome me as the Nazis went down in front of me. I thought, OK you bastards, this is where you belong."

All at once, the morning light reached the summit of Mount Belvedere, and Morley and Charlie Company stood on top of the mountain, knowing they had done it. They had stolen the high ground. The ridgetop assault toward the Po Valley had begun in earnest. Now the 10th had to hold the position and continue to push to the north.

"While we prepared for the counter-attack, we searched for a defendable position on the forward side of the mountain," Morley says. "In doing so, we came across a big bunker hidden under the snow on the north slope, and three of my men stepped out, Sergeant Ureslen, Sergeant Kennewick and Sergeant Stolen. As they came out, all at once two Germans came out of a deep bunker in the snow. 'Nix, boom comrade,' the Germans said, with their hands over their heads, as if they were going to surrender, and the minute they got out of the hole, they stepped to the side, and German storm troopers behind them opened fire with machine guns.

"Sergeant Stolen was looking off to the side, and he was hit immediately. A bullet took out the ridge of his nose and both eyes. Sergeant Stolen screamed and staggered over to the south side of the ridge, trying to get away. My men shot all the Germans, we ran over to assist Sergeant Stolen, and we saw one of the most horrible sights any man will ever see. Sergeant Stolen was trying to go down through the snow with blood coming out of his head and face, and he'd fall down and get up and go some more, and the men were so horrified by his injury that they couldn't gather up the strength to help out. Sergeant Stolen fell down several more times before a few men grabbed him and called the medics, who gave him a sedative so we could take him to the hospital for treatment.

"This whole episode resulted in a story going through the company that no one would ever forgive the trickery that had been used by the Germans. Every man in Company C agreed that the next time a Kraut said, "Nix, boom comrade," they would take careful aim and shoot him between the eyes to be sure that there wasn't another German behind them, ready to fire at our troops. So from that day on, for the rest of the war, Company C never took a prisoner."

The German daylight counter attack came quickly. But the 10th Mountain and Morley's Company C were ready to defend the top of the mountain and the ridge. Morley had time to make a reconnaissance trip around the top of the mountain, and he placed men with machine guns in a position where they could shoot down the slope several hundred yards. Therefore, it would be almost impossible for any Germans to pierce the American position, especially in daylight.

"This came as a great surprise to the Germans," Morley says. "When the attack started, the Germans came running up the slope with a lot of fanfare and bravado, but when my men opened up the machine gun fire, the Germans had to break

Nelson family collection

Captain Morlan Nelson in the Italian Alps, 1945.

ranks and they never even came close to making an attack with bayonets, which they hoped to use to retake Mt. Belvedere. We felt a tremendous emotional high, knowing we had easily stopped the attack. We had control of the high ground, at least for the moment."

Reinforcements came in to hold the position secured by the 10th, and the company commanders got together to plot their strategy for the next move. Morley and the other commanders met with the Air Corps to work out a plan that would help eliminate German tanks, which were well hidden in the Italian countryside, in a position to fire on the ski troopers in the mountains.

After taking Mt. Belvedere, the 10th continued to push north, over the top of Mt. Gorgolesco, Mt. Della Torraccia and other peaks and ridgetops in the Apennine Mountains. The strategy was to attack German fortified positions at night, using small arms, hand grenades and fixed bayonets. As the American troops advanced, the Germans counter-attacked with Tiger tanks from the west flank of the mountains. The tanks had deadly 88 mm guns that could shoot shells at 4,000 feet per second. Plus, the Germans knew the exact coordinates of the American positions, because the Nazis had held those positions for several years.

"The Germans had extremely accurate knowledge of the mountain range," Morley says. "You could hear the blast from one of those tanks, and a good many seconds later you would hear the explosion on the mountain. The shells hit with such velocity that it jarred your body right down to a point where you had to shake and get a hold of yourself, and then push on."

Tenth Mountain commanders stayed in communication with the Air Corps as they fought on the ridgetop. "When we got close to the tanks, I'd call the Air Force with their position, and the fighter planes would come down and knock 'em out. That was a wonderful thing," Morley says.

The casualties were heavy at times, when the German tanks scored a direct hit on the American troops. It wasn't uncommon for Morley and his men to see their friends blown to bits by the fire or shot up by a machine gun. He and his fellow soldiers found ways to rationalize how they felt with death all around them. Morley remembers talking about that issue with Sergeant Ureslen as they sat under a tree, eating lunch.

"I stopped eating for a second, took a sip of water, and looked up into the tree, and damned if I didn't see a hand hanging there with a sweater on it," Morley says. "We got up and looked closer at the tree, and it was obvious that somebody had been hit by an eighty-eight right there and blown to bits, scattering body parts throughout the tree. We sat down and continued to eat our lunch and said, well, he doesn't have to worry anymore.

"That was the way we rationalized losing our friends. There would be no tears coming to your eyes because you're still facing the crisis of whether you're going to live or die every second. But the thing that went through everyone's mind was, OK, he doesn't have to worry anymore. There was a kind of relief, knowing the man's spirit was going to rise above it all, and this was his last salute."

After several weeks of fighting, Charlie Company received a break, a little rest and relaxation. At the same time, Morley received a surprise visit from a boyhood friend from Fargo, Chuck Putney, a first sergeant in the Air Corps. Putney was the same age as Morley's late brother, Norman, and they had played together as kids.

Company C took its rest period in a fortified farm house near the top of a mountain. Putney arrived at the house before Morley, so he had a chance to ask the men how they liked Lieutenant Nelson The men told Putney, "Oh he's a great combat officer, but he's not very good at meeting all the requirements of being an executive officer, he's not a spit-and-polish type of guy."

That description fit Morley to a T, Putney said. Morley was always a lot more comfortable in jeans and a flannel shirt, than in his Sunday dress.

When Morley arrived at the farm house, he was quite surprised. "My god, Chuck, what the hell are you doing up here?"

"I heard you were on R&R, and I was able to get a command car, and I had the time, so here I am. How the hell are you?" Putney replied.

The two men went over to a quiet spot to speak privately, and they talked about the loss of Norm and old times. Putney had been stationed in North Africa, and had never been directly in the line of fire, so his visit to the Italian farm house made quite an impression, especially when German artillery shells began to shake the place. Some of the shells landed within forty feet of Putney's jeep, but fortunately, it was still driveable. Morley sent him off after nightfall and instructed Putney to go very slow with his lights off. He made it out just fine. Later, he sent a letter to his wife about visiting Morley, and the letter was printed in the *Fargo Forum.*

Company C continued to keep the Germans on the run. Morley and his company held Hill 913, as it was called, during a German counterattack, just to the northwest of Mount Della Spe. During the night, Morley checked on his men in their foxholes and was pleased to find a high level of morale.

"It was one of the most endearing things that I had ever experienced. When I showed some concern for how they were doing, trying to live through such a stressful situation, the men assured me that they were proud of being there, and yes, they were frightened, but they were willing to go on. We're going to be with you, Captain, all the way, they said. It was such a beautiful shining thing to see the men coming through under such difficult conditions."

As Company C pushed north, they ran across a family living in a farm house in the mountains. Morley went inside and checked for any occupants, and a man and his family were huddled inside one of the rooms of the house. They were in dire need of food.

"I told the man, well, I'll share my mountain ration with you," Morley says. "And he gratefully accepted it. So we sat down and ate our meal together, and when we got through, he offered me an apple. In my poor Italiano, I said no, I don't need an apple when you need it so bad. But he insisted that I eat it, so we both ate an apple. When he finished his apple, I noticed that he ate the whole thing, including the core. And I thought, well, I can't throw the core away after watching him do that, so I ate the core. It's kind of an odd thing, but ever since that day, I always eat the apple core when I eat an apple, and I think of that man and his family."

By mid-April, the 10th Mountain was poised to take out the last high-mountain outposts from the Germans, just above the Po River Valley. The spring snowmelt had begun, and the mountains were muddy under the snow.

Morley joined an advance party of commanders to scope out a plan of attack for taking the high ridge. The Germans were firing on the mountain as Morley and an officer from Great Britain huddled in a conference, trying to figure out what to do.

"Boom!" A shell exploded directly behind Morley and the English officer, sending them flying in the air. Two pieces of shrapnel went through Morley's arm at the same time.

"Terrible stinging wounds went through my body as many big fragments cut my right leg and went into my back," Morley says. "I landed on top of the other officer, who had been hit in the neck and was bleeding profusely. I couldn't move my left arm because I was lying on it, and my right arm seemed to be paralyzed."

A soldier rolled Morley off the English officer and called for medics. However, none were available because there had been so many men wounded in the German counter-attack. So an American officer ordered German prisoners to place Morley in a sleeping bag, roll him up in a tent, and carry him down the mountain to an aid station. An American officer gave Morley a shot of morphine to dull the pain.

At the aid station, a doctor removed some large pieces of steel from Morley's back, sewed up the wounds and placed bandages over them to stabilize him until he reached a hospital. They also gave the patient a couple ounces of Scotch whiskey to dull the pain.

The only road leading out of the area was still under heavy enemy fire. It was a risky drive for the ambulance. The driver and his assistant placed Morley in a stretcher in the back of a Jeep next to a man who had lost his leg and was in a lot of pain. They kept the Jeep lights low to keep from attracting the attention of German gunners.

"About half way up the hill, there was a tremendous barrage of artillery fire," Morley says. "One shell hit near the front wheel of the Jeep, and the front wheel bounced through a cut in the side of the road, but then the rear wheel hit a big hole in the road, and we were thrown from the jeep. We slid down the mountain in our litters, flipping end over end, until we landed in some brush and trees, not too far from the road. The man who had been next to me was screaming in pain and yelling for help."

The driver and his assistant immediately ran down the slope to retrieve the men. They dug the Jeep out of the hole and continued on to the field hospital. When they arrived, there were about twenty men lying on cots outside, waiting for their turn to go into the operating room. Doctors were constantly performing triage, moving the men around according to the severity of their wounds. A few men died waiting their turn. Morley had so many wounds and big metal fragments in his body that he was moved toward the front.

Morley was injured on the bloodiest day of battle experienced by the 10th Mountain Division since the fighting in Italy began. More than 500 men were wounded or killed. The Germans tried everything to stop the Americans from taking the last high stronghold above the Po River Valley, but they failed. Lieutenant Robert Dole, a member of the 10th who later would serve as Senate Majority Leader and was a candidate for president of the United States, was injured that same day.

After surgery, Morley faced mixed news. The good news was the shell fragments that hit his left arm passed through muscle tissue below the bone and exited the skin. They were considered "superficial" wounds. But he also had four deep cuts in his back, about 32 inches of incisions, to remove most of the shrapnel. Some shrapnel penetrated his back so deeply the doctors decided it was best to leave it in place. Morley's right leg wounds required skin grafts—so many that the doctors openly discussed amputating Morley's leg below the knee.

"First they said they might have to cut it off, then about four days later, they said no, the skin grafts were going to work and they could save my leg," he says. "I did have permanent nerve damage in the bottom of my right foot. I couldn't put my right heel on the ground, and the sensory nerve had been damaged to the point where I couldn't feel the bottom of my foot."

About three weeks later, Morley began physical therapy in hopes of getting his right foot to work properly again. As he walked into the therapy room, he noticed all kinds of pictures of smiling veterans on the wall, with thank you notes to the therapy nurse. She must be a really nice woman, Morley thought.

"All right, Captain Nelson, stand up, let me have your crutches, you are a big 10th Mountain soldier. Now I want you to walk across this room," she said.

"But I can't put any weight on my right leg," he replied.

"Well, you know you have to get over this, so give me your crutches and let's see what you can do."

Morley took two or three steps, much to his surprise, but by the fourth step, the pain was too great and he fell on his face, catching himself with his arms as he hit the padded floor.

"OK, Captain, don't take it too hard. Get up and try it again," the nurse said. "Come on, let's see you walk."

Morley tried to walk again, and the nurse could see that he

U.S. Army photo

Members of the 87th Infantry Regiment march through Rome after winning their battle with the Germans in Italy.

couldn't get his right heel on the ground. So she placed him on a therapy table, flat on his stomach.

"Well, now, I'm going to take your foot and I'm going to push your toes up as if you're walking on the heel. This might hurt a little bit."

She yanked his toes upward, causing a fair amount of blood and fluid to shoot out of his heel. She repeated the procedure several more times, and said, "OK, that's fine. We're going to work on this for several more weeks until we can get that heel down to where you can walk with a cane instead of crutches."

The nurse had inflicted a great deal of pain on Morley, cranking his foot up and down. But after losing the fluid, his foot felt better. He looked back at the photos on the wall and realized that the nurse was tough-as-nails for a reason. Instead of showering the American soldiers with sympathy, as the bedside nurses had done, she had to be tough and force the soldiers to push through the pain of physical therapy. The first few days were the hardest part of all. Morley had survived Day One.

As he left the therapy room, he said to her, "You're as much a psychologist as a physical therapist, aren't you?"

"Shhhhh. That's right, that's part of the game here."

And Morley realized that if he was going to get back to leading an active life in the outdoors, as he loved to do, he would have to stay with the therapy program and work hard at it.

Several days after Morley was injured, he urged the medical authorities to send a telegram to Betty Ann and his parents and inform them on his condition. He was going to be OK, and would be home soon. Actually, Morley thought that he might be able to recover to the point where he could return to the front lines of battle in three weeks. The doctor told him he'd be lucky if he could walk in three months.

As it turned out, the 10th Mountain Division captured the last of the mountain positions in the Apennines, and marched across the Po River Valley to the edge of the river by April 22, a week after Morley was wounded. The troops crossed the river in wooden assault boats and kept the Germans on a fast retreat to the north.

On April 26, the 10th marched into the city of Verona, which had been finally liberated from Nazi occupation. This was a key victory.

"It was important to take this city quickly to seal off the main escape route for all of the Germans who had been bypassed in the recent actions, and also to trap most of the enemy units located all over northwest Italy," wrote Doctor Albert H. Meinke Jr., in his book *Mountain Troops and Medics: Wartime Stories of a frontline Surgeon in the U.S. Ski Troops.*

Just over a week later, on May 2, the Germans surrendered in Italy. The relentless pursuit of the Germans by the 10th in the Apennine Mountains had broken one of the last strongholds of the Nazi troops.

But the real killer blow was dealt by Adolf Hitler himself. As the Russians approached Berlin, Hitler hid in an underground bunker and married his longtime mistress, Eva Braun, on April 29. They committed suicide together the next day. On May 7, the Germans surrendered. Hitler was dead, and the war in Europe was over.

Morley, meanwhile, was still undergoing therapy at the 45th General Hospital in Naples. He was very thin, weighing only 140 pounds for a muscular guy who stood 5-foot-11. But he was getting better. He was walking with only one crutch now, and he started to get antsy. So Morley was quite excited when his buddy Sergeant Chuck Putney showed up at his bedside in mid-May. Putney had a car, and he wanted to take Morley for a drive.

Up to that point, Morley and other soldiers in the hospital who were mobile could go down to a little town square several blocks away to listen to Italians sing and play music. They weren't supposed to be gone for more than a few hours at a time.

But Morley didn't care at this point. He told Putney, "I'll put my Eisenhower jacket over my pajamas and my kimono, put on my overseas cap, and we'll put my crutch in the trunk and go for a drive."

No one stopped them in the hospital. When they drove past the guard by the hospital entrance, they gave a salute, and they cruised down the road. Putney didn't have a map, so they just followed the road signs and toured around the country. One of the first places they went was to Pompei, where they discovered a most peculiar museum of sexually explicit paintings called the House of Vetty.

Morley was quite taken by the art in the museum, which depicted the sexual behavior of animals and people in mythical times. "We were more amazed each time that we went to a new room because they depicted sexual scenes between lions and donkeys and all the positions of humanity on each one of these walls, all of them very beautifully done, but very much a surprise to anyone from the United States who had never made a special point of looking into all the peculiarities of history," Morley says.

Putney drove Morley back to the hospital and promised to come back a week later to take another drive. Putney, who served in the Air Corps in ordnance and aviation supply, took Morley up to his base camp, in the town of Civitavecchia, about fifty miles north of Rome. They stayed up there for a couple days and went back to Naples.

So far, Morley hadn't gotten into any trouble for leaving the hospital, so they went on a third foray to a cliff that Morley had spied out the hospital window. He thought he could see a hawk nest in the cliff.

Morley brought a climbing rope that he had in his personal bag in the hospital, and he placed it in a small pack. They drove to the cliff, and Morley spotted a nest in the cliff wall from the heavy presence of white bird droppings. Morley decided to climb up to the top of the cliff and rappel into the nest area, even though he still had a heavy bandage on his right leg, covering the wounds.

"I threw the rope around a tree at the top of the cliff, tied it off, and rappelled down the face of the cliff, using a quick rappel rope technique (no sling around the legs and waist). The trip down the cliff was very easy, because I didn't need to use my injured leg at all. Inside the nest, I saw several European kestrels. They are a much bigger bird than the American kestrel, and a most interesting bird. I reached into the shallow nest and placed one of the juveniles in my rucksack. Chuck joined me at the bottom of the cliff, and took a picture.

"Just as I sat the bird back into the pack, I was so happy about having this bird that I wasn't paying attention to my footing, and when I stepped back on my right leg, it didn't hit level ground and I fell over backwards, and a rock hit my leg from underneath and broke the bandage on my knee."

Fortunately, Morley was not badly injured and Putney took him back to the hospital, along with the kestrel. Morley made arrangements with a hospital employee to keep the bird in his office, and they found some fresh meat in the cafeteria for the bird to eat.

Morley expected he would have a harder time explaining the black burn on his soiled pajamas to the nurses. "They said, Captain Nelson, where have you been, the house of ill repute? They were flabbergasted to see these peculiar burn marks in the groin area of my pajamas, but they didn't really want to know what had happened. So I just laughed and told them I've been doing a little climbing on a nearby cliff to check out a

Nelson family collection

Recovering from bullet and shrapnel wounds, Morley sticks out his tongue at the photographer before going AWOL with Sergeant Chuck Putnam to search for raptor nests in Yugoslavia. The episode almost cost Morley his Silver Star.

hawk. They could tell that my bandages had been torn up, so they fixed me up with some new dressings, and brought me a new set of pajamas."

A week went by and Putney arrived for the next adventure. They decided to go back to the Po Valley and check out Brenner Pass, at the Yugoslavian border. They were gone several days. When the commanding officer of the hospital inquired about Nelson's whereabouts, the nurses said Morley had been spending a lot of time at the library, but they really had no idea where he was. So the officer put out an order for Nelson to be picked up.

Since the war was over in Europe, no lines of communication were available for searching for a missing soldier. Plus, they had no idea that he was in the northern end of the country, a long ways from any existing military outposts. Morley and

Putney had a good time cruising around the countryside, taking pictures and swapping stories.

When they returned to the 45th Hospital, Morley was summoned to the office of the commanding officer. "We have never heard of an officer or a soldier going AWOL from a hospital for as long as you've been gone. You are way out of line," the officer said.

"I've fought your damn war and I just don't give a damn what you think," Morley shot back. "I'm a captain in the 10th Mountain Division, I fought on the front lines of the war, and I'm not going to take any shit from a rear-echelon hero like you."

"Well, you should be court marshaled for breaking all the rules and causing everyone to worry about your welfare," the officer said.

Putney says neither he nor Morley thought they were doing anything wrong. "Morley just wanted to go for a ride," he says. "As far as I was concerned, I wasn't thinking much about the authorities. The war was over. The military has lots of rules and regulations that aren't convenient, and this was one of them."

The real reason the hospital commander had been looking for Morley was that a special ceremony had been planned for Morley to receive a Silver Star for his leadership and bravery in combat. The ceremony was scheduled to take place in five days, and the officer finally cooled down enough to forget the AWOL business and let Morley receive his Silver Star. However, he told Morley that he would transfer him after the ceremony was over, and he could finish his rehabilitation at a hospital in Marina de Piza.

On the day he received the Silver Star, Morley learned that Sergeant Ureslen had nominated him for the commendation. Ureslen had been killed the same day that Morley had been wounded during the German counterattack. Morley wept during the presentation, thinking of his friend and so many other men who had been wounded or killed in the Apennine Mountains.

During 114 days of combat, 992 men from the 10th Mountain Division died and 4,154 were wounded, according to Hal Burton, author of *The Ski Troops*..

After the ceremony was over, Morley went to visit other men from the 10th who were in the hospital, recovering from serious wounds and injuries.

"I went back to these men with tears in my eyes and let them know that it was all right to receive the Silver Star but it was almost too much to bear, knowing the wondrous strength of character that all the men showed in every battle," Morley says.

As one soldier from Company C told Morley, "Well, Captain, you know you have to accept that award for all of us."

The leadership and bravery Morley displayed in Italy was quite typical of the company commanders and the men who fought in the 10th Mountain. Perhaps the greatest compliments came from German and English generals. Generalleutnant von Senger, who signed the surrender in Italy, said "The 10th Mountain Division had been my most dangerous opponent in the past fighting," according to Ellis in *See Naples and Die.*

In William Johnson's article in *Sports Illustrated,* he quoted British Field Marshal Harold Alexander. "The only trouble with the 10th Mountain Division was that the officers and men did not realize that they were attempting something which couldn't be done, and after they got started they had too much intestinal fortitude to quit. The result was that they accomplished the impossible."

On August 16, 1945, Morley finally was released to return to the United States. He boarded the Navy hospital ship *S.S. Exchange* in Marseilles, France. The ship contained doctors, nurses and orderlies to take care of the 2,500 wounded men onboard.

Morley found his mind spinning at the thought of returning to Betty Ann and—quite possibly—a new child while he reflected on the horrors of the war, the success of the 10th Mountain Division, and the loyalty of his men. For the pervious eight months, he'd been living a dark existence in a totally foreign place. There hadn't been room in his mind to focus beyond the immediate task in front of him, the responsibility of leading Company C, killing Germans and trying to keep American casualties to a minimum. In the back of his mind, he knew he could be killed at any moment.

The reality of it was, in the heat of battle, he had almost completely forgotton everything else. Now that he was headed home, he had space in his mind to start thinking about Betty Ann again. Would he have a boy or a girl? Would his right leg heal to the point where he could climb mountains and ski and fly his birds, and do all of the things he wanted to do in life? Morley stared at giant swells in the foaming gray seas, as the ship sailed across the Atlantic, and the realities of life in America flowed back into his mind.

Chapter Five

Back to work
1945-1948

The *S.S. Exchange* arrived in New York Harbor August 25, 1945, on a sunny and breezy morning. Standing on the deck of the ship, Morley was struck by his first exposure to the turquoise Statue of Liberty, standing proud on a concrete platform, amid all of the ships in the harbor against the backdrop of high-rise buildings in New York City.

"I was very impressed," Morley says. "To see the Statue of Liberty there in the harbor made me think back to all of the men I'd lost in the war, and the loss of my brother. It also made me think about our country, and all of the Allied forces that stood for liberty and freedom and decency for all of the world in the war. It was a wonderful feeling to see it and be a part of it."

Within hours of arriving in the United States, Morley received a telephone call from his wife, Betty Ann, who was nine months pregnant. Her due date was in five days, and she was 1,000 miles away in the Midwest!

Initially, the Army refused to release Morley, but Betty Ann had a plan: "Tell the superiors that I have uremic poisoning," she advised, a deadly malady. It worked. Morley found a seat on a TWA flight, and flew to O'Hare International Airport that afternoon. Betty Ann's family had moved to Oak Park, Ill., a suburb of Chicago. Since she didn't know how to drive very well,

Betty Ann's mother insisted that the family's gardener take her to the airport to meet Morley.

"I remember that I couldn't find him in the airport, so I went to call my family, and then all of a sudden, Morley saw me in the phone booth," Betty Ann recalls.

They gave each other a big, long hug. They were overcome with emotion to see each other, especially in light of Betty Ann's condition and Morley's injuries.

At Betty Ann's home in Oak Park, her mother served a big dinner. They all were pleased to see Morley, who regaled everyone with stories about the war.

Right after dinner, Betty Ann started to feel labor pains. Morley jumped in his old car and fired it up. Betty Ann had taken care of their Chevy sedan while Morley was away. "I had the car oiled and greased, but I never put water in the radiator—I didn't know you were supposed to do that," she says, laughing.

Before they pulled out of the driveway, the car was overheating, so they transferred to Betty Ann's stepfather's car, and hurried to West Suburban Hospital in Oak Park. Upon arrival, hospital attendants placed Betty Ann on a wheeled stretcher and whisked her into a delivery room. Back in those days, fathers weren't allowed inside.

"Nobody told me it was going to hurt," Betty Ann says. "They all sat outside and listened to me scream."

It was a healthy boy. Norman Wendell Nelson, five pounds, three-fourths ounce and eighteen inches long. He was named after Morley's late brother, and he took Morley's middle name. The doctor predicted that Norman would be a lean lad, which proved to be absolutely correct. Betty Ann nicknamed him "Eager Beaver" because he had been so active in the womb, kicking and rolling around. The nickname, shortened to "Beav," stuck.

"I felt so lucky to have a son right after I got home from the war. He looked like a Nelson," Morley says. "It was a wonderful feeling to experience the joy of childbirth."

While Betty Ann remained in the hospital for 10 days, Morley drove to North Dakota to go bird hunting with some childhood buddies. With his bum right leg, he wasn't able to walk very briskly in the farm fields chasing pheasants, but Morley didn't care. It was fun to get reunited with a few of his hometown buddies like Chuck Putney and Dick Brown.

Morley arranged to return to his job with the Soil Conservation Service in Salt Lake City. Army officers tried to convince Morley to consider a career with the military, but he wouldn't have any part of it.

"I just had to say no," Morley says. "I had my old job waiting for me, and I didn't want to have anything to do with the military. I wanted my old life back."

Five weeks after Norman Nelson was born, Morley and Betty Ann filled their blue Chevy sedan with all of their belongings and a crib for the baby. They drove for several days to the Salt Lake City area. Betty Ann was breast-feeding little Norman, so the baby did well on the drive, either sleeping or eating.

The Nelsons rented a small house in Salt Lake City and began hunting for a home to buy. Almost immediately, Morley got into a conflict with their landlord, and Betty Ann discovered how the war had changed her husband.

The landlord was a Swiss-German man who had an accent that sounded a lot like the Nazi soldiers that Morley fought in Italy. Without notice, the landlord knocked on the door and told Betty Ann that he had a buyer for the rental, and he wanted them to move out in two or three days. She told him they had bought a house, but they weren't going to be ready to move for two weeks. By law, a landlord was required to give the Nelsons two weeks.

When Morley came home from work the next day, he saw the landlord trimming some trees with an ax in their front yard.

"So I hear you're asking us to leave in a matter of two or three days, is that right?" Morley asked.

"Yep, that's absolutely right," the landlord replied in a heavy German brogue accent. "By god, you will get out of here or I'll force you out."

Suddenly, Morley felt like he was back in the war. He grabbed a shovel on the front deck, and pointed the blade upward toward the man's neck, as if it were a bayonet on the end of a rifle.

"All right, damn it, it won't make one bit of difference if there's one less Krout in the world," Morley said, gritting his teeth as he rushed toward the man. The landlord dropped the ax, leaped over the wooden railing of the porch, and ran to his car. "If I see you in less than two weeks, you had better be prepared for action," Morley yelled at him.

As it turned out, they didn't see the man for nearly a month, and they moved to a small two-bedroom home in a Mormon neighborhood—the first home that they owned together. It had a dirt floor in the basement, and no tiles in the bathrooms. The house cost $4,000 in 1945 dollars.

Morley went back to his job, doing soil surveys in an area west of the Great Salt Lake to determine if the land was suitable for irrigated farming. Locally, the dry, unpopulated area was known as "Death Valley." "It was surprisingly good land if you could get water on it," Morley says. "If you used calcium phosphate on the land, a powder-like fertilizer, you could leach the sodium into the soil."

In mid-January, cold weather gripped the Wasatch Front and the Salt Lake City area, sending temperatures well below zero at night. On a weekday evening at 5 p.m., Morley came home from work, kissed Betty Ann and baby Norman, and settled into an easy chair to read the afternoon newspaper.

Suddenly, they heard a knock on the door.

"Who would that be?" Morley asked, getting up to answer it.

It was a man from the natural gas company, and he wanted to inspect the furnace and flue pipe. He took one look at a jerry-rigged metal flue pipe that ran from the downstairs furnace into the brick chimney. "Oh my goodness," the repair man said, "that pipe doesn't meet the city code. I'm going to have to shut down your furnace right now."

Morley looked at the man and replied, "Well, now wait a minute, you can't turn our furnace off tonight, it's cold as hell outside. Let's wait until it warms up in a few months and we can get it fixed then. The pipe has obviously been set up this way for several years, and it hasn't caused a problem."

"No, I'm sorry, it's not up to code, and I must turn it off," the man said. He moved toward a door on the floor that opened to the cellar, and Morley leaped out of the chair and blocked his way.

"You're not going to turn it off in here," he said, staring at the man in the eyes.

"All right, I'll go outside and turn it off." The man went out the door and headed for the valve by the street.

Morley went into his bedroom and grabbed his Colt .357 magnum pistol from its holster hanging on the wall and ran outside. Pointing the gun at the utility man, Morley said, "Mister, if you turn that off it'll be the last thing you ever do."

The utility man stared at Morley's clenched teeth and bulging eyeballs. He looked down at the silver pistol, and realized that this guy was dead serious. He inched away from the confrontation and walked to his truck, looking back at Morley the whole way. "You'll hear about this!" he yelled as he drove away.

Morley shook his head. He couldn't believe how a man could be so ignorant. God almighty, he thought, when can I quit fighting the war? Why would a man from my own country threaten my family during a bitter cold spell just to follow the rules?

The following morning, Morley was summoned by his superior officer in the Army Reserves in Salt Lake City. Technically, he was still under military control because he was receiving therapy at the Veterans Administration Hospital, and a decision was pending on what type of disability payments he would receive for the partial loss of his leg.

"Now Lieutenant Nelson, you obviously have some feelings about the war that could be solved by visiting a therapist," the officer said. "We understand your concern about your son and your family, and it was wrong for the man to try to turn off the gas in your house without advancing warning. But you can't go around threatening people with a gun. If that happens again, you will be arrested."

"I don't need a therapist," Morley replied. "The man was wrong, and I wasn't going to let him endanger my family. That's all there is to it."

The commanding officer let Morley go, repeating the admonition that he didn't want to hear about Lieutenant Nelson pulling a gun on anyone again. After he left the military post, Morley crumpled up the name and phone number of the therapist and threw it on the floor of his car. He didn't need any damn therapist, he thought to himself, but he had to admit that this business of war had changed him when it came to dealing with confrontation in every day life. As a company commander, he was used to "playing God"—making snap decisions to attack and kill. He had developed sound instincts that had served his company well. If someone tried to face him down, like the landlord and utility man had done, he would stand up for his position, ready to pay with his life if it came to that. *Johnny be quick or dead.* But now, the rules were different. He would have to avoid trouble if at all possible.

Betty Ann, who had witnessed both incidents, says it was spooky to see Morley threatening to kill people. "It was scary, and it was weird," she says. "But over time, he mellowed out as he started to get out and fly some birds and have some fun."

Springtime came early to the Salt Lake Valley in 1946. Morley had been counting the days until he could climb into the nest of a prairie falcon or a peregrine and start training his own brood. On a sunny Saturday in early April, he took off in his Chevy sedan to check out a few nest sites in Ogden Canyon, a narrow slot with steep cliffs, and many orange-red rock outcroppings that made for excellent nesting platforms.

A few miles east of Ogden, he stopped the car in a pullout and gazed at the cliffs with a pair of bulky green 8x30 military-issue binoculars. The best way to detect a falcon nest was to look for white streaks caused by the falcons' droppings on the cliff wall. *There!* He spotted a nest site that appeared to be active, but he was too low to confirm it was occupied.

He threw a climbing rope, gloves, a small cloth bag and his lunch into an Army knapsack, and headed for a spot where he could walk up the side of the cliff without ropes. It had been nearly a year since he was injured, and his right leg was quite strong now. He walked with a slight limp, but the leg worked reasonably well in a climbing situation. Morley scaled the edge of the cliff on a rocky ridge, occasionally stopping to catch his breath. He raised the binoculars again when he had a better view, and he could see a couple white downy heads sticking out of the nest and one adult prairie falcon, likely the mother. *Fantastic!*

Now he had to find a way to approach the nest from above. Morley climbed about fifty yards above the aerie, and he tied his rope around a large rock, wrapping several lengths around it and tying it off with a bowline knot. He didn't bother with a sling for a short rappel. He wrapped the rope around his shoulder, back and leg, and leaned back into the rope. It felt good to be on a cliff again.

He dropped slowly. As he got closer, he could hear the young birds chirping, and the adult began to issue and repeat a high-pitched distress call. Morley kept descending, and the adult female flushed from the nest, while making quite a fuss. The bird flew up toward Morley as a warning sign and then banked away into the wind. Morley quickly rappelled to a point where he was even with the nest, and in his low baritone voice, he spoke calmly to the juvenile birds and whistled in their own language.

Whoosh! Morley felt the breeze from mother prairie falcon dive-bombing his head and instinctively ducked. He reached into his knapsack, grabbed the cloth bag, and sized up the brood of four juveniles. Morley knew from experience that he wanted a female because they are larger birds, broader across the chest, and can take bigger quarry than a male. He quickly tried to determine which of the two females would be best. In very soft tones, Morley said a few words to the birds and studied the look in their eyes, trying to sense their attitude. He whistled to them like a falcon to check their reaction. He settled on the female that didn't appear to be as spooked by Morley as the rest. That meant the bird might be easier and quicker to train than the others.

Very slowly, he reached around the back of the nest and immobilized the bird by gently folding its wings to the body and placed her in the bag. Wasting no time, he stuffed the bag into his pack and climbed back up the face. He reached the top of the rock where he tied off the rope, and rested for a moment. Morley smiled at the thought of capturing his first falcon in four years. He couldn't wait to fly it.

Now the training began. Back at the house, Morley put on a large leather glove—known as a gauntlet in falconry parlance—and he placed the prairie falcon on his fist. The first step in training the bird is called *manning*. In essence, the falconer carries the bird around on his fist, indoors and outdoors. The bird can't fly away because it has leather jesses around its ankles, and a leash tied to the jesses.

Initially, the falcon stares at Morley with its black eyes, and breathes heavily, sometimes with its beak open. Over a period of several days, the falcon begins to tame down and recognize its master. "You carry her, and talk to her, and you move very, very slowly," Morley says. "After a while, she starts to trust you."

What happens, if all goes well, is the falcon imprints on its master, meaning she recognizes Morley as mother falcon. From the time the bird leaves the nest and begins a new life with Morley, she puts all of her trust in him, from super-slow movements and enticements of fresh meat. By now, Morley has shown that he's capable of gaining its trust.

Morley puts it this way: "It gives them the understanding to trust you far enough to think you're a falcon, talk to you in falcon language, and they expect you to conduct yourself like a falcon. It doesn't change their wild instincts or their ability to hunt and fly like a wild bird. They do the same things they do in the wild, except you are the mate. The trust is so deep that they trust you with their life."

Today, many falconers learn how to train a bird by starting with a kestrel or a red-tailed hawk, which are comparatively easy to train. First-timers can refer to a number of falconry books that describe the training process in detail. But when Morley learned to train his first prairie falcons in 1938, and again now in 1946, there weren't any books that described how to train prairies. More than thirty-five falconry books had been published on how to train primarily peregrine falcons, dating back several hundred years.

Nelson family photo
Morley with haggard falcon "Tlanvwa," July 1946, Salt Lake City.

Unbeknownst to Morley at the time, prairie falcons were considered one of the most difficult birds to train. In the book *North American Falconry & Hunting Hawks* by Frank L. Beebe and Harold M. Webster (Beebe and Webster, 1964), Webster says this about training prairie falcons: "At best, the prairie falcon is a problem child. Their temperament is very difficult to predict. It is not uncommon to have the tamest hawk turn suddenly into a raging demon before your very eyes. On the other hand, the wildest prairie can just as suddenly turn into a pillar of virtue and behave better than the most trusted peregrine."

Prior to the war, Morley had trained two juvenile prairie falcons in Utah, so he already knew that they took a while to tame down. From his upbringing on the North Dakota farm, he knew how to handle dogs, horses, cattle and bulls in a gentle way that built trust. He knew that patience was critical in developing a strong bond with a falcon.

How much patience? William F. Russell Jr., author of *Falconry: A Handbook for Hunters* (Charles Scribner's Sons, 1940) notes that while training a bird, if it "becomes frightened or disturbed, everything will have to be done over again."

If the bird is not trained properly in the early stages, it will not perform well in the field, Russell says. "Neither anger nor haste will prevail, and if the falconer plans to fly his wild-caught falcons loose later on with certainty that they will return, he must be patient and do this first big job very carefully."

After a full day of carrying the prairie falcon, it was time for Morley to check and see if the bird trusted him. Time to feed her. He placed fresh bits of uncooked chicken between the thumb and index finger of his gloved hand, and watched the young falcon size up the situation with its ink-black eyes. After several minutes, she leaned over and began to nip at the meat with her dark-blue black beak. She was taming down quite nicely.

Now, the next step: hooding the bird. A falcon must be trained to accept a hood to prevent it from becoming nervous and flying away while carrying the bird on foot or horseback, or transporting it in a car or airplane. Morley already had fashioned a new hood for the bird by sewing together strips of lightweight leather. He preferred to use an Indian-style hood (developed by falconers from India) as opposed to tight-fitting Dutch hoods used by European royalty. From trial and error, Morley found that his birds preferred the Indian hoods because they did not touch the bird's eyes or the corners of its mouth. He learned to trim the hoods so they fit perfectly, without aggravating the bird's sensitive cere, the tissue around the beak.

First, however, he had to indoctrinate the bird to the hood. In *North American Falconry,* Beebe and Webster point out that this is no simple matter. "Hooding is an indignity at any time, even for the best of falcons, and the least the falconry can do is to adopt the lightest, easiest-fitting hood possible," Beebe wrote.

"It is not unusual for an intelligent hawk to bluff a falconer clear out of his own shoes and I mean this quite literally," adds Webster. "A big scene is usually the starter and the pupil absolutely refuses to have anything to do with the hood . . . ."

Morley learned from his falconry books and previous experience that hooding a bird in the dark is a good technique. Plus, it made sense. "It was just an obvious thing to me," he says. "When you're in the dark, they can't see a window or any light, so they're not going to try to fly."

The technique of placing a hood on a bird is critical—very slow and deliberate. Morley takes the hood by the fancy leather tassels on the top, and being careful not to show the bird the hood, he raises it from her chest toward the beak, and in one steady motion, he places the hood over the falcon's head. Then, he pulls the drawstrings on the back of the hood with his right hand and his teeth.

"Everything has to be super slow—slow and steady," he says. "If you move too fast, she'll think you're attacking her and she'll bate."

That means the bird could try to smack its master with its wings, bite him in the face or grab him with its razor-sharp talons. He knew he had to be careful.

After three days of carrying the bird and feeding it, Morley had the falcon accustomed to being hooded. She also stopped staring at her master and began to watch other things around her. That was a telltale sign that she was becoming comfortable with her master, and overcoming her wild instincts to shun a human.

The next step would be to introduce her to the lure, and train her to fly to the lure in the back yard. First, Morley had to let several days go by without feeding the bird to ensure it was hungry.

During the week, when Morley was gone to work all day, Betty Ann would bring the falcon upstairs from its block in the dirt-floor basement, and place it on a block outside. Morley taught her how to carry the bird, and how to hood it and unhood it. She quickly developed the right touch.

On a Thursday evening, Morley came home after work, ready to introduce his falcon to the lure. He came into the house and gave Betty Ann a kiss, squeezed baby Norman and hoisted him high in the air, and then collected a number of items that he'd need for this phase of the training. The lure consisted of an eight-foot long, one-fourth-inch diameter nylon rope, with two duck wings sewn to a leather piece in the center attached to one end. A quarter-pound weight was placed on the other end of the lure to prevent the bird from leaving with the meat and lure. At feeding time, Morley attached several pieces of chicken or beef to the duck wings.

Morley walked out the back door and approached his falcon, perched on the block with its hood on. "Hello there, big hawk,"

Morley said, reaching for the leather strands on the back of the hood to remove it. Standing proud and alert, the falcon looked around with its piercing black eyes. Late-evening sun cast a platinum glow on the bird's white and brown-speckled breast. Morley could tell that she was hungry. The bird extended its wings and started to flap them on Morley's fist, as if she were perched on the edge of its nest, getting ready to test her wings for the first time. Morley smiled. "OK, now, big hawk, let's simmer down a little and see if you'd like something to eat."

To start with, Morley placed the lure about four to five feet away on the grass. The falcon studied the lure and flapped its wings. "You can go anytime you want," he said. The falcon flapped its wings and took off directly for the lure. It landed on the duck wings, and gripped the lure with one of its yellow feet, standing proud.

"That's good," Morley said, letting the bird eat a morsel of chicken. "Now step up, you big hawk," he said, reaching under its feet with another piece of fresh meat in his glove. The bird stepped up. Next, Morley carried the lure thirty feet away to the other side of the yard. Then he walked across the yard and faced toward the lure. The falcon flapped its wings with excitement and issued a high-pitched shriek. *Kak Kak Kak.*

"Take it easy, now," Morley said, quietly. "Go anytime you want."

The falcon took off and flew to the lure, landing on the wings. It dug into the feathers to find more fresh chunks of chicken below. It began to rip off pieces of meat with its sharp beak, while looking up and around to instinctively protect its catch. Morley went over to fetch the bird with a morsel of fresh meat in his gloved hand. He smiled. The falcon was making good progress.

Over the weekend, Morley took the falcon into the field so it could fly a greater distance to the lure. He reduced the amount of meat he fed her to make sure the bird would be hungry enough to pursue the lure. "You have to lean them down," he says. "They've got to be hungry to do this."

Morley used a scale to ensure the bird was at the right weight for flying—about thirty-five ounces. He took the falcon to some farm fields near Tremonton where he had permission to fly his birds. He didn't want to go to a place where they might run into a lot of other raptors or shorebirds, because the prairie falcon might try to pursue wild prey before she was ready.

For this exercise, Morley tied a much longer leash, a creance—a ten-pound test nylon fish line—to the falcon's jesses. He walked across the field with his falcon on his fist and placed the lure on the ground, about fifty to sixty yards away. Then he walked back toward the car, turned around, and stood facing the lure. A steady breeze blew directly toward the falcon—perfect flying conditions. The bird looked around, tested its wings and called. *Kak Kak Kak.*

It was hungry and ready to go.

"You go anytime, now you big hawk. Just go when you're ready," Morley said quietly, next to the bird's ear.

The falcon took off and flew low, directly to the lure. Like before, Morley let the bird eat some fresh meat from the lure as he walked over to retrieve the bird with a fresh chunk of meat in his gloved hand. He repeated the same procedure for several weeks, while adding a little more distance to the flight each time. "You need to repeat it so she gets it in her head that she's got to catch the lure, and she'll get fed when she does," Morley says.

Two weeks went by, and on a breezy Saturday morning, Morley took the falcon into the field for the next stage of training. The bird was ready to fly without a leash, and he'd see if she returned to the lure—this time, swung by the master. Morley drove to Tremonton again, and walked into a field, facing a stiff breeze. He unhooded the falcon, and she snapped her brown and white head around, sizing up the surroundings.

"Whoa up!" Morley said, lifting his arm up slowly, casting off the falcon. She flapped her wings rapidly and rose up quickly in the wind currents. Morley felt the newfound freedom that the bird sensed at that moment, and marveled at the way she instinctively flew into the wind, gaining altitude all the way. She circled around and looked for the lure, but Morley didn't produce it, as yet. He wanted to let the bird fly for fifteen minutes or so, and gain some confidence in the air.

Then, he pulled his lure out of his falconry bag, a leather satchel, grabbed the quarter-pound lead weight on the end of the lure string, and began to swing the lure in a counter-clockwise direction. Without knowing it, the falcon was learning how to "wait-on" by flying high and waiting for her master to produce the lure before she dove for her morning meal. Seeing the lure, she folded in her wings and dove in a nearly vertical position toward the lure. Morley let her grab it on the first pass. The bird clenched the lure with her sharp talons, landed on the ground and pecked at fresh beef in the duck wings.

"You're learning to be quite a hero, you smart hawk," Morley said to her as he crouched low, and offered fresh meat in his gloved fist. The bird stepped up on the fist and released the lure.

It never ceased to amaze him how interesting it was, during the training procedure, to watch a falcon mature and employ its natural-born talents to fly like a wild bird. For Morley, it was a treat to watch a bird's intelligence and ability increase with every flight. Plus, the bond between him and the bird grew stronger by the day.

Morley just about had her trained in. He'd fly her to the lure for another week or so to give her more practice and confidence flying in the air. The following week, Morley took the falcon out to a nearby golf course. He released the bird into the wind, and she took off with zeal. He watched her fly higher and higher until she was a speck in the partly cloudy sky. After about 15 minutes, Morley began to swing the lure. She was getting quite independent now, and he couldn't see the bird anywhere. This was the first time that she didn't return immediately to the lure, while waiting-on high in the sky. Perhaps she wasn't hungry enough today.

As it got dark, Morley had two thoughts: either the falcon had decided to fly off into the hinterlands to live as a wild bird; or she would likely return the following day, looking for a meal. He went back home for dinner, quite concerned about the quandary. The biggest problem was he had to go out of town on assignment for the SCS for several days. Over dinner, he explained the dilemma to Betty Ann, and she vowed to keep a lookout for the falcon.

She was pregnant with their second child now, so she had her hands full taking care of Norman, keeping house, grocery shopping, cooking and getting enough rest for herself. Of course, she never got much rest.

A day went by, and Betty Ann didn't see the falcon. She started to wonder if it would ever come back. Then, the paperboy knocked on the door the next afternoon, and mentioned that a bird was flying over the house. Betty Ann pulled the lapels of her pink chenille bathrobe closer together and walked out on the front step. From one glance, she could tell that it was Morley's hawk.

"Oh my god," she says. "I had this fear that if I didn't catch the bird, I'd be hung by my teeth until death."

So she told the paperboy, "Stay here and watch my baby."

She went into the kitchen, pounded some chicken heads, grabbed Morley's glove and lure, and stuffed the meat in his falconry bag. She went out the front door, jumped on the newspaper carrier's bike and headed for the golf course. She dropped the bike by the edge of the golf course, and climbed over an unwieldy hurricane fence with an armful of falconry gear twisted around her pink robe with Morley's Army jacket over the top to reach the fairways.

"Twenty-Seventh Street South was a big road so all of these truck drivers were honking their horns at this weird lady in a pink bathrobe," Betty Ann says, laughing at the thought. "I went out in the middle of the fairway and swung the lure, and the bird came down to get it. I guess she was finally hungry enough to come home."

Betty Ann decided that she'd walk out of the front entrance of the golf course instead of climbing over the awkward fence with the bird. "I thought hell, I've got the bird. I'm walking out the front door."

When Morley returned home, he was relieved that Betty Ann found the bird. He chuckled at all the details of the recovery mission. "You're getting to be quite the falconer," he said.

Now, the falcon was ready for the last major element of the training regime—he would *enter* or introduce the bird to live quarry—a pigeon, quail, duck or pheasant. Instead of coaxing her back by swinging the lure, Morley would release a live bird.

She flew off from Morley's fist into the wind, and quickly gained altitude, her wings pumping all the way. Morley let her fly for 15 minutes or so, and she was more than 1,000 feet in the sky, surveying the scene, while keeping an eye on her master. It was time: Morley released the pigeon and looked up to see the falcon dive toward the pigeon like a guided missile.

She smacked the bird with a balled up talon and doubled back to catch her quarry in the air, just before it hit the ground. Then Morley, the proud master, walked over to her with a fresh piece of meat in his fist, and she stepped up on the glove.

"It's a wonderful thing to train a falcon to the point where she can go hunting with you as her partner," Morley says. "You're dealing with nature's fastest characters. They can outclimb the eagles and the hawks, and it's so spectacular to watch them learn how to use their own natural instincts to fly in such an acrobatic fashion in the air, and take other swift-flying characters as if they're standing still. They are masters of the air."

At times like this, when Morley has successfully trained a new falcon, he could enjoy the payoff of more than a month of tenacious, hard work. Now, he could take the falcon hunting by the northeast shore of the Great Salt Lake and watch the bird chase a wide variety of quarry in the air. Inevitably, he felt an urge to catch another bird, and expand his brood of trained hawks.

As the days grew warmer, however, spring turned to summer, and it wasn't a good time to catch any juvenile birds from a nest—too late for that—and too early to trap adult falcons migrating through the Salt Lake region in the fall. It was time to take a break from falconry and have some fun with the family.

Morley and Betty Ann took baby Norm to the mountains or the Southern Utah canyonlands country for camping. They went car-camping by some high mountain lakes in the Uintah Mountains, and enjoyed the cooler weather in an alpine setting. Morley and Betty Ann enjoyed the wonder of visiting the canyonlands, a unique landscape of orange-red sandstone arches, bluffs, monoliths and canyons.

"We camped a lot," Betty Ann recalls. "I'd gather up all of the baby stuff, and Norman would go with us out into the boonies. I still remember heating up bottles on the radiator of the car."

The Nelsons had two dogs now. They were large collies, one named Dak, short for North Dakota, and Toby, the paper boy's dog that was adopted by Morley and Betty Ann. When baby Norman started to walk, he'd grab the fur of each dog and walk in between them. "One time he walked four blocks away from the house, I couldn't see him between the dogs," Betty Ann says.

Morley's treatments at the Veterans Hospital were over now, and the medical authorities listed him as being one-third disabled, for which he would receive small monthly payments for the rest of his life. He had permanent numbness in his right foot, nine pieces of shrapnel lodged in his back, and his hearing was partially impaired. The good news was that the bullet-related injuries had healed. He walked with a slight limp, but he could rock-climb and ski without feeling impaired.

During the summer months, Morley got a call from the Salt Lake City Zoo, which had been accepting a number of ailing peregrine falcons that had been turned in by citizens who found them by the shore of the Great Salt Lake. The birds suffered from a malady called "ring-neck," caused by excessive ingestion of salt from ducks they ate by the lake, near the mouth of the Bear River. The disease caused them to twist their neck around incessantly, and they couldn't fly.

Since Morley had a permit from the Utah Department of Fish and Game for taking wild game with his falcon, the zookeepers knew that Morley was a falconer, and would know how to handle peregrines. They asked if he would accept five of them and see if he could nurse them back to health. When he got the call, he was quite excited because he'd never had a peregrine—the preeminent game-hawk in the world.

"It was a tremendous experience," he says. "They were marvelous birds—big Anatum-type peregrines, with very dark plumage. And after they had a chance to recover, it was wonderful to take them hunting."

Once again, Morley took the birds through all of the steps of training for falconry. Betty Ann had her hands full as the brood increased and she had to bring the birds out of the basement to the back yard every day. As adult birds, the peregrines already knew how to fly and take wild game, so that part of the training would go much faster. When the birds became healthy again, Morley made an effort to wean them of any food from their master, knowing that they would eventually fly back into the wilds and survive on their own.

One day he was out flying one of the peregrines, and the bird decided to take a rest stop on top of a powerpole. Morley didn't think much of that. But then, a pickup truck drove by and the driver slammed on the brakes. Morley saw the gunbarrel of a double-barrel shotgun emerge from the passenger window.

"Boom!" the gun went off, killing the peregrine. The bird dropped to the ground with a thud.

Morley fired his .357 pistol in the air, swearing under his breath. But the pickup sped off down the road before he could do anything about it. There was no use getting the license plate number because it wasn't illegal to kill birds of prey in Utah at the time. Morley went over to pick up the dead peregrine and placed it in the car in disgust.

This wasn't the first time that someone had taken a shot at one of his trained hawks, but it was the first time that someone had killed one. He was angry. Yet, he knew that most people didn't know any better. Birds of prey were known as "chicken hawks," "duck hawks," "sparrow hawks" and "bullet hawks." No one knew the value that they represented to nature and society. "Everybody shot them—it was very common," he says.

He'd pick up dead birds on the side of the road when he was out hawking or working as a soil scientist. He found nearly all of the commonly found raptors in the area—red-tailed hawks, Cooper's hawks, Swainson's hawks, goshawks, peregrine falcons and prairie falcons. He took the carcasses to the University of Utah, where wildlife biology professors stuffed the birds and used them for identification and education.

In hopes of sending out a more enlightened message about the value of birds of prey, Morley went to the editorial office of the *Salt Lake Tribune* and talked to a reporter about the frequent shootings he'd witnessed. "I said it was wrong for people to shoot birds of prey, and that they were valuable to farmers and society because they ate mice, gophers and snakes," Morley says.

Public attitudes and state and federal laws would have to change before people would stop shooting birds of prey at will, but at least Morley made an attempt to present the other side of the issue. As time went on, he would focus more and more on educating the public about the need to understand, appreciate and protect all birds of prey.

In the fall of 1947, a well-known falconer from the East Coast, Jim Fox, sent Morley a young peregrine falcon named "Black" as a gift to use in falconry. Very few falconers existed in the United States at the time, and they all knew each other from reading British falconry journals or from word of mouth. Fox had heard about Morley's high praise for prairie falcons, and he wanted Morley to experience the thrill of flying a peregrine.

It was a gift that Morley appreciated greatly as Blacky became an excellent flier and hunter. "She was a very large peregrine, known as a rock peregrine in the East," he says. "She weighed over forty ounces when she was hunting, and more than forty-four ounces when she was in the moulting stage. She was very, very big and very healthy. She was a tremendous flier. Once I had her trained, she made tremendous stoops to the lure, and eventually, she learned how to take just about anything in the air."

Kent Carnie, a longtime friend of Morley's, knows the story of how Blacky was found. "A bunch of Boy Scouts were walking along the Appalachian Trail, and they found a peregrine nest," Carnie says. "It was easy enough to reach, so they took three baby hawks out of it. And they proceeded to roast and eat two of them. Boy Scouts living off the land. There you go. But the third one, they were going to make a pet out of it, so they took it home. And they found that there was a lot more to raising a hawk than they thought. So they gave it to Jim Fox. And Fox sent it to Morley."

That winter, Morley spent a fair amount of time skiing the world's finest light powder snow at Alta Ski Area in the Wasatch Mountains. He hooked up with a fellow ski trooper named Monty Atwater, who was working on developing avalanche-control techniques with explosives at Alta in Little Cottonwood Canyon as an employee of the U.S. Forest Service. Morley helped Atwater test Alta's steep slopes for stability on skis, and it was interesting to try to predict whether slopes would slide after they threw explosives.

Atwater's project was important due to the incessant slides that occurred in Little Cottonwood Canyon every winter. A number of people had died in avalanches since the area opened in the mid-1930s. Atwater's work with explosives would later spread to many ski areas in the West, where slopes were steep enough to require blasting and preventative avalanche control.

On January 21, 1948, Betty Ann delivered their second child, a boy. They named him Timothy Morlan Nelson.

Betty Ann remembers that it was a struggle to deliver Tim because she had a bad cold, and she was in labor longer than she had been with Norman. Fortunately, she had asked her mother to come out to Utah to help take care of Norman and the new baby.

"I can remember that I thought I'd die if my mother didn't come out," she says. "They wouldn't give me anything for my cold, since I was about to deliver the baby, and it was bad."

After Tim was born, the doctors discovered that he had a blood blister on the back of his head, near the nape of his neck. He had to be treated with radium to remove it. "I had to take him into the hospital two or three times a week, and each treatment took seven hours," Betty Ann says. "I remember that they wrapped me in a protective apron, and the doctor who did the radium treatments had only three fingers. Eventually, the treatment fixed the problem, and Tim had a pink spot where the blood blister had been, and his hair didn't grow back."

In the spring of 1948, Morley resumed flying his birds. He had Monty, his prairie falcon, Blacky, the peregrine, and he had released the rest of the peregrines that had suffered from ringneck. He decided to search for a juvenile peregrine falcon in the canyons and cliffs in the Salt Lake region.

Typically, when he searched for a falcon aerie, it would be occupied by prairie falcons. But Morley remembered from the days before the war, when he lived in the Salt Lake area from 1939-1941, he had found a number of peregrine aeries. Nearly ten years later, he returned to the same nest locations and found them occupied by prairie falcons.

He scratched the top of his flat-top crewcut in wonderment. What could be causing the peregrines to surrender their nest sites to prairie falcons? Certainly, peregrines were capable of guarding their own territories. There had to be another explanation. From his soil survey work, he knew that the level of the Great Salt Lake was in a period of steady decline, the obvious results of a prolonged drought. Many of the small ponds adjoining the lake were now dry. Previously, they harbored large numbers of small birds that peregrines depended on to exist in the area.

So the problem of the disappearing peregrines could have been drought-related in terms of a loss of prey base, or it could have been something else. Morley recorded his observations in his field journal, notes that would become very valuable in the future, when falconers began to sound the alarm in the 1960s that peregrines were disappearing throughout the United States.

Monty the prairie falcon, meanwhile, and Blacky the peregrine were an impressive aerial force when Morley took them flying in the field. Being a large female prairie falcon, Monty had been going after the largest prey possible—even Canada geese.

In June, the Soil Conservation Service made Morley an offer to take a new position working on snow surveys in the Columbia River Basin, based in Boise, Idaho. It would be a promotion in terms of pay, and Morley found the prospect intriguing. The federal government was concerned about unpredictable spring floods in the Pacific Northwest, which afflicted towns near the big rivers such as the Snake and Columbia. By hiring experts to survey snowpack in the mountains, federal officials hoped that they could do a better job of predicting floods and streamflow runoff for farmers and ranchers, power companies, municipalities and federal agencies.

At the same time, the Army Corps of Engineers and the U.S. Bureau of Reclamation were engaged in a dueling dam-building era on major rivers in the West. New dams would increase flood-control, produce hydroelectric power, and expand the amount of mountain runoff that could be stored for irrigation.

Before Morley decided to take the job, he and Betty Ann took a weekend reconnaissance trip to Boise with the two boys. They drove across the undulating sagebrush-dotted Snake River Plain toward Boise—the same route followed and cursed by Oregon Trail emigrants. When they made the final descent into the Boise Valley, they were impressed. They marveled at the cottonwood-lined Boise River flowing through the valley, and the mile-high mountain backdrop behind the state Capitol. They drove a few miles up the Boise River along an area known as the Black Cliffs, and Morley's keen eye instantly spied a golden eagle nest, several red-tailed hawk nests and numerous prairie falcon aeries on both sides of the river.

"I was so excited, I could hardly stand it," Betty Ann says. "It was a beautiful town with a nice river flowing through the center of it, and there were lots of trees. It was a much smaller town than Salt Lake, which I liked, because it seemed more like the town I grew up in. And the people were friendly. I remember we needed some cash when we arrived, and I went into the bank and asked them if I could cash an out-of-town check, and she said, "Oh sure, honey, that's OK." And I thought, hey, this town is for me."

Morley felt the same way, and so it was an easy decision. They would move the family to Boise in a matter of weeks, along with Monty and Blacky and the dogs.

Chapter Six

The mews
1948-1959

On a Saturday morning in early June, Morley woke early. He lifted a corner of the white cotton curtains in the kitchen window, and marveled at the azure sky. It was going to be a beautiful day. Finally, spring was beginning to surrender to summer.

Today, Morley planned to do a little exploration in the Snake River Canyon, thirty miles south of Boise. He made a quick lunch, and hopped into his Chevy sedan before any of the kids got out of bed.

Purple lupine, red Indian paintbrush, yellow buttercups and scarlet gilia twirled in the breeze as Morley followed a rutted, single-lane dirt road that runs dead south from Kuna, a little farm town fifteen miles southwest of Boise, to the Snake River Canyon. The dashboard chattered as he cruised along at thirty mph. Morley's blue eyes focused on the rolling fields of sagebrush in front of him. Nearly everywhere he looked, he saw dark-brown ground squirrels standing on their hind feet, sounding off with a high-pitched one-note whistle. In between the sage, he saw green strands of bluebunch wheatgrass and Idaho fescue bending in the wind. Occasionally, a jackrabbit raced across the road in front of the car.

Not more than three miles south of Kuna, Morley saw the

first raptor—a red-tailed hawk perched atop a three-strand powerpole. Then he saw a pair of marsh hawks circling overhead. Off in the distance, Morley saw a ferruginous hawk swooping across the tops of sagebrush in search of prey.

Ah, he thought to himself, scratching the unshaven stubble on his chin, this is looking good. But he hadn't seen anything yet. The dirt road took a ninety-degree bend to the left, near the edge of the Snake River Canyon. Morley pulled over and walked to the edge, jumping over brown volcanic basalt rock smothered with yellow, green and red lichen.

At the cliff's edge, the grandeur of the canyon came into view. It was a sheer 500-foot dropoff below his feet. He could hear the white, thundering Snake River rushing below the powerhouse at Swan Falls Dam. He scanned the opposite cliff for any signs of raptor nests, and saw none in the layer-cake blend of orange, yellow and rouge on the rock wall.

And then, he looked directly below his feet, only to see a golden eagle riding the current of the upstream breeze with its giant wings fully extended. In seconds, the eagle rose 150 feet to the point where it was almost even with the canyon rim, not more than twenty yards away. Morley stared into its golden eyes, and marveled at the majestic bird.

If the eagle only knew how important it was that this particular fellow had arrived in Idaho. Driven by an insatiable curiosity, Morley was in the midst of discovering a treasure here in the Snake River Canyon that would surpass all of his dreams. Maybe the eagle was saying "welcome."

Morley had to see more. He leaped over the sharp basalt rocks as if he were walking on air and jumped into the car. He followed the dirt road down a steep grade and parked by the Idaho Power Company hydro plant. He grabbed his 8x30 binoculars and walked through the sagebrush at the base of the cliffs. First, he took in the canyon view without magnification. Almost immediately, he noticed a male prairie falcon fold its wings and go into a steep, 500-foot dive from the rim to the canyon floor. Morley understood that the tiercel was trying to display its acrobatic prowess to attract a female.

"Boy those tiercels are something else," he whispered.

Against the north rim of the canyon, he watched prairie

falcons chase large flocks of pigeons that flowed like a sheet blowing in the wind against the brown wall. Then, he brought the binoculars to his eyes and spotted eagle and prairie falcon aeries on the same cliff.

By the end of the day, he had seen more active aeries in closer proximity than any other nesting area he had previously seen. The possibilities of adding a new prairie falcon to his hawk house quickly came to his mind, even the prospect of working with a golden eagle.

"I was excited, and I was amazed," Morley says. "I'd see a nest, and then I'd walk less than a quarter-mile and see two or three more. Normally, I had never seen a situation where the falcons and the eagles were living so close together. It wasn't long before I was thinking, my god, I've found something here that may be unique in the world."

Morley and Betty Ann bought a nice two-bedroom house in Boise's North End at 326 Ada Street. It was a cozy abode—about 1,000 square feet—with a red exterior and white shutters. In 1948 dollars, the house cost $6,000. It had a one-car garage for the Chevy sedan, and in the backyard, there was a small shed that became the "mews," a falconry term for the hawk house.

Just days after the Nelson family moved into their new home, Morley had to leave town on snow survey business for several weeks.

"By the time he came home, I knew all the neighbors, and they were all really nice people," Betty Ann says. "Most of them had young kids as well, so all of the women, we just sat out in the front yard in lawn chairs, smoked cigarettes and watched the kids play."

Once the summer weather kicked in, Boise residents typically enjoyed a long string of clear blue days from July through September, afforded by a strong ridge of high pressure that hung over the Inland Northwest. It also got hot during that time, with temperatures in the nineties. So Betty Ann and her friends frequently took the kids swimming in the Boise River to cool off.

To live on the far eastern end of Ada Street, adjacent to the Boise foothills, was an ideal location for the Nelson family. The kids had immediate access to the foothills, a series of gulches and grassy ridges which rise 3,000 vertical feet from the edge of town to the tree-cloaked Boise Ridge. Near their home, Norman and Tim could play with trucks in the sand next to a blond cutaway cliff. They played army while hiding from each other in the sagebrush, and they explored Cougar Cave, an actual cavern with bats flying in the ceiling.

Morley discovered several areas within a mile of their house where he could fly his falcons. For quick exercise, he'd take them down to a large public lawn in front of the Veterans Administration Hospital on Fort Street, and fly the birds to the lure. In the east side of Boise, a flat-top butte aptly called Table Rock provided a perfect place to fly the birds in a more wild setting, where they could take quail, Hungarian partridge and chukar partridge. Almost always, the winds were favorable on Table Rock for flying falcons because of the butte's prominent location on the east side of the foothills. No matter if the wind was blowing from the north, south, east or west, Morley could find a good place for casting off his birds into the wind from the circular butte.

After a long, hot summer, Morley looked forward to the fall when Blacky the peregrine and Monty the prairie falcon would be ready to fly again. During the heat of the summer, mature falcons go through a moulting phase, in which they shed their feathers and grow new ones in preparation for the fall migration. Morley didn't fly his birds during this period because "they're fat and lethargic, and they have blood flowing through their new feathers. It's best to let them rest during the summer," he says.

On weekends in the fall, Morley took Blacky and Monty to nearby farms to go hawking. One location was called McManamy's Farm, a large cattle feedlot, surrounded by corn and hay fields. The farm had a number of ponds that attracted scores of ducks. Today, the Boise Towne Square—the city's largest mall with five anchor stores and 200 other shops—occupies the same tract of land.

Back in the late 1940s, the farm teemed with pigeons, pheasants, quail, Hungarian partridge and ducks. It was a perfect place to go hawking. Morley had permission to hunt his falcons there anytime he wished, and he'd return the favor by offering the McManamy family a duck or a pheasant after he was through for the day.

Morley says Monty and Blacky were so well-trained, "they could take anything in the air."

Morley had a knob installed on the steering wheel of his Chevy so he could drive with his right hand and hold the hooded falcon on his left, gloved hand. On alternating days or every third day, he'd take one of the birds to McManamy's Farm and catch quite a quarry.

Just for kicks, Morley trained his falcons to catch crows that fall. "The crows are intelligent, and they dodge and they fight back," Morley says. "And they're hard to catch. My falcons could dive out of the air and take a pheasant or a duck real easy. But a crow dives and turns, so the falcons have to grab them underneath. They have to be real quick to get a crow."

Often times, crows gathered in a large locust tree on McManamy's Farm. Sometimes there would be more than thirty in the tree at once. Morley would cast off Blacky or Monty nearby, and if the crows were flying above the tree, they'd quickly hunker down in the tree for protective cover. Then Morley picked up a pellet gun and fired into the crown of the tree, forcing the crows to flush.

The chase was on.

Blacky, meanwhile, was flying 1,000 feet above, waiting on, and when she saw the crows flush, she settled into a dead-vertical dive. Falconers call this *the stoop.* Hurtling toward the earth at speeds approaching 200 mph, she isolated her victim, swooped underneath the crow and gave it a lethal blow with her foot. The action happens so fast, the kill occurs in a blur, dropping the limp bird into corn stubble below.

Morley had to act fast now because the rest of the crows would fly toward Blacky and her victim to protect their own. If he didn't get there quickly, the crows could peck Blacky's eyes out. He approached the flock with a warning call of his own, causing the crows to scatter. Then, he offered a small piece of meat on his glove to Blacky and said, "step up big hawk," and she'd jump up on the glove.

When he had time, on weekends or evenings, Morley liked to fly his birds at Lake Lowell, a large irrigation reservoir on the western outskirts of Nampa, about thirty-five miles west of Boise. Lake Lowell had an abundance of waterfowl, particularly ducks and geese, and pheasants. He had permission to go hawking on several farms next to the reservoir, and his birds always had a field day.

On a cool and breezy Saturday in November, the air was scrubbed clear by the wind, making the sky appear true blue, a shade afforded by the low angle of the sun.

"Up you go," Morley said and Blacky took off like the strong peregrine that she was. You could hear her strong wings go "whoosh whoosh" in the wind before she rose up in the sky, and rapidly gained altitude. Morley wanted her to wait-on until the dog, Kiska, flushed a duck or a pheasant. But Blacky had ideas of her own. She went after a flock of geese that flushed from a wheat stubble field.

Morley watched with awe as Blacky went into an inverted dive and knifed through the sky, on a direct bead for a goose. What a remarkable sight to see 2 1/2-half-pound Blacky take down a full-grown, ten-pound Canada goose. The honker fell to the ground, and Blacky was on it in seconds with her right foot clamped around its neck.

At the moment of the kill Morley felt a visceral tug inside, watching Blacky, the peregrine, a bird he had bonded with and trained to perfection, perform the instinctual act of taking prey just as effectively and artfully as a fully wild peregrine falcon. That singular, spectacular moment of the dive and the kill evoked the same thrill, the same jolt he felt as a twelve-year-old kid on Grandpa Steen Nelson's farm.

"Hey, what the hell are you doing!" a man yelled from out of no where.

Morley ran over to Blacky lording over the goose.

"Well, my falcon gets a little carried away once in a while," Morley said. "Step up, big hawk." Blacky jumped up on his fist.

The goose wasn't dead, but it was dazed. It slowly got up on its feet, shook its head and waddled away.

"That isn't legal for a falcon to kill a goose is it?" he said.

"Oh, yeah, it's legal for them to take any damn thing in the air," Morley said. "It's just pretty rare for a falcon to fly so strong and powerful that it could take down a big bird like a goose."

"Yeah, well, I'm going to call Fish and Game and report you," the man said.

"Oh, that's not necessary. They don't have any problem with me and my falcons," Morley said.

"We'll see about that!" he said, storming off.

Two days later, Morley saw an article in *The Idaho Statesman* in which a man from Nampa complained about a falconer whose trained birds killed geese and deer out by Lake Lowell.

Idaho Fish and Game authorities called Morley and inquired, "What is that man talking about? Does your falcon attack deer?"

"Heh, heh, heh, no, I don't know what that man is talking about," Morley said. "I did catch some geese out there, but not a deer for Christ's sake. That's the silliest damn thing I've ever heard."

"Well, we're glad to hear that," the Fish and Game officer said. "As you probably know, you don't need a permit to hunt with a falcon, but be sure that you don't violate the bag limits on ducks and geese during the hunting season. That's probably what set that fellow off. He's probably a waterfowl hunter, and he didn't like seeing a falcon taking down some birds that he could have shot himself."

"Yeah, you're probably right," Morley said. "Don't worry. I'm not going to take more than my share. My birds take pheasants and crows as well as ducks and an occasional goose. And often times, I fly them to the lure, and they don't kill anything."

"Very good, sir. Thank you for your cooperation."

During the 1948 hunting season, daily bag limits were five a day for ducks (ten in possession), two a day for geese (four in possession), and three a day for cock pheasants. So Morley limited the take of his falcons to match the bag limits. There were a few times when his falcons took five ducks in a day. For a falconer to see that much action on a breezy afternoon is about as good as it gets.

Far too often, when Morley went out to fly his falcons, he'd run across the bullet-shot carcass of a red-tailed hawk, a golden eagle or other raptors lying in a ditch. Most of the time, the birds had buckshot in the breast or a bullet hole in the head. The birds were victims of a "frontier" mentality in the West, a pervasively bad attitude that had been passed down through the generations by ranchers, farmers and hunters. The prevailing attitude was that hawks were bad because they killed chickens. Golden eagles should be shot because they killed and carried away newborn lambs. Falcons were evil because they killed wild game birds. Bald eagles were nasty because they ate salmon.

Having grown up at a time when most Americans had no understanding or respect for birds of prey, Morley expected to see bullet-shot hawks and eagles along the roadways, but the extent to which it occurred sickened him. It made him angry every time he saw a dead raptor, and it strengthened his resolve to do something about it.

On the rare occasion that he ran across a live raptor that had been shot, he brought them home to the mews to see if he could repair an injured wing or nurse a bird back to health. The seeds planted in Utah, where Morley had rehabilitated more than ten peregrine falcons, took on a new form now. From this point forward, the Nelson household and the mews doubled as a raptor hospital.

Ever the savvy media hound, Morley called up the newspaper, radio and TV stations after he found a live red-tailed hawk with its wing shot off. He put the bird on his gloved fist, so the cameras could get close-up pictures, and talked about the value of red-tailed hawks to farmers because they ate all of the unwanted rodents digging up the fields, and they killed rattlesnakes.

"If it weren't for the red-tails, we'd be overrun with mice and rattlesnakes," Morley said. "They do a great service to all of humanity. People should stop shooting them for mere kicks."

Morley added that if anyone found an injured hawk, eagle or falcon, they should call him or bring the bird to his house so he could try to save it. It wasn't long before people started calling, and the mews filled up with quite an assortment of birds.

Quite innocently, out of the goodness of his heart, Morley had taken two steps forward in a new role as a conservationist—he started to make stump speeches with a bird on his fist to educate the public about the benefits of birds of prey, and he opened up one of the first raptor-rehabilitation clinics in the United States.

Early in the winter of 1949, Morley and Betty Ann left the kids with Morley's parents, Norris and Agnes, and went skiing at Bogus Basin Ski Area, atop 7,650-foot Shafer Butte, the highest point on Boise's northeastern skyline. Bogus Basin had a rope tow and a T-bar for shuttling skiers to the top of the mountain. Morley was a volunteer ski patrolman. It was a rare occasion that Morley and Betty Ann got to go off on an outing without their kids, and they had a nice day together on the slopes.

Norris was retired now, and so he and Agnes liked to visit Morley—now their only surviving son—and the grandchildren in the winter for a short time before heading to Arizona for the rest of the cold months. Then they'd come back through Boise on their way back to North Dakota.

When Morley and Betty Ann got home after their day on the slopes, they could sense that something had gone terribly wrong.

"After we walked into the house, I went into the bedroom and Morley's dad said, "There's something wrong with Agnes. She won't wake up!" Betty Ann says. "She had gotten sick, thrown up on the bed sheets. I don't know why I realized it, but I took one look at her, and I knew it was really serious. We hadn't lived in Boise for that long, so I didn't know any doctors. So I walked across the street and asked my neighbor if they knew of anyone to call. She recommended Doctor Ralph Jones. I called him on the phone and he came over. He checked her pulse and said she was dead. She had had two or three strokes and died."

"It was a total shock," Morley says. "No one had any idea that my mother had any health problems at that point in her life. And it was very sad for us to lose my mother at such an early age."

Betty Ann agrees. Agnus was only fifty-eight at the time of her death, and Norris was sixty-five. "Norris was devastated. It never occurred to him that something like that could happen at that point in his life," she says.

Doctor Jones said later that Agnes had clogged arteries. She had the circulatory system of a person in their seventies, not late fifties. Growing up on the farm and eating a diet heavy in dairy products, fat and grease—a diet that was so typical in those days– had taken its toll.

Now Norris was on his own.

In the spring of 1950, Morley noticed that his four-year-old female peregrine falcon, Blacky, was starting to get "clucky," as the farm hands at Steen Nelson's farm used to say, when describing chickens anxious to mate. "It was obvious that she wanted a family as she scraped nest depressions in the mews and went through all the sounds and actions of an adult falcon at the aerie in March or April," Morley says.

Morley had read about a man in Germany who had raised peregrine falcons in captivity, so he thought he'd try the same thing. He wrote several falconers that he knew and asked them to send him a male peregrine to give Blacky a proper mate. But since she was an imprinted bird—meaning, she thought Morley was her mate—the male peregrines (tiercels) did not fare particularly well, trying to share a nest with Blacky in the mews.

"The first tiercel I got was an adult and Black thought he was to eat only," Morley dead-panned. "The second went the same way and I barely saved the lives of both birds. The third tiercel seemed to make out slightly better but only for a short time. After the second day, while I watched things went fine. However, it was not long until I heard the cries of the tiercel again and was lucky to save his life."

In the meantime, the neighbors were getting quite upset about all of the commotion in the Nelson's mews. "Our neighbors, Bill and Pat, lived next door, and she had a new baby, too," says Betty Ann. "And she came over one morning, and said, "All of that bird squawking noise is driving me crazy!" And I thought, the poor dear. But it didn't matter to Morley. He was conducting this great experiment. He just told them it would be over in a couple days. But I remember that the birds just fought constantly. It's like your kids right before they go to college—you'd just like to ring their neck."

After the third male failed to mate with Blacky, Morley was ready to give up.

Much to his surprise, however, Blacky laid an egg on April 24, 1950. In the next few days, she laid a total of six eggs. She broke one of the eggs by mistake, and laid another one two days later to replace it.

Morley had a hunch that the eggs were infertile, given the cool reception that the male peregrines received in the nest with Blacky. He wrote to several falconers in the United States, and they all agreed that the eggs couldn't be fertile. This turned out to be correct: the eggs didn't hatch after more than thirty days of incubation.

"This seemed like a rough situation for Blacky, so I decided to go out to a local prairie falcon aerie and take two young prairie falcons to see if she would raise them," Morley says. "After lengthy discussions with my wife, I decided to put them under Black that night. The minute the little falcons touched Black, they began to peep and she became very excited. There was no doubt in my mind, Black thought she had hatched her own eggs."

After the birds settled down for the night, Morley went to bed and set his alarm for 4:30 a.m. He wanted to see if Blacky would feed the young prairie falcons. He took some beef heart and gave it to Black in the mews.

"Much to my surprise, she immediately began to feed them. The little falcons were not stuffed as is so typical of other young birds. They went after the small bits of meat and took it from Black's beak as she held it above them. It was a very interesting experience for me and I was lucky enough to get around 400 feet of movies of the whole situation."

Morley let the two young prairie falcons grow up to the point where they were ready to fly. But he couldn't set up an outdoor hack site, an artificial nest, from which they could take off as they would in the wild, because Morley's house was located too close to other homes where young children were present. The nest in the mews was in a protected area inside the garage—not outside on a high perch, like a hack site would need to be.

"I tried it for a few days but the young birds flew down to children and got into all sort of trouble, and so did I," Morley says.

And so he released the birds to the wild after they were strong enough to fly and hunt on their own.

Morley's captive-breeding experiment caught the attention of a handful of falconers in the United States after he wrote an article for an early falconry journal, published by Luff Meredith. "The whole process did not prove anything although there is no doubt in my mind about the possibility of husbandry with any of the falcons," Morley wrote. "Certainly the peregrine falcon, and possibly others, could be raised in captivity. It would be possible to raise ten to fifteen birds with each pair in captivity. The eggs could be taken to hatch in an incubator and the old birds would breed again to lay another set of eggs. Artificial insemination should work, too, which would give the falconer a chance at a good many fertile eggs."

By anyone's yard stick, Morley's sky-blue thinking on the subject of captive-breeding and artificial insemination laid an early foundation for more serious experiments on the subject twenty years later at Cornell University by Tom Cade and his associates. By looking ahead to the next possible steps—artificial insemination and removing eggs to increase productivity—Morley articulated the general method that The Peregrine Fund would use to rescue peregrine falcons from the brink of extinction. By writing it up for the falconry journal in 1950, Morley showed just how knowledgeable and far-sighted he was about the subject.

"Morley's experiment with Blacky was the first indication that imprinted birds could be used as surrogate parents to raise

Nelson family collection
The Nelson family prepares for an outing in the late 1950s.

young in captivity," says Cade. "He gave us some good ideas that we used in our own experiments much later."

The article about the peregrine and its prairie falcon chicks caught the full attention of many falconers, triggering a number of letters to Morley from all over the country. "I saw the picture of your bird feeding the prairies. At first it seemed to me quite incredible," said Johnny McCabe, in a letter to Morley dated July 6, 1950. "Your write up of this business is very interesting and unusual indeed."

In the fall of 1950, the Nelsons welcomed their third child into the world, Suzanne Lee Nelson. Susie was born on November 20, 1950, the day before Thanksgiving. Betty Ann's good friend, Dorothy Beyerle, brought her Thanksgiving dinner in the hospital, and Morley and the two boys were outside the window of the old red-brick St. Alphonsus Hospital in downtown Boise, jumping up and down in the snow.

Now the Nelson family would have three children sleeping in the kids' bedroom on 326 Ada Street.

In the early 1950s, Morley began to dream about taking a

major trip to Alaska to search for the king of all falcons, the gyrfalcon. He had to find a way to pay for a trip to get a gyr.

To any accomplished falconer, a peregrine falcon is considered a trophy, and a gyrfalcon is the ultimate prize. Gyrfalcons are the largest birds in the falcon family. Found only in remote areas north of the Arctic Circle—Alaska, Canada, Greenland, Iceland, Scandinavia and Russia – they are fierce hunters that can do everything a peregrine can do and more, and their large wings propel them to great speeds, during the stoop or a thrilling tail-chase.

Now that Morley had been practicing falconry for more than 15 years, he was ready for the ultimate challenge—capturing and training a gyrfalcon. For centuries, gyrfalcons were the most desired bird of medieval falconers, such as Emperor Frederick II of Hohenstaufen, Holy Roman Emperor, King of Sicily and Jerusalem, (1194-1250). Morley had Frederick II's 637-page treatese on *The Art of Falconry* (English translation published by Stanford University Press, 1943), on his bookshelf. Frederick II writes for pages in great detail about how to train "gerfalcons" and hunt cranes, herons and even gazelles. Clearly, the emperor pushed the envelope as far as possible taking big forms of game with the most powerful falcon on earth.

In a narrative in which he compares the value of different falcons for hunting cranes, Frederick II said, "The peregrine falcon resembles the gerfalcon in that she can be taught to take cranes, and in her flight she approaches that of the larger falcon in swiftness more nearly than any other species. The (gyrfalcon), however, holds pride of place over even the peregrine in strength, speed, courage, and indifference to stormy weather."

Morley wanted a gyrfalcon for all of those reasons and more. During the ill-fated attack on Kiska Island, he had observed gyrfalcons and Peale's peregrine falcons, and he was impressed by the size and the coloration of gyrfalcons. In the Alaska region, they have a white dark-speckled breast, and dark blue feathers covering their wings and head. Their eyes are pure black, just like the peregrine and the prairie.

Ultimately, Morley wanted a gyrfalcon for the sheer thrill of watching it hunt and no doubt, for the prestige. "I had read so much about gyrfalcons in the old books from England and

Persia and all of the other books that I read about falconry," he says. "They're the most powerful falcons on earth, and the biggest ones, too. The Al-Can Highway was just opening up, so I was very interested in finding an aerie that I could reach from a car or a boat."

In the winter of 1950-1951, Morley's correspondence file was thick with letters from a handful of falconers around the United States who gave Morley hints about where to find the birds in Alaska. The best clues came from his snow survey boss, Arch Work of Ashland, Oregon, who had known Morley for several years now, and understood his passion for finding the king of all falcons.

Work told Morley that he had talked to several agents with the U.S. Fish and Wildlife Services, including a Mr. Scott. He also identified the proper contact for applying for a permit to remove a gyrfalcon from Alaska.

Says Work, "Scott had the dope as follows: Tom Cade, a senior student or so at the University of Alaska is believed to be studying the gyrfalcon and Scott thinks he is writing up something on his studies. Cade's address is University of Alaska, College, Alaska. Scott feels sure that (Cade) would gladly answer any inquiry you might address to him."

What a nifty tip that would prove to be for both Morley and Cade! Talk about two birds of a feather.

Morley wasted no time writing Cade, inquiring about his research in Alaska, and Cade wrote back to Morley on May 12, 1951.

Dear Mr. Nelson,

I received your letter about gyrfalcons today, and I want to assure you that I am very much interested in your project. I am afraid, however, that I am not in a position to be of much help to you right now. Mr. Scott was a little bit confused about the bird I am studying; it is the peregrine falcon, not the gyrfalcon.

But I am interested in the gyrs and hope to do a study on them sometime in the future. The only gyrfalcon aeries that I know about in Alaska are as follows: (1) Last summer in August I made a trip into the interior of Seward Peninsula and found two miners, Mr. Sterling Montague and Mr. Frank Whaley, who own and operate the Rainbow Mining Camp, who had a pet gyrfalcon that they had taken from a nearby aerie about five miles from their camp on the Nuxapaga River. They showed me the aerie, and at that time, the male gyr was still on the bluff. Mr. Montague unfortunately had killed the female at the time the young bird was taken. They had raised the gyr on fish, of all things! – but despite this, the bird was one of the most handsome, well-proportioned falcons that it has been my experience to handle. I worked with her at the camp for about a week, hardly enough time to train her properly, and then I had to leave. Unfortunately, I could not persuade them to give the bird up, and I do not know what finally happened to her. This is the only aerie that I am personally acquainted with; however, I know that there are two gyr aeries on the Savage River in McKinley Park. They have been under observation by Dr. Adolph Murie, the park naturalist, for several years, and I am sure he would be more than happy to have you film the birds in the park, but I doubt that you could take any of the young birds for training. If you are interested in this possibility, I would suggest that you write directly to Dr. Murie for more specific details. He is not at the park right now, but will be there in a few weeks. Then, Dr. Arthur A. Allen, when he was in Alaska looking for the Bristle-thighed curlew's nest, found three aeries in the Askinuk Mountains, south of Mountain Village on the Yukon River. I do not know the exact location of these aeries; Dr. Allen said they were on what he calls the "castle rocks." There is one other lead, and you may take it for what is may be worth. Mr. Montague, who I believe is a fairly accurate observer, told me that on a trip that he took in 1948 down the Noatak River he and an Eskimo killed over ten gyrfalcons, a lamentable slaughter admittedly, but it does give a good idea of how many falcons must nest along that river. Mr. Scott has suggested the Kobuk River also as a good possibility. He has flown over that country and reports that both rivers have extensive bluffs along their upper courses. I myself have tentative plans for visiting these rivers in the future to check on the gyrfalcon population in that region.

This summer I will be on the Porcupine and Upper Yukon Rivers studying the breeding peregrine population of that region. There is a possibility that I might come across some gyrs on the Porcupine. (I believe none have ever been recorded from the upper Yukon). In the event that I do find one or more gyr aeries you and Mr. Work might like to know about their location, and if we could make some arrangements for keeping in contact with each other this summer, it might prove most profitable for both of us. I personally would be very much interested in seeing any films that you might get of gyrfalcons. . .

I hope this information will be of some use to you, and here's wishing you a most successful season in Alaska.

Sincerely yours,
Tom Cade

Thus began an exchange of letters that allowed these two men to get to know each other better as Morley planned his trip to Alaska in the summer of 1952. It was the beginning of a life-long friendship involving two visionaries who would take leadership roles in many aspects of the raptor-conservation movement in the years to come.

At the time, Cade was twenty-one years old. He had just graduated from the University of Alaska at Fairbanks with a bachelor's degree in biology. Cade had been practicing falconry since he was a teen-ager in Texas. "I had absolutely no contact with any other falconers until I went to southern California," Cade says. "I just learned it from books, the *National Geographic* article and the Craigheads' book, *Hawks in the Hand.* That book was quite an inspiration for many people."

Morley sensed from Cade's letter that the budding student was quite knowledgeable about birds of prey. Cade's studies on the Yukon and Porcupine rivers almost made Morley drool with envy.

Morley fired off a letter back to Cade, asking him to send him a gyrfalcon if he ran across one during the summer research studies. Morley also gave Cade a lot of advice about making hoods, how to use them, and how to set up a dho-gazza trap. Cade wrote back that he wanted Morley to send him a hood and a bell, and asked about the best way to ship birds in a box without harming them. Cade also talked about visiting Morley at the end of the summer of 1951 when he arrived back in the continental United States. But as things turned out, Cade had to wait until the fall of 1952 to visit Morley in Boise.

In the fall of 1951, Cade sent Morley two male peregrines, an eyass and a haggard from the Yukon River.

Over the winter of 1951-52, Morley learned that all of his schemes for bankrolling a gyrfalcon expedition to Alaska fell through. The SCS rejected the proposal to do a joint snow survey-gyrfalcon trip. None of Morley's inquiries with magazines and book publishers bore any fruit.

Cade fared better. He obtained a grant from the Office of Naval Research to conduct the first ornithological investigation of the Colville River in northern Alaska. There, he discovered a major concentration of cliff-nesting raptors—comparable to densities that Morley had observed in the Snake River Canyon in Idaho—including both peregrines and gyrfalcons.

With no other options available, Morley put together a personal trip to Alaska with his friend and fellow falconer, George Bradshaw, and his dad, Norris. They would leave in late June and drive up the Alaska Highway to some areas where they might find gyrfalcon aeries on the east slope of Mount McKinley. Norris offered to drive his car, a four-door DeSoto sedan. The trip lasted nearly a month. Morley had enough work time logged in the winter months that he took all of the time off as paid vacation.

Morley, Norris and George Bradshaw loaded up the dark-blue DeSoto at 5 a.m. on June 15, 1952, and headed north. They towed a small trailer with their camping gear and climbing equipment packed inside. They crossed the Canadian border near Glacier National Park at the Port of Roosville, and passed by the magnificent peaks of the Canadian Rockies in the vicinity of Banff and Lake Louise, heading north for Edmonton. From there, it would be a two day's drive—475 miles on paved and gravel roads—to the beginning of the Alaska Highway at Dawson Creek, just inside British Columbia.

From Dawson Creek to Fairbanks, Alaska, the gravel road covers 1,523 miles, an enormous distance, the equivalent of driving from Denver, Colorado, to New York City, on dirt. The trio camped out along the way, and with Bradshaw's help, caught fresh salmon for a hearty meal whenever possible. They camped on the rocks to avoid grizzly bears, and of course, they had Morley's .357 pistol in case they needed it.

"The trip up through British Columbia and the Yukon Territory was wonderful, and it had some very interesting moments," Morley says. "We had a grizzly bear come into our camp one night and check out all of our equipment, making all kinds of racket, and I was sure glad that I hadn't brought my German shepherd with us at that point. Fortunately, the bear didn't try to break into our tent and eat us. When we woke up the next morning, we saw these monstrous tracks in the sand and gravel around our camp. That was a wake-up call, that's for sure."

After ten days of continuous travel, the men rolled into Fairbanks on a rainy afternoon, and they decided to get a hotel room, take showers and get cleaned up. They had reports from Cade and park service officials that a good spot to look for gyrfalcons would be near Summit Lake on the Richardson Highway.

A half-day's drive south of Fairbanks brought Morley, Norris and Bradshaw to Summit Lake. The setting was breath-taking: Enormous snow-bleached peaks of the Alaska Range rose dramatically above the circular emerald lake. Morley fetched his field glasses and began to survey steep cliffs above the lake for a gyrfalcon aerie. With the naked eye, Bradshaw immediately saw several of the largest falcons he'd ever seen circling above.

"Whoa, Morley, can you see those big birds flying above the cliffs?"

"Yep, yep, I see 'em," Morley said. "I can see one aerie in the cliff, and it's got a couple of young birds inside. Heh, heh, we are a couple of lucky dogs!"

Before Morley set up ropes to scale the cliff nest, known as The Hoodoos, he and Bradshaw set up a dho-gazza trap, basically two nets standing side by side suspended between wooden poles. They placed a pigeon, tied down with a leash, to the ground in front of the nets. If one of the juvenile birds tried to fly away, it might be attracted to the pigeon and get entangled in the net.

Morley solo climbed the cliff face with his rope in the rucksack. He climbed above the nest, and then slowly rappelled down to the rock face. Bradshaw waited near the trap to catch a bird if it flew in that direction.

"One of the most exciting things that I'd seen for many, many years was to get on top of the cliff, and peer down on two young gyrfalcons almost ready to fly," Morley says. "I couldn't see the adult birds, but I expected that they might come after me at terrific speed. About that time, the mother gyrfalcon, a pearly gray marvelous bird, dove out of a cloud at tremendous velocity down toward me, and then with this most powerful bass voice, an entirely different call than I'd ever heard before, it went, *gak gak gak gak gak*, and her mate returned the call, *gak gak gak gak gak gak*, and there was a great feeling of authority in this call. The bird was literally saying, I will really take you, however strange as I may sound.

"But the bird did not attack me in the violent way that some prairie falcons and some peregrines have done in the past. She came by me, and she had an unusually powerful wing beat, and my, she was big. I went over the cliff, and the first juvenile falcon flushed from the nest, as I rappelled to a point where I was even with the nest. When I got there, the second bird flushed and flew directly to the pigeon by the dho-gazza trap. Apparently the gyr was hungry, and it decided that now was the time to give it a try, so it made an attack on the pigeon and got caught in the net, just as I hit the base of the cliff. I ran over to the bird and got it untangled in the net and quickly put a hood on it."

Now, you're mine, Morley thought, marveling at the beauty of the young gyrfalcon. There was something special about the appearance and the attitude of the hawk that reflected the extreme elements in which it lived.

On their way to Summit Lake, Norris drove the DeSoto down the same draw they came up, called Gun Creek. They had left their small trailer near the Alaska Highway due to the rough terrain. As they cruised down the draw, pausing to miss small boulders and holes in the dirt road, two grizzly bears ran past them and cut in front of the car.

"I looked over at the speedometer, and it indicated the car was going about thirty mph," Morley says. "Now it's true that there was a great deal of slippage on the gravel road, so the car

Nelson family collection

Morley and Betty Ann with kids, dog and birds near their Boise home.

was not going thirty mph ground speed, but those two grizzly bears for some reason cut right in front of the car, and put more distance between us as they ran down the road in this incredibly smooth gait. You could see their golden fur ruffle around their powerful shoulders, and I suddenly had a profound respect for the grizzly bear.

"Everyone had told us that you couldn't protect yourself with a .357 magnum pistol, and I didn't believe it. I felt that there wasn't anything that could come in and bind to you if you had a .357 magnum single-action Colt to fire and you waited until the animal was within a few feet and then dodged it like you would if a bull was charging at you, and you kept firing. I figured there was no way that a grizzly bear could really hurt you when you were armed that way, but after seeing those bears charging down the road so effortlessly at thirty mph, I had to change my opinion and eat my words. From that point on, we made even more of an effort to stay away from the bears."

As the group drove back through the Yukon Territory and Alberta to Idaho, Morley worked on manning and training the gyrfalcon. "The bird took to domestication beautifully. She would stand on the fist and eat on the fist, but her wild drive for life was certainly beyond anything that I had seen before in any of the falcons I'd trained. There was a critical intensity of action displayed by the gyrfalcon that you would not see in a peregrine, but it was very similar to the prairie falcon. The bird was very powerful and very active and it understood that in quite a dignified way."

When the men rolled back into Boise, Betty Ann and the family were overjoyed to see them, and it was a wonderful reunion. The men's cheeks glowed a ruby red color from being outdoors for nearly a month.

But then, a tricky question arose—who would get to keep the gyrfalcon? George Bradshaw was interested in training the bird, and Morley remembered what a fine job Bradshaw had done with training and flying a prairie falcon when he lived in Trementon before the war. But of course, Morley wanted the gyrfalcon, too, and if he hadn't been on the expedition, no one would have had the rock-climbing skills to get it.

"So we had a big discussion about, well, who was going to keep the falcon? Obviously, we both wanted it. So we decided to flip a coin, and I lost. I honestly never thought I'd lose the coin flip, but I did. That was tough to watch George drive away with the bird."

Bradshaw took the bird to his house in Boise and began training it. On a professional level, Bradshaw was a civil engineer, and he was in the midst of changing jobs from the U.S. Forest Service to Morrison Knudsen. He told Morley that he'd try to train the bird until he had to leave Boise for an overseas assignment with MK. That seemed reasonable to Morley.

"Well, the way it turned out, I was a little eager to see the bird fly, and I tried to convince George to let it fly without a leash. This was a different situation because I had not trained the bird all the way from the beginning, and I was not certain that the bird could be trusted to fly on its own, but I was anxious to see the bird fly up into the sky and wait-on, and then come down and take some wild quarry. George agreed, and he cast off the bird into the wind, and it flew up into the sky with its powerful wings. Then it must have seen something way off

Photo courtesy Tom Cade

Tom Cade prepares to release a shrike at Peters Lake, Alaska.

in the distance, which is typical of a gyrfalcon, and it went after it, and never returned. I was devastated."

So that was the beginning and the end of Morley's first gyrfalcon expedition. It had been a wonderful trip, he thought, and no one expected it would end this way.

In the fall of 1952, Cade promised to visit Morley in Boise on his way back to Southern California after a busy summer research field season in Alaska. Cade was married now, and his wife was expecting their first child in late October. Morley and Cade corresponded throughout the summer, so Cade knew that Bradshaw had lost the gyrfalcon, and he told Morley that he would bring him another gyrfalcon. That certainly got Morley's attention because Cade already had sent him two peregrines. He knew that Cade would be in a position to get a bird for him, and he'd be true to his word. So he sent Cade enough money for a bus ticket to Boise from Seattle.

On a fine Indian summer morning, Cade arrived in Boise after an all-night bus ride from Seattle with the gyrfalcon. Morley picked him up at the bus station. The two had a lot in common, and heaps of mutual respect for each other. Morley was considered one of the finest falconers in North America—along with his other accomplishments to date—and Cade had spent months in the remote wilds of Alaska, banding peregrines and rough-legged hawks, and studying the behavior of peregrines in particular. At the end of the 1952 field season he had banded twenty-seven peregrines and forty-one rough-legged hawks.

"I was really impressed with his work in Alaska," Morley says. "He took many solo trips down the Yukon River for over a month. You've got to be a real man to be able to pull that off and survive."

"Morley was really a gung-ho guy. He was going 100 miles per hour all the time," Cade says. "Boise was just a little itty-bitty place back then. Morley liked Boise because it was located on the western edge of the time zone, and he had lots of daylight hours in the evening for spending time with the family and flying his hawks. After work, we'd go swimming in the Boise River, have a nice family affair, and then we'd go hawking. There were lots of farms around here then that attracted tremendous numbers of ducks. Half a dozen species or more of ducks would flock to these feedlots, so we'd fly our hawks at them."

Cade had a chance to see Blacky perform. "She was a fantastic bird," he says.

And Morley had a chance to see the gyrfalcon fly for Cade. The bird had learned to fly to the lure, but he had not been able to train it to fly at wild game or to wait-on. Morley figured the bird would rise to the occasion, after it had some more training.

But unfortunately, he never got a chance. Soon after Cade left town, the bird got aspergillosis, a deadly fungus, and died. Now Morley was 0-2 with gyrfalcons, and more than a little depressed about the fate of his first two gyrs.

But there was a happy ending. Morley's good friend, Luff Meredith, just about cried when he heard that Bradshaw had lost the gyrfalcon, and he enlisted a friend of his in Alaska, Vern Siefert, to go back to Summit Lake in the spring of 1953 and see if he could trap another member of the brood for Morley. In a matter of weeks, Morley opened a shipping package to find a large, healthy gyrfalcon inside with blue-gray plumage. He named it Tundra. It was likely the first gyrfalcon to be trained successfully to capture wild quarry in North America, Cade says.

When Tundra arrived the following June, however, she was in good health. Morley was extra careful as he went through the training regimen with Tundra. He didn't want anything to happen to this gyr, and he certainly didn't want to let it fly loose before it was trustworthy. He took about six weeks to man the bird and train it to fly to the lure.

Once Tundra was ready, Morley enjoyed flying her at wild game more than any other bird. "She became one of those inspirational things in life that comes from developing an intimate relationship with a very special wild bird or a dog or a horse," he says. "'Tundra personified the power and strength and feeling of singularity that she had to have to survive in probably the harshest environment that any bird of prey experiences anywhere in the world. She was a good eater, and she even ate egg yolks from a spoon with a vengeance, which was good for her health. And when I flew her at wild game, man, she ruled the air. If any other birds tried to attack her, she would climb above them and make a dramatic vertical stoop at them, telling the others, you'd better not mess with me. She'd do that to golden eagles and ravens and falcons of any size.

"The effect of her aggressive behavior was quite impressive to me. When the other birds tried to attack her, she would make a frightening stoop on peregrines migrating through the area, and it looked as though she was going to take them down. The peregrines had to turn upside down and hold their feet up to keep from being seriously hurt by Tundra's attacks. I had never seen a peregrine endure such a dominating adversary before."

After Morley had met Tom Cade in person, they began to correspond with each other more frequently about falconry, trips, family including Tom's research on peregrines in Alaska and

Morley's new connection with the Walt Disney Company on several movie projects.

In the spring of 1959, Cade invited Morley to accompany him on the Colville River in Alaska to assist with an annual raptor survey. The Colville was located in an uncommonly biologically rich area in the extreme northern portion of Alaska, above the Arctic Circle. It contained many pairs of gyrfalcons, peregrines and rough-legged hawks, not to mention wolves, caribou, grizzly bears and more.

"For the Arctic, it's the richest area in Alaska," Cade says.

Morley arranged to take some vacation time. He was not only anxious to see all of the gyrs and peregrines with his own eyes, but he also planned to film as much of the trip as possible for a new personal film that would be called *Nature's Birds of Prey.*

Cade had been down the river several times before. He arranged all of the trip logistics. They would fly to a launch point on the upper Colville River, and float about 100 miles to the airstrip and research-staging outpost at Umiat. Cade arranged for a collapsible Kalamazoo twenty-foot boat, a green canvas-covered craft with a square stern. When collapsed, the boat could fit into a small aircraft. They would sleep in a large Army surplus tent. They packed mosquito hoods for the swarms of biting bugs they knew they would encounter on shore. They also had all of Morley's camera gear, a tripod, and ropes for rappelling into nests.

They would spend about a month on the river, banding juvenile birds at every nest they encountered, and filming whatever action they could find. Cade packed enough food for two weeks, and he arranged for an aircraft to resupply them about half way through.

When they launched the boat on the banks of the Colville, Morley felt a wonderful feeling of calmness settle in as they drifted into the northern Alaska wilderness. The early morning sun created a metallic glow on the river's surface. Already, they could see all kinds of bird life flying overhead. The river was broad, about 200 feet wide, and it was bordered by 300-foot cliffs on both sides. They could look forward to seeing a peregrine or a gyrfalcon aerie around nearly every bend. Morley almost felt like an early explorer, the countryside seemed so pure, wild and untouched, and the falcon populations were robust. This was the domain of nature at its finest.

He was so happy to be there. "Tom always had this wonderful sense of humor, and he was very knowledgeable about the whole area," he says. "My respect for Tom grew even more as I could see how he had this wonderful academic ability for assimilating his knowledge of falconry with his sophisticated biological knowledge of birds of prey from studying at the universities and doing his field work in Alaska. His depth of knowledge and understanding was certainly going to be needed in our efforts to improve public awareness about the virtues of birds of prey."

As they floated downriver, Morley got lucky more than once with some great scenes for his movie project.

"I remember one particular time, when we were camped across the river from a gyrfalcon nest," Cade says. "Morley was set up with a telephoto lens, and while he was shooting the female gyrfalcon, the male came back to the nest, and landed on top of a rock with a ptarmigan in its claws. Then, right at the same time, a grizzly bear casually walked by in the background. Morley got the whole thing, the bird landing, and the bear walking by, all on cue. That was neat."

Sometimes they'd set up a blind close to a nest, or they'd see if they got lucky without one. "We got several really good shots without a blind," Cade says. "Morley set up next to a gyrfalcon nest, and one of the adult birds flew in with a ground squirrel in its talons, not more than twenty feet away. I mean we were close. We got the whole sequence with the young eating the ground squirrel.

"And then another time, Morley got a shot of a female peregrine sheltering her brood in a pretty heavy snow squall. She came in and landed, and spread out her wings to protect the young, and Morley got all of that."

Morley and Cade took turns climbing into the nests to band the young. In some cases, Morley asked his partner to do the climbing so he could get the sequence on film. In one instance, though, Morley stayed in the boat while Cade climbed up to an old raven nest that was occupied by peregrines on a rocky point in the face of a near-vertical cliff. There was a ledge in the center of the cliff that Cade planned to use to traverse over to the nest. He attempted the climb without a rope.

"Tom started out on the traverse and got around on to the point, but just as he was stepping out and around to where he could reach around the point and have a hand on both sides, the exfoliated rock began to slide and the entire traverse ledge broke off and slid down," Morley says. "Tom grabbed onto a rock hold on the cliff with both hands and one foot, saving himself from going down with the rock slide, which crashed down to the bottom of the cliff, next to the boat. Tom needed help fast. He was in a gnarly position, with no where to go, and no possible way of going down. It was obvious to me that I would have to go around the cliff, where there was a couloir that I could climb and get over the top, and lower a rope to Tom."

Morley leaped into action, grabbed the climbing rope and his rucksack, and started up the slope.

"It took me a half hour to get up there, and Tom hung in there in that most precarious position on the cliff. I tied off the rope, and proceeded down the cliff to a point where I could throw the rope down to Tom. It was a difficult descent because I was trying to avoid breaking any more rocks loose and hitting Tom with debris. I drove in a number of pitons along the way as a safety precaution. When I let the rope go, it hung only 10 feet above Tom, so he was going to have to climb up the cliff to get out of there.

"By the time I got down to him, I could see that look in Tom's brown eyes that I'd seen in battle so many times—the strength and courage and the wonderful feeling of come on, let's go give her all we've got, we're going to make it. It's one of the most beautiful things that a man can ever see in another man—to see that kind of strength and courage in the face of adversity."

Cade scaled the cliff successfully, and Morley gave him a big slap on the back when they got to the top. "Thanks, Morley," he said. "That was an awful spot. You saved my ass."

Morley and Cade had been on the river for about two weeks now, and they expected to see an airplane with their food drop anytime. All they had left was bacon and pancake flour. On an early morning, the two were working on reaching another peregrine nest to band the young. They heard an airplane fly over, but it was mostly cloudy, and the aircraft apparently didn't see their greenish-gray boat tied up on shore. They shot a flare into the sky, but the pilot didn't see it. They assumed that the pilot might come back and try to find them again, but he didn't. They would have to get by on whatever food they could shoot with a shotgun, and the bacon and pancake mix.

Nelson family photo collection

A hitchhiker eyes Tom Cade as he and Morley Nelson prepare for another day of their 1959 float trip down Alaska's Colville River.

"We tried to kill each other with our cooking," Cade joked. "Morley's pancakes left a lot to be desired, and I wasn't any better at it. We both lost a fair amount of weight on the trip."

The next morning, the two woke up, made coffee, and noticed that their boat was gone. "What the hell?" Morley said. "Where's our boat?"

During the middle of the night, a big windstorm jolted the bow line loose and blew the boat into the river.

"I thought maybe it had sunk," Cade says. "But we walked downriver about a mile or so, and there it was, washed up on a gravel bar. So we had to swim across the river to get it, and then drag it back upriver to camp. That was miserable and cold, but hey, we didn't lose the boat. If we had, we would have had to wait until the pilot found us, if he ever came back to look for us."

Ah, the rigors of a true wilderness adventure.

Morley recalled that toward the end of the trip, Cade had increasing trouble stomaching Morley's pancakes. "The problem of eating was really getting difficult, because all we had was a slab of old bacon, and we didn't have any eggs or butter left for adding to the pancake flour. So I had to use bacon grease to mix with the pancake mix and a little water, and after many days of the same thing, Tom could barely bring himself to lift his fork."

"Oh man, this is pretty bad," he said, grimacing with each bite.

"Yep, it is," Morley freely admitted. "I'm such a lousy cook that you're not going to hurt my feelings."

"It's a good thing we're almost to Umiat."

Indeed, they were a couple days away, and they knew the end was in sight. Morley had captured two juvenile gyrfalcons that he would bring back to Boise for use in falconry and films. They had visited forty-one peregrine aeries and twelve gyrfalcon aeries. They had seen wolves, moose and grizzly bears, and many other species of birds. All in all, it had been a very successful trip, despite the paucity of food.

Finally, they pulled into Umiat. Cade gunned the motor, and they beached the boat on shore at the take-out. They walked over to an area where four men were busily loading food and material into an airplane with pontoons.

"They seemed to be in such a hurry that they didn't notice a couple of scruffy looking guys with beards walk up to say hello," Morley says. "And we overheard them saying, 'We'd better help those fellows who got lost on the Colville River.' And then I said, 'well, you guys look awfully busy, where are you going with this big load of food? And they said, "Well, we have a big problem. Tom Cade and Morley Nelson have been out there on the Colville River for almost a month, and we're getting really concerned that they have been hurt or they are out of food . . . .'

"And I smiled and said, 'well, now wait a minute, this is Tom Cade and I'm Morley Nelson, and we'd like to help you load the food, but I think it's meant for us.' And then we all had a big laugh, and the men invited us to join them for a meal in the mess hall. We had a big dinner, and several stiff drinks, and Christ, we were very happy to eat some real food."

They ate a tremendous amount for dinner—it seemed as though they had a bucket-sized hole in their stomachs. "Suddenly, we both had to step outside because there was no way our systems could take on so much alcohol and food in that quantity, and we both got sick," Morley says. "It was a strange feeling to be so hungry, and not be able to hold the food down, so we realized that we would have to take it a little bit easy coming back to eating normal food again."

The next morning, they flew out of Umiat to Fairbanks, and caught flights back to Boise and southern California. Their wonderful trip together cemented a lifelong friendship.

Cade was ready to finish his doctoral dissertation on peregrines and gyrfalcons and start teaching. In the fall of 1959, he took his first job as an assistant professor at Syracuse University in New York. Morley was certain that he'd accomplish great things.

"I read his dissertation with interest, and the paper had very outstanding academic strength and it was very complete," Morley says. "He had discovered more about the ecology of Alaska falcons than anyone knew at the time. It was obvious to me that Tom Cade was going to be one of the finest men that ever worked for the benefit of raptors in the history of the world."

Chapter Seven

Minding the snow
1948-1971

Morley embraced his new job as snow survey supervisor of the Columbia River Basin with uncommon zeal. In short, he hit the snow with both skis pointed straight downhill, freshly waxed for maximum speed. Here's a guy who loved being out in the mountains as much as possible, and was fascinated by the interrelationship between snow, soil, water, rivers, dams and crops. He had landed the perfect "day job."

From the Soil Conservation Service perspective, the government had found the ideal person for the job. Morley had taught skiing and snow survival as a company commander in the Army's 10th Mountain Division. He was an instant asset because he could teach his fellow surveyors how to survive in the snow—that is, how to build snow caves and lean-tos to survive a cold night in the mountains, in the event of an emergency, without a tent. And from his association with fellow ski trooper Monty Atwater at Alta Ski Area in Utah, he knew how to recognize and cope with avalanche hazards.

Plus, the job *forced* Morley to scout the mountains for new snow course sites and, at the same time, raptor territories. He always had his eyes peeled for birds of prey—in the sky, on a cliff, or in a giant nest atop a tall ponderosa pine tree.

As Morley traveled throughout the backcountry, he documented raptor nesting areas for future reference.

"I found peregrines, prairie falcons, golden eagles, red-tails, ferruginous hawks and goshawks, and every kind of bird there was because I had to go to all of these places," Morley recalled. "And I loved it. The principle of the work was so important to everybody that we had lots of money in the budget for travel and setting up snow courses. We didn't spend very much, though, because we camped out most of the time."

For Morley, the first order of business was to establish as many snow course sites as possible in all of the major river basins in Idaho, Wyoming and Montana on the west side of the Continental Divide. From 10,000-foot snow-streaked peaks butting up against the Teton Range and the west side of Yellowstone National Park, all waters flowed to the Pacific Ocean. Snow depth and snow-water data in the mountains provided the basis for forecasting stream flow runoff on the Snake and Columbia rivers, the two largest arteries in the Pacific Northwest. As new snow course sites were added to a skeleton of sites in Idaho and elsewhere, the accuracy of forecasts would improve. Snow course sites had to be located near the headwaters of major streams, as well as in scattered locations in lower elevations to get the most accurate data about snow depth and snow-water content in each river basin.

Along major rivers such as the Snake, as well as in subbasins of the Snake, farmers, power company officials and municipal leaders were all extremely interested in the streamflow forecasts because they provided advance warning about the potential of big floods running in the springtime, as well as the potential for dramatically low flows that could harm irrigated agriculture. Farmers were especially interested in the forecasts because about seventy-five percent of their irrigation water originated as snow in the mountains. In the arid Great Basin, rainfall didn't amount to much in the summer months. Periods of zero appreciable rainfall could occur for sixty to ninety days from July 15 to September 1. Hence, farmers relied on deep snowpack in the mountains for their summer water supply. An elaborate network of twelve major dams and reservoirs on the Snake River caught the runoff and a labyrinth of irriga-

U. S. Department of Agriculture Soil Conservation Service photo
Morley, right and another Soil Conservation Service snow surveyor, pose for a publicity photo demonstrating their winter survival skills.

tion canals delivered enough water to irrigate more than three million acres of cropland throughout the banana-shaped Snake River Plain, from the Wyoming border to Oregon.

"Watermasters, irrigation district managers, reservoir managers and many others in the West have come to rely on a new tool in water supply management. It is a tool that enables measurement in advance of seasonal water supply; to determine in advance how large a check may be drawn on the bank account, that is, how much water can be drawn from reservoirs or will come from watersheds. This new tool is the snow survey," wrote R. A. "Arch" Work in Oregon, and W. T. Frost in the November 1948 issue of *Agricultural Engineering.*

The business of measuring snow on a West-wide coordinated basis was just beginning to develop when Morley joined the Boise Snow Survey office in 1948. The first snow course sites were set up by Dr. James Church, a University of Nevada hydrologist, in 1906 in the Lake Tahoe area. A number of other enterprising individuals and utility officials launched snow surveys in California, Washington and Utah in the 1920s. By 1935, Congress was convinced that snow surveys should become a nationally sponsored program. Four years later, the Soil Conservation Service's Division of Irrigation and Water Conservation was charged with overseeing and implementing the program.

In the winter of 1948, approximately 1,000 men from a variety of federal and state agencies made 2,300 snow surveys on 950 snow courses in twelve western states and British Columbia, according to Work and Frost. The snow surveyors traveled 21,000 miles on skis, snowshoes or mechanical oversnow vehicles to collect the snow data.

In Morley's first summer on the job, he spent a great deal of time in the mountains, setting up new snow course sites with the assistance of officials from the Forest Service, Bureau of Reclamation, Geological Survey, National Park Service and the Idaho Water Board.

Typically, Morley and his associates in the Forest Service drove on dirt roads to a mountain pass closest to a potential new snow course site, and then they'd haul in the equipment from there on foot. They'd search for locations in the headwaters of a stream that were sheltered from the wind. Big snowdrifts would render a distorted picture of the snow depth.

"Because I'd been in the mountains so often, I knew the types of areas that we were looking for to get the most accurate forecasts," Morley says. "Finding places out of the wind, protected by trees, was important for the best accuracy possible."

Beyond the task of setting up new snow course sites, one of Morley's first responsibilities was to travel around the Northwest and recruit other people to assist in measuring snow course sites in the wintertime. The SCS reached out to rangers with the Forest Service, and farmers who were members of a Soil Conservation District, to help out with the job. They also recruited a few "freelance" pilots to assist with snow course readings.

In the Soil Conservation Service Snow Survey office in Boise, Morley worked under the leadership of James C. Marr, an irrigation engineer who was involved in establishing many of the first snow course sites in Idaho. When the SCS first jumped into the Snow Survey business with both feet in the mid-1930s, very few snow course sites were established anywhere in the Snake River Basin. Much like Morley, Marr was a man of action.

In August 1935, for example, Marr covered 2,300 miles on primitive roads and trails, setting up new snow courses in Yellowstone National Park and western Wyoming. Marr's enthusiasm for the job caught the attention of the national chief of the SCS Bureau of Agricultural Engineering, Henry McCrory. Concerned that Marr might be working too hard, McCrory suggested that his superiors should tell the man to slow down. "He is working so hard that I am afraid he faces a nervous breakdown if he does not ease off somewhat," McCrory said, according to a history of the SCS.

At the end of the 1935 field season, Marr had established one-fourth of the 1,000 snow course sites that the SCS expected it would need for the most accurate streamflow forecasts possible.

Marr was happy to have Morley on the Snow Survey staff in Boise. Morley was just as energetic about setting up new snow courses, he understood the importance of making the most precise water forecasts possible, and he had the built-in expertise of skiing and snow-survival skills from the war. Morley also had established ties with the media. He made sure that when new snow measurements were made, he publicized the information as widely as possible so farmers, ranchers, power company officials and the general public had the latest data to plan accordingly.

During the winter, snow surveys occurred on the first of each month, beginning in January. Toward the end of December, Morley and Forest Service officials would head into the mountains to measure the snow depth and snow-water content at established snow course sites throughout the upper Columbia River Basin. The measurements were taken by two-person teams, for reasons of safety and accuracy. They could either ski or snowshoe into the snow course site, or they might take a Tucker snowcat, depending on access considerations.

Once they arrived, they assembled a cylinder-shaped 1.5-inch diameter aluminum tube and drove it into the snowpack until it reached the ground. The bottom end of the tube was ringed with a set of sharp teeth for cutting through the snow. When the tube reached the ground, the snow surveyor lifted the tube out of the snow to verify that he had found the ground layer.

The amount of snow inside the tube indicated the snow depth, measured in inches. Then the surveyors weighed the tube to determine the snow-water content of the snow, calculated in inches of water. The snow surveyors would take at least five measurements at a snow course and calculate the average snowdepth and snow-water content for the site. The same procedure was repeated at hundreds of sites throughout the Columbia Basin.

After collecting snowpack data, Morley and his comrades had to develop streamflow forecasts for all the major streams in the Columbia River Basin. Because about eighty percent of the water used for irrigation arrived in the form of snow, streamflow forecasts from the SCS Snow Survey office gave potato and sugar beet farmers advance warning about what type of runoff was expected. The forecasts also provided information on whether reservoirs would fill to a full depth during the summer, and what kind of late-season flows could be expected.

The Army Corps of Engineers and the Bureau of Reclamation watched the Snow Survey's water forecasts with interest because they provided early warning about whether reservoirs would need to be evacuated in the spring to accommodate heavy runoff.

Power companies kept a keen eye on forecasts to determine how much electricity could be generated from dams in the Snake River Basin and the Columbia River Basin, based on natural runoff. The same was true for the Colorado River Basin, which provided irrigation water and drinking water for southern California.

Many important aspects of the economy depended on the forecasts. In water-rich states like Idaho, government officials wanted to make sure that every drop was put to work before the snowmelt flowed in Washington and Oregon. Idaho Republican Governor Robert Smylie described the state's water strategy in 1955, when Congress was considering the construction of a high dam in Hells Canyon. "We are seeking—we always have been and we always will—the ways and means of developing, comprehensively, every drop of water that tumbles from the snow packs of the Snake River watershed," Smylie told Congress. "And when we have used that water, whether to help grow a potato or turn a turbine, or both, then and only then, will we willingly send it flowing into the canyons below Weiser to help develop still another empire farther west."

U.S. Dept. of Agriculture
Soil Conservation Service photo
Morley, right, uses a probe to check the depth and water content of the snowpack.

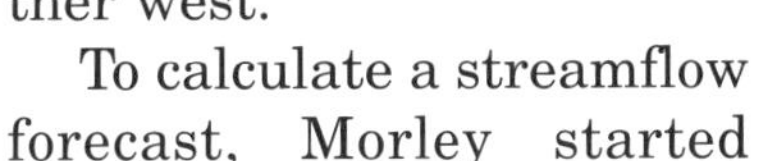

To calculate a streamflow forecast, Morley started with snow-water information, and combined that with base flow information, soil-moisture, ground water levels, long-term weather forecasts and historical flow information to make a coordinated forecast. For example, the Geological Survey had a number of stream gauges set up on streams to provide base flow information on an historic and real-time basis.

Soon after Morley joined the Boise Snow Survey office, the number of streamflow forecasts increased markedly each year, as new snow course stations were installed and more data were available. In 1948, the first year of Morley's service, the Boise Snow Survey office produced thirteen streamflow forecasts for Idaho. The following year, thirty forecasts were issued, and in 1951, sixty-six forecasts were produced. In 1956, more than 100 forecasts were made, with an accuracy level of eighty-nine percent, according to SCS records.

In the surrounding western states, a similar trend occurred. More than 100 forecasts were developed for the states of Montana, Colorado and California. By the early 1960s, the SCS

Snow Survey was hitting its stride, pumping out highly accurate forecasts in the mountain states, benefitting farmers, power companies, dam-regulators, recreationists and many others. At the end of each water year, snow surveyors from eleven-western states congregated at the annual Western Snow Conference, held in a different location each year, to share new insights about improving water-supply forecasts, new oversnow vehicles and equipment, the benefits of skis vs. snowshoes, and more.

The water supply forecasts in 1960 and 1961 were particularly helpful because of a shortfall in snow during the winter months. Below-average snow survey measurements gave farmers and others advance warning that their irrigation supplies would be reduced by as much as fifty percent of normal.

"In the spring of 1960, southern Idaho farmers saved close to a half-million dollars by adjusting their operations to the predicted irrigation water supply," Morley says. "There wasn't enough water to irrigate their usual plantings so they kept some land out of production."

In 1961, the winter drought became more pronounced, and farmers were duly warned by the snow survey offices throughout the Northwest. In a speech to the Idaho Reclamation Association in September 1961, Morley gave a review of the leanest water year in more than two decades, and he was quite candid about the grim prospects for the coming year.

"The forecasts for 1961 were probably more helpful to individual farm and ranch operators than they have ever been," he says. "Almost everyone was aware of the light snow pack and low water supply outlook and made the best possible use of water supplies. We have many quotes from members of your Reclamation Association stating that they dropped the use of water from six acre-feet to 3.5 acre-feet per acre and had, if anything, slightly better crops.

"It is interesting to note that the data from our mountains also forecast the drought in the Midwest in North Dakota, South Dakota, Montana and parts of other states.

"I know all of you are concerned about 1962, and I have brought my worn charts from last year to make a preliminary estimate of the situation as it stands to date. Soil moisture conditions as of this time are extremely dry to unusually great depths. Reservoir-stored water is critically low throughout southern Idaho excepting on the Payette River. For this coming season, we are almost entirely at the mercy of the natural flow of rivers. Unless there is drastic change in the present trend of precipitation and snowfall, we face a situation as bad or worse than this year.

"The first real chance of detecting the trend for the coming year will be through the snow survey measurements made as of January 1, 1962 . . . .

"In an effort to increase the precision of forecasts, all of the basic data, snow-water content, precipitation, base flow and soil priming factors have been analyzed with an electronic computer. I am very happy to say that all of our equations have been improved through this detailed analysis which would have been impossible with engineers and calculators alone. The analysis indicates the best snow courses and other data to make the most accurate forecasts for 1962 and future uses."

To give farmers and ranchers as much advance warning as possible, Morley coordinated seven water-supply forecast meetings in the spring of 1960 and seventeen such meetings in the spring of 1961. Information was spread by the Bureau of Reclamation, the University of Idaho Extension Service, the Idaho Reclamation Association, various irrigation companies, Soil Conservation Districts and the Soil Conservation Service.

As part of the forecast meetings, Morley and others suggested that long-term water conservation methods could squeeze more drops of water out of a lean water year. He recommended leveling farm fields, lining ditches with concrete, installing water pipelines to reduce evaporation, water-control structures, water-measuring devices and more efficient distribution systems.

Looking ahead, Morley suggested that the drought of the early 1960s would not persist forever. Showing a graph of a snow course in the upper watershed of the Snake River, and historic streamflow information on the Snake River at Heise, just east of Idaho Falls, he told farmers, "We must remember that before another five years have passed, we probably will face the possibility of floods stemming from a heavy snowpack which has developed in our mountains. All scientific evidence supports this contention, and I am sure we will all look forward to that change."

Morley received a big hand from the farmers in the room. Even though the situation was grim, Morley provided accurate and candid information about the true water supply. Everyone appreciated the honest truth.

By this time in the evolution of the snow survey program, most every western state had enough snow courses established to provide highly accurate streamflow forecasts for the April-September irrigation season. Morley's own track record in 1961 showed an average standard of error in the range of three percent to twelve percent, based on April 1 snowpack information.

"It doesn't get any better than that, still to this day," Morley says with pride.

In an SCS oral history interview with R. Neil Irving, a farmer, extension agent and former SCS employee, Irving pointed out the magnitude of the work that Arch Work and Morley Nelson supervised in the Northwest.

"Well, we had a boy in Boise by the name of Nelson that was very active in this when he was with the Service. He was a corker. Morley Nelson was our snow survey supervisor, and Arch Work was in charge of the thing from Portland, Oregon, for these western states. They got the information, they figured out where they'd have to make the measurements over the years, and they can't predict it down to the gallon of water that's coming, but it seemed like they could get pretty close.

"They were the gospel."

On a good day, Morley could be coaxed into exalting a little humor, and in 1949, Morley's boss Arch Work laid out a challenge:

The loggers have their legends about Paul Bunyan and the rock drillers have their yarns about John Henry, and now comes a tall tale from the snow surveyors. The first that your forum editor heard of it was through a copy of a letter from Arch Work of Medford, Oregon, to Morley Nelson of Boise, Idaho, which read in part:

Is it true that you have trained your falcon "Blacky" to take snow surveys?

I find it hard to believe that Blacky, with his plunging dive of 400 mph can drive a thirty-foot sampler clear through twenty feet of snow and on down through ten feet of soil, thus securing combined snow and watershed soil moisture samples. Our informant stated that Blacky makes fifteen round trips a day to courses 800 miles away from Boise. Why don't you teach him to weigh the sampler to avoid so much flying back and forth to laboratory scales?

Yours doubtfully,
Arch Work

As a result of Arch Work's inquiry and also to put the quietus to the many wild rumors which by that time had gained wide, and growing, circulation, Mr. Nelson was moved to make public a full report of a fantastic research project which he enthused with the true snow surveyor's ardor for bigger and better snow surveys, had undertaken during the past winter. His report is herewith reproduced in full:

The Snow Surveying Falcon
By Morley Nelson

For the past few months, what was thought to be a secret work was carried on in training a hunting Falcon to make snow surveys. However, some way the startling initial results seeped through the iron curtain of the Boise office and a communication was received from Mr. Arch Work stating the exact nature and capability of the Falcon in this study. Since the results were startling and at first promising, then a complete failure, I will describe them.

A Falcon which had been used as a game hawk for two years was selected for the study. Her name is "Blacky" and she is one of the noble family of peregrines. The long association of her ancestors with the Aristocracy and Medieval Nobility of Europe as a sporting Falcon is what eventually caused her to be dangerous in snow surveying.

She was trained to carry the snow tubes as if they were the "lure" and scoop (or dive) on the snow course, which was first shown to her by going in on skis. From then on, she could be sent out to the snow course as often as necessary, at any distance, and she would return to her perch in my front yard with the results. All that was necessary was to go home with the scales and weigh the samples which might come from any desired course in the Columbia Basin. The samples were doubly useful because of the Falcon's speedy, diving approach to the course. As is traditional with all Falcons, Blacky climbed thousands of feet above the course. Folding her wings she stooped at the course and dropped the tubes with such velocity that they would penetrate 20 feet of snow and 10 inches of soil, returning to me with a composite snow and soil sample, which was very useful.

The remote courses first studied in this manner gave very promising results. However, upon attempting this procedure with a course very close to a recreation area I ran into insurmountable obstacles.

Blacky, in her first flight over Bogus Basin Recreation Area, could not resist her old sporting blood's urge and made her speedy attack on a skier's alpine hat with its pheasant feather. The feather made this fair game and Blacky, in attempting to strike her game, konked the skier with the snow tube. She returned to me in great humor with a few pheasant feathers and no snow and soil sample. This did not at first appear serious to me until I heard of the strange incidents at Bogus Basin. There was talk of guided missiles and all sorts of fantastic things among the skiers.

In desperation, I tried one more flight which was near Sun Valley. It happened to be a bright day and most skiers had on their goggles. Again, Blacky's fighting blood was aroused and she confused the skiers' goggles with the large eyes of the owl family, her natural enemy. Every skier on the hill with goggles, as well as those with feathers or fur on their head, were rolled down the mountain by Blacky's speedy stoop. She came back in great spirits with an assortment of feathers and bits of goggles instead of samples. After news got out of incidents at Sun Valley, it was obvious that such a sporting bird is dangerous in areas dotted by recreational skiers such as the western United States.

As a result, Blacky has been reassigned to her traditional work of hunting crows, rabbits and game in the ancient methods of Falconry, which fit only remotely with the civilized world, skiers and snow surveying."

When Morley traveled around the Columbia River Basin to present water-supply forecasts for the coming irrigation season, he often brought a falcon or a golden eagle with him to educate the public about birds of prey. He knew that farmers, ranchers and hunters frequently shot hawks, falcons and eagles, fearing that they preyed on livestock and game birds. But few of the ranchers and farmers that attended water-supply meetings had ever seen a falcon or eagle up close, and Morley knew that a falconry demonstration would capture the fancy of the agricultural community and the media.

On April 6, 1958, for example, Morley traveled to Twin Falls, Idaho, to give a water-supply forecast on the Salmon Falls area to local farmers. After the water meeting concluded, Morley gave a demonstration on the art of falconry with two of his two prairie falcons and a couple of pheasants that he secured at a local game farm. The demonstration was covered by the Twin Falls *Times News,* resulting in a front-page article in the Sunday edition. Now the entire Magic Valley region would read about Morley's fascination with falconry, and his sermon on wildlife conservation.

Noting that he had just finished work on the Walt Disney movie, *Rusty and the Falcon,* Morley dazzled the audience with an exhibition. Standing on the snow-covered sagebrush plain adjacent to the incised Snake River Canyon nearby, he cast off "Whizzer" into the wind, and she climbed the brisk air currents of a spring day, rising higher and higher in the sky with each rapid wing beat. Before long, Whizzer was waiting on, a speck in the sky now, keeping a watchful eye on her master.

Morley grabbed a colorful ring-necked pheasant and threw it into the air, causing Whizzer to immediately tuck in her wings and begin the vertical stoop. The crowd gasped as the falcon sped toward the ground at high velocity, approaching more than 150 mph.

"Poof!" The brown and black feathers of the pheasant went flying when Whizzer nailed the bird with a major body punch with its foot. The lifeless pheasant hit the ground with a thud, and Whizzer stood atop its prey with pride. Morley let the bird have a snack before putting it back on his fist. "Gawd, these falcons are so courageous, it is fascinating," Morley says. "They will tackle anything, and can even be trained to hunt eagles. Both falcons and eagles will kill and eat rattlesnakes, bull snakes, lizards, squirrels or rabbits. They are strictly meat eaters."

Raptors protect farmers by eating lots of mice, he said. "Mice have done nearly $5 million dollars damage to crops in Oregon in the past year. If the raptorial birds were allowed to thrive, the mice population would not be so high."

Mice are a favorite meal for falcons, he said, pointing out that it takes thirty-five mice to keep a single family of falcons fed during the springtime.

These were precisely the right things to say to farmers, who were not only impressed by the falcon's hunting prowess, but also with the information about what the birds ate on a daily basis. Morley's credibility was unassailable when it came to predicting mountain streamflow, and by extension, his authoritative words about raptors and falconry carried more weight with the farmers than they might have otherwise.

Former Senator James A. McClure, R-Idaho, was the executive director of the Idaho Reclamation Association in the 1950s. The reclamation association was a politically powerful umbrella group representing all of the farmers and ranchers in southern Idaho. Morley's dual role as a snow surveyor and falconry expert opened doors for the wildlife conservation message that wouldn't have been opened otherwise, McClure says.

"I think the birds of prey had a constituency, they always did, but it was small at the beginning," he says. "The interests in irrigation water and water supply was a much broader interest, with a much different group of people, and I think Morley's credibility in the water user community greatly enhanced his ability later on when he was working on the protection of birds of prey. Because, you see, the water users were a different group of people who would ordinarily not listen to a Morley Nelson, but since he was the snow survey man, they listened to him with keen interest."

Neil Sampson, who worked alongside Morley for the SCS in the 1960s, says Morley was pushing the limits of tolerance inside the Dept. of Agriculture when he showed up on national TV, wrote papers for journals or starred in movies.

Nelson photo collection

Film legend Gary Cooper poses with Morley's eagle, "Clyde," in the early 1960s at Sun Valley.

"Morley always stuck out of the organization, and big government organizations don't take kindly to that," Sampson says. "They much preferred to have employees who don't rock the boat and stay out of the public eye. I know that Morley was always at the ragged edge of stuff because he was always doing something that was too creative or too bizarre for the administrators, but nobody was able to put the screws to him. His golden reputation with the Service—I mean even though he was a renegade, he was always extremely loyal to the Snow Survey program—and his reputation with the public meant that administrators could fuss and fume, but not much else."

Plus, the SCS was open to Morley's conservation message regarding birds of prey because he placed the value of wildlife conservation on an equal footing with soil and water conservation. The SCS carried an article by Morley on the virtues of birds of prey in the April 1957 issue of *Soil Conservation*. For the most part, the SCS supported his crusade.

After more than twenty years on the job, the national office of the SCS embraced Morley's celebrity status and sent him to several national Boy Scout jamborees in the late 1960s. Sampson and Morley were sent to a week-long event at Camp Asaayi in New Mexico in the summer of 1970 to give falconry demonstrations. Morley tied his educational pitch about birds of prey to the need to conserve habitat, and protect the soil—the backbone of all land-dwelling critters on earth.

"The Service finally recognized that the birds could be used as a major conservation tool," Sampson says.

When Morley struck out into the snowy backcountry to make snow measurements, he usually drove a Tucker Snowcat, a Frandee oversnow machine or a Sno Ball Industries snowcat. The heavy-duty snow machines predated the invention of snowmobiles, so they were the best machines available for the job at the time. Some of the models had been used by the U.S. military in World War II.

In steep country with deep snow, it was not feasible to use a large oversnow machine. Morley had to use skis or snowshoes to take the snow measurements.

Driving the snowcats around in the backcountry allowed the snow surveyors to reach snow courses quicker than they could on foot, and they were subject to less exposure to the outside elements—that is, unless breakdowns occurred. The big snow machines did break down fairly frequently, requiring the surveyors to know a fair amount about trouble-shooting engine trouble.

"The snowcats were kind of a half-way rig, and they broke down a lot," Sampson says. "They were usually a pickup truck or a jeep mounted with tracks, and nothing was engineered to match."

But sometimes, the breakdown had something to do with operator error. Morley wasn't bashful about admitting that he'd screwed up more than once, and flipped a snowcat. He wrote about one of the flip episodes in the 1953 edition of the "Snow Surveyors' Forum," presented at the Western Snow Conference.

High Center of Gravity

By Morlan W. Nelson
Snow Survey Leader, SCS
Boise, Idaho

A couple of years ago, the Boise office received a Frandee over-snow machine for use in Southern Idaho. It was, and still is, an excellent vehicle. But one fault was brought to my attention and sort of burned into my memory.

It was in April and the season melt had started. Two extra men were going along for training purposes, on a trip which involved rather long over-snow movement. Wayne Criddle, of the Boise office at that time, Phil Whiting and Mel Williams of the local office were the other members of the party.

We started out in the new Frandee. The trip went well as there was a crust on the snow, and we averaged unusually fast speed for about 20 miles. Then it came my turn to drive. In this model, two men sit side by side in the front, with the battery in between, and two more can ride in back. I was amazed at the way the machine maintained its speed and the power of the light motor. Tires driving a rubber belting and wooden cleat track make very little noise compared with Tucker Sno-Cats. There were many melt holes alongside the big trees but I did not have occasion to sidehill the machine until it was too late.

As the trail began to get a little rougher, I maintained approximately the same speed. Suddenly, we came to a ridge, formed by two large trees melting the snow almost to the ground and leaving a high curving ridge between them. Well, before we got to the ridge, I knew we were going too fast and that I could not ride it out. There was only one choice. To try and go down into the hole, which would tip us uphill, rather than downhill. As we slid off the ridge it was obvious that we were going to turn over so I called, "Look out, she's going to go over!" Then in sort of slow motion we slid off the ridge, went about 20 feet further, and the machine started rolling over. With loud and complex noises caused by bouncing saws, snow tubes, gas cans, lunch boxes and breaking glass, we stopped – upside down. The oil was gurgling out of the motor through the oil cap. It occurred to me that I should catch that oil, but somehow I could not get organized to do so. There seemed to be a lot of weight on top of me but no pain.

After a moment of silence, Wayne called out, "Anyone hurt?" We all answered no and then someone started to laugh.

However, there was a strange feeling along my right leg, a sort of burn! I could not move but Mel began to break out of the machine. In a very short time, it became imperative that I move, and fast, as there was no longer any doubt about my problem.

As soon as Mel got out, I found I was upside down and directly below the battery. The battery acid was slowly trickling down my leg and had gone too far anyway. With a very unusual burst of speed and energy, I broke loose from the wreck and jumped into a nearby stream to dilute the acid. Of course, this caused considerable laughter but my passengers were not too happy with my driving and the humor came much later.

By using the two ropes, we were able to right the machine, get one track back on and after adding two quarts of oil, we were ready to go again. The machine looked sad but it would still run. Mel and I went on to make the last snow survey and we all returned home late at night, in relatively good shape.

When the time came for a report on the new machine, I recommended lowering the center of gravity, along with several other points. As things have turned out, they also moved the battery but I did not have nerve enough to make that recommendation, although I heartily concurred.

In the late 1960s, Morley took some dignitaries to a high-mountain snow course, including former Idaho Republican Governor Don Samuelson, his wife and Betty Ann. By some stroke of good luck, Morley asked the governor to drive.

"We all went out in the Tucker Snocat, and I'm sitting behind the governor for some reason, and he goes up on this hill and flips the damn thing!" Betty Ann says. "And everyone inside is going, ho ho ho, and I thought, I don't know if this is all that funny. How are we going to get out of this mess now?"

Well, Morley had two ropes in the back of the cab, and they righted the machine and went on their way.

Sometimes, the Snocat couldn't be fixed, and the snow survey crew had to spend the night out in the chilly winter air. Morley remembers that he flipped another snowcat during a field trip at the Western Snow Conference near Jackson Hole, Wyo. The cat couldn't be fixed before dark, and the men had to hastily dig a snow cave for their sleeping quarters. They had extra sleeping bags in the snow machine.

"When that happened, the men who were with me said, "Oh gawd, we can't understand this. This is a real problem, yada yada yada. And I said, well, you've got to remember, we've got to walk thirty or forty miles back to get out of here, and we're better off sleeping here in the snow. And they're saying, this guy is crazy as hell. But I said, look, it's not that dangerous because there's no Germans out there shooting at us. All we've got to do is sleep in the snow. In other words, I had to remind them that I've been in much worse situations. I had lived so much in the mountains that I didn't look at it as a great big disaster."

And so the naysayers had to zip it, and camp in the snow. Morley, the master snow survival instructor, would make sure that everyone was warm and comfortable for the night.

Because of Morley's winter survival skills, Betty Ann says she didn't worry that much about him when he was gone on snow survey duty, often for days at a time. "I didn't worry about him because he knew so much about survival. I really didn't worry about him being able to take care of himself because he was good at that," she says.

"But I used to get more worried when he'd go out and look for lost skiers. They'd always call Morley because he had access to a snowcat. And he'd go out there in the middle of the night and look for them up by Bogus Basin. I thought that was scary. They were always running out of gas in those machines. Anything could happen."

Morley's snow survival workshops, and the SCS's commitment to safety made a difference in saving lives, Sampson says.

"In large part, due to Morley, we never lost a person in the mountains. It was a pretty darn remote operation. Lots of things could go wrong. But there was a constant safety push by the Service, and Morley taught everyone a constant respect for the wild country in the winter."

New technology for oversnow machines hit a breakthrough

U. S. Department of Agriculture Soil Conservation Service photo
A commercially-produced snow machine demonstrates its ability to ford a stream near Sun Valley in 1950. The early "Sno Cats" tended to be top heavy and sometimes tipped over or got stuck.

in the early 1960s, when several manufacturers produced one-person "snowmobiles." A number of innovators had created one-person sleds before that time, but none of them performed well enough in the deep snow to entice a manufacturer to produce them on a major scale.

In 1964, Ray Brandt of Boise talked to Morley about ground-testing some snowmobiles manufactured by Polaris. Brandt wanted to put the Polaris sleds through a rigorous examination to see if they could travel long distances in deep powder snow. Morley welcomed the challenge. He put together a high-powered snowmobile expedition, from Mackay, Idaho, across the Pioneer Mountains to the west, and over Trail Creek Summit to Sun Valley. The five-day expedition would feature a climb of more than 3,000 vertical feet, take them through deep snow in dense forests in rugged terrain, and pass by one of the worst avalanche areas in Idaho.

Morley recruited nineteen men for the trip, including doctors, mechanics, Idaho Fish and Game wardens, a ski instructor from Sun Valley, an official from the New Hampshire Fish and Game Department, and more. Polaris provided eleven ruby red snowmobiles for the trip, and the company paid for all of the trip's provisions. The trip logistics required Morley to cache food and fuel along the route in the fall before the expedition began, before the snow came.

Morley wrote up the trip for *Snowmobiling* magazine in November 1970. Here are some excerpts from the article, called "Pioneer Safari."

It all started when Ray Brandt of Boise, working with Polaris Industries, suggested that we undertake a difficult trip that would really test the new light machines. Jack Wilson and I, working the snow survey program for the U.S. Department of Agriculture's Soil Conservation Service, had been working closely with Ray as the new machines were being developed, so he came to us for a suitable location. . . .

"We were more than willing, as there was much more at stake than merely an exciting trip. The snow survey job in Idaho was, at that time, growing faster than our ability to keep up. New demands for water forecasts were springing up almost daily with the rapid development of Idaho's land and water resources, and new forecasts meant more snow data must be gathered. The old methods of travel into the mountains, either on foot or on large oversnow "snowcats" were rapidly becoming too slow and unreliable. We were getting desperate for a faster, better way to get men in and out of the high mountains in the winter, and the small "snowmobile" looked like just the ticket – if it worked.

The accomplishment of two primary objectives received foremost consideration in planning the expedition. First, the capability of snow machines to perform in powder snow at high elevations. Secondly, the use of over-snow machines for a new form of mountain recreational activity would be evaluated. If the machines proved practical and maneuverable under the severe test conditions, winter work such as snow surveying, forest and game management, powerline maintenance, etc., would be advanced significantly. Also under the second objective, if they proved feasible for general recreational use, a whole new field of exciting winter sport would be born . . . .

During the night, the temperature dropped to 20 degrees below zero. In the morning, the hoarfrost on the machines was an inch thick and it was a severe test to get the machines started and keep them running. Many ideas for the new developments in the machines were conceived during this trip. Principles learned and considered at that time have since been adopted, resulting in the production of a more versatile, capable snow machine for use in all snow conditions. . .

The following day, we planned to make our final run to Sun Valley. Sometime during the night, a great storm gathered. When we woke up snow was falling rapidly and the snow plumes and banners flying from the mountain peaks signaled increasing avalanche danger. It was now a race against time, and the same question was in all our minds, could we cut the ledge and cross the avalanche path before the new snowfall started another slide? We hurried to the slide path and began cutting a ledge in the snow for the machines. The job proceeded slowly as it was necessary to post outlooks and to allow only one party on the face of the slide at one time. Dick Stauber expressed his concern about the increasingly high avalanche hazard as the new snowfall continued to raise the danger level significantly.

Finally, the ledge completed, the machines began their one-by-one advance over the perilous slope. Idling across to keep excess vibration from triggering the dreaded slide, the operators kept one eye on the trail and other cocked warily on the precariously balanced snow above. As the last machine crossed safely, I breathed a deep sigh of relief and turned to the final part of the trip—the relatively easy run downhill to Sun Valley . . . .

Near the bottom of the trip, we experienced the most serious machine breakdown of the entire trip as Gene Stoker's motor developed an ominous knock and, finally, with continued driving, literally flew apart. We were able to tow him the last mile with another machine and even the prospect of buying a new motor failed to dim Gene's happy spirits. It was a triumphant group that breezed into Sun Valley to be welcomed by Louie Shadduck, then head of the Idaho Department of Commerce and Development, and the resort executives. A pioneer feat had been accomplished and I, for one, felt both relief and exhilaration.

As we look back at this trip, it is interesting to note how the

Nelson family collection

Morley's eagle "Clyde" lands on a Larson Snomobile in a promotion photo for the machine. Clyde was actually going after a piece of meat attached to the snomobile.

winter recreation business has exploded today. What was a very questionable trip only six years ago is now a common feat for anyone with a modern snowmobile and the necessary experience. The dangers still exist, but the proven practicability of these small machines is now history. What was once an unreachable part of Idaho's magnificent winter landscape is now a popular recreation area, where families enjoy outdoor thrills formerly restricted to the hardy few.

In the mid-1960s, Polaris improved its line of snowmobiles, and Ski-Doo, Yamaha and many other manufacturers leaped into the growing business to satisfy customer demand. The Boise office of the Snow Survey and other stations in the West acquired a fleet of snowmobiles to quicken trips to snow survey sites. Now people like Morley would become technically proficient in the operation of snowmobiles and learn how to fix them in the field. Morley was sold on the technology, and he and his boys used to go snowmobiling on weekends for fun.

"Dad got a little bit crazy on those things," says his third son,

Tyler Nelson. "He'd show us how to jump the snow drifts over the sagebrush in the Owyhee Desert, and he liked to burn down the plowed roads at blazing speeds."

Morley wasn't ready to stow his skis in the garage forever, but after more than twenty years of scaling mountains the hard way, he was ready to get there by snowmobile. Plus, he was almost fifty years old, and the boyish energy and athleticism that had kept him in good shape over the years was beginning to diminish—just a bit.

At the same time that Morley was experimenting with snowmobiles in the mid-1960s, he and a number of other snow surveyors began discussions with nuclear scientists and NASA officials about the possibility of measuring snow with space-age technology. The discussions at that time would lay the groundwork for automated snow survey sites that would relay snow depth and snow-water information by satellite to SCS Snow Survey offices throughout the nation. The new sites would transmit snow data in a matter of seconds, accelerating the process of obtaining measurements and saving tremendous costs in field-collection work.

Use of the new space-age technology was just the beginning of a number of new techniques that would be tried in the late 1960s to gather snow course data electronically. Another method, using nuclear technology, measured snow using an isotopic snow gauge. Having studied nuclear engineering in college, Morley was intrigued by the technology, and as a snow survey supervisor, he had to be on the front lines of that advancement as well.

In the winter of 1963, Sampson remembers an official experiment with a cesium nuclear snow gauge on Hemlock Butte, near Pierce, Idaho, that went awry. The government was hoping to find "peaceful" uses of the atom, and was often a little cavalier in pushing new technology before it was completely safe, he says.

"I got a call from the Army Corps of Engineers in Walla Walla, saying that we had just lost two feet of snow overnight at Hemlock Butte," Sampson says, "and I knew that couldn't be right. We were in the middle of winter, and there was snow everywhere up there. So I went up there to see what went wrong. I had an ammunition box full of radioactive monitoring equipment, and we'd been trained to use it. When I got up to Hemlock Butte, and checked the radioactivity of the snow gauge, it lit up like a church bell. Then I took another reading on the mountain top, and the woods up there were just glowing with radioactivity. I mean the readings were just pegging the meter.

"So that winter, the woods were full of radioactive snow, and when the runoff came, it went all over the Pacific Northwest. I called the Army Corps about the Geiger counter results, and they told me to keep it a secret. The public never knew about that radioactive snow."

The cesium gauge was discontinued for a few years and sent back to the research lab for more experimentation.

On October 28, 1970, Morley joined a number of nuclear scientists and engineers in Sun Valley for a special informational meeting on an isotopic snow gauge. The meeting was co-sponsored by the Idaho Nuclear Energy Commission and the Soil Conservation Service.

As one of the lead-off speakers at the meeting, Morley tried to convince the participants that a better way of measuring snow course data was needed by remote means. The isotopic snow gauge seemed to be the perfect answer.

"In all of the years of snow surveying, we have not been serving every interest with our traditional snow tube or pressure pillow measurements," Morley said in his speech. "The structural characteristics of the snowpack itself and its depth are impossible to measure with pressure pillows. Those measurements made by snow tube and men on foot are not frequent enough to produce significant data for skiers, recreationists or the detail of the operation of multiple-purpose reservoirs in a great system such as the Columbia Basin."

Morley and the nuclear scientists proposed that a test pilot site for the isotopic snow gauge should be established on top of Bald Mountain, the 9,151-foot summit of Sun Valley Ski Resort. Morley and two other fellows from the Forest Service had established the snow course on Baldy in 1949. The proposal was approved by the SCS and put in place in the winter of 1971.

The main benefits of the isotopic snow gauge were that it could provide information on total snow depth, snow density, total water content, water content increases and decreases (trends), the amount of snow that had fallen since the last measurement, rainfall and rain intensity and snowmelt rates between measurements.

How did it work? The gauge used a radioactive source, Cesium-137, along with a detector to provide a more accurate profile of the snowpack. Using a specially engineered instrument, a probe would move up and down the profile of the snowpack to measure snow density and water content in half-inch increments—a much more detailed measurement than ever before.

"Sun Valley is an internationally known ski resort. Therefore, the use of an isotope at this site would have the potential of increasing the understanding of the peaceful uses of isotopes in general," Morley said. "The data from this isotopic gauge in some ways would affect most of the people in Idaho, and in a general way, many throughout the nation and in some foreign countries."

In the winter of 1971, an isotopic snow gauge manufactured by Aerojet Nuclear Company was installed on the top of Bald Mountain. The gauge did provide high-quality results, but after five years, the SCS determined that it was too expensive to operate, compared to other types of remote-sensing gauges.

In the meantime, research was underway to refine electronic snow course gauges that could beam information to a SCS Snow Survey office in milliseconds. In 1977, the SCS settled on automated radiotelemetry data sites, called SNOTEL sites. The technology for SNOTEL sites is similar to the "space-age" site installed at Bogus Basin Ski Resort in 1965, except snow data is transmitted by bouncing electronic signals off of tiny meteorites as they enter the earth's atmosphere.

Morley retired from the Soil Conservation Service Snow Survey office on December 3, 1971, well before hundreds of new SNOTEL sites were installed throughout the West. But he was on the leading edge of technological advances up to the day he retired. He always searched for ways to improve water-supply forecasts as long as he worked for the Snow Survey.

By the time Morley retired, the science of snow-water forecasting had been perfected throughout the United States to the point where all interested parties could expect highly accurate streamflow forecasts every year. Through the course of Morley's career, 144 manual snow-course sites had been established, and after he retired, a full array of 112 SNOTEL automated sites were added.

"It's still the most important thing we have going today," he says.

Chapter Eight

In the bull's eye

1951-1972

On a hot Saturday morning in May 1964, Morley drove his blue-and-white International Scout up a boulder-strewn four-wheel-drive road, near Melba, Idaho. He and his sons, Norm and Tim, were bumping along the jeep trail to check out a golden eagle aerie near the top of a brown, cone-shaped lava-rock butte on the north side of the Snake River Canyon. But as they neared the nest, all Morley could see was the unmistakable black barrel of a high-caliber hunting rifle sticking out of the driver's side window of a new Chevy pickup.

Morley stopped the car immediately. "I'm going to put a stop to this," he said, bolting out of the car.

He marched up to the driver's side window of the pickup, and stood between the gun barrel and the eagle aerie.

"Now, gentlemen, what's going on here?" Morley said, staring at the driver with his steel blue eyes. "You must be the kind of sportsmen who needs a little education if you're taking aim at those young defenseless eagles over there. To shoot at those young eagles for target practice is the lowest form of poaching that I can imagine."

"You son of a bitch," the driver said, charging out of his truck.

Morley took several steps back, planted his feet on the ground and lifted his chin. He was ready to fight the big hoss of a man, if it came to that. The shooter was about six-foot-five, 250 pounds, and his face was all twisted with anger. Morley was only five-foot-nine on a good day, but he was a whole lot tougher than that. He'd experienced hand-to-hand combat in World War II; he knew what it took to survive. The smartest thing to do, at this moment, was to lay it on thick.

"Now, look here. My name is Morley Nelson. I'm a falconer, and I take these eagles hunting in the field, and I've made Walt Disney movies with them, and they're one of the most intelligent birds in the world. I grew up on a farm, and I used to shoot hawks, too, but I know better now. If you're a hunter, you'd have to respect a bird like the golden eagle because it can take a wolf's head and crush it like nothing flat. They've got long sharp talons that'll bore a hole into the skull. If one of those eagles grabbed onto your arms, he could drive his talons all the way through, right to the bone."

"You've made Disney movies with eagles?"

"Yep. That's right."

"Well, we were just up here looking to test out my new rifle. We weren't going to shoot any eagles up there, we didn't even see them. We were just waiting for something to move. . . ."

"All right then," Morley said, reaching over to shake the man's hand. "you guys carry on. My sons and I are going to check out this eagle aerie."

The man returned the handshake, grasping Morley's hand like a vice. "Nice to meet you. I'll watch for your movies."

Norm stood outside the car, watching the confrontation dissolve into a reconciliation. "When I saw my dad's chin go up like that, I thought there was going to be a fight. Whew!"

From the time that Morley began flying hawks in North Dakota in the 1930s, he had many close encounters with people who shot at birds of prey. Remarkably, as many times as it happened, he never got into a fistfight or a gun battle. When people shot at his trained falcons in the air, Morley often fired a warning shot in the air or in the dirt, in front of the shooter's toes. But after he'd been home from the war for about five years, Morley was more likely to confront the shooters and give them a quick education about why they shouldn't shoot birds of prey. It was dangerous, but his silver tongue worked every time.

The sad truth was that most "sportsmen," ranchers and farmers had the same bad attitude towards birds of prey, not only in Idaho and the West, but throughout the United States. The only good eagle was a dead eagle—the familiar mantra that outdoorsmen learned from dealing with rattlesnakes. Somehow, Morley had to change that way of thinking. Being a rare falconer in the West, he knew that if he didn't do something, the situation wouldn't get any better. It was up to him to teach farmers, ranchers and sportsmen about the rich tradition and history of falconry. He had to share his wisdom of wild. He had to convince the public that birds of prey deserved absolute and permanent protection from wanton shooting.

Knowing what he knew about the history of falconry, it was a shame to realize how public attitudes toward falcons and eagles had changed so radically, from the time when the birds were treated like royalty, to a situation where they were shot on sight. What had gone wrong?

Following the confrontation with the hunter, Morley and the boys climbed to the top of the lava rock pinnacle, and had lunch. They observed three eagle chicks in the nest below. They sat on the lava summit and listened to the cool, north wind swirl through the rocks. Morley's mind faded into thoughts of kings, Crusaders and Mongols, who carried falcons and eagles with them during distant travels for food and sport. How could so many people be so ignorant of the rich tradition of falconry? Why were people so self-possessed when it came to shooting their guns?

In the midst of a daydream, Morley's mind wandered to a time 2000 years before Christ, in Central Asia, where the nomadic Mongols used falcons and eagles to hunt for food. Birds of prey could be more valuable than arrows or spears because they could travel great distances for prey, and they often made the kill. For centuries, trained falcons and eagles were a popular, effective weapon for gathering food until the advent of guns and firearms in the 1600s.

By the beginning of the Christian era, falconry had spread throughout most of Asia, the Middle East and Europe. It continued to grow and expand, becoming a favorite sport of the aristocracy. Morley looked at his boys, Norman and Tim, munching on their sandwiches, making idle talk on the pinnacle rock. He knew they could have been expert bird-handlers in the Middle Ages even as teen-agers. Marco Polo (1254-1324) wrote about his observations of participating with enormous hunting parties in central Asia in the 13th century in "The Book of Ser Marco Polo—The Venetian—Concerning the Kingdoms and Marvels of the East," edited by George B. Parks (MacMillan, 1927).

"After he has stopped at his capital city those three months that I mentioned, to wit, December, January, February, he starts off on the 1st day of March, and travels eastward toward the Ocean Sea, a journey of two days. He takes with him full 10,000 falconers, and some 500 gerfalcons besides peregrines, sakers and other hawks in great numbers; and goshawks also to fly at the water-fowl."

Roger Tory Peterson in *Birds Over America* (Dodd, Mead & Co., 1948) amplifies the scene.

"Marco Polo, dazzled by what he saw, reported that the Great Khan of Cathay journeyed once a year to the eastern part of his kingdom, to within two days of the eastern sea. He rode on a pavilion borne on the backs of four elephants . . . . Men on horseback gave the signal when cranes or other birds appeared overhead. Then the curtains of the pavilion were drawn while the Khan, in sensual splendor, watched the sport from his couch."

Frederick II (1194-1250), the Holy Roman Emporer and king of Sicily and Jerusalem, wrote an exhaustive treatise on falconry that revealed his intimate knowledge of many species of hunting falcons, especially the prized gyrfalcon. "Frederick the II was clearly one of the most amazing men of the Middle Ages, known as a marvel to the world and wonderful innovator," said Bill Mattox, a longtime friend of Morley's and president of the Conservation Research Foundation, who researched the origins of falconry for the Institute of Current World Affairs from his home in Denmark in 1967. "Whole books have been devoted to Frederick, who was, in addition to a falconer, a mathematician, architect, naturalist, and a man of great burning inquisitiveness. He founded the University of Naples in 1224 and has been variously described as the "first modern man upon a throne" and the "real beginning of the Italian Renaissance," Mattox says.

Not surprisingly, gyrfalcons were considered the most valuable birds. The birds were captured in Greenland and Iceland and traded for diplomatic favors. In a dramatic illustration of the birds' value, Mattox noted, "During the siege of Acre in Syria in 1104 A.D., a white falcon of the crusading King Philip August of France strayed and was later captured by the city. The King offered Sultan Saladin the enormous sum of 1,000 gold ducats for return of the falcon, but the bid was refused."

The penalties for stealing a king's falcon were correspondingly stiff. In sixteenth century England, during the rule of Henry the Eighth, anyone who tried to take a hawk from the royal nest, take a hawk egg from the nest, or steal the king's falcon would be subject to the "paynes of death," according to Kent Carnie, a longtime falconer who oversees the falconry archives at The Peregrine Fund in Boise.

If only these slob hunters realized the storied tradition of birds of prey, Morley thought to himself. I'd love to put one of these guys on a rack, and stretch their bones, and see what happens.

With the advent of gunpowder, rifles and shotguns, falcons and eagles were seen as direct competition between man and wild game. By the late 1600s and early 1700s, falconry had been all but eliminated in Western Europe as the aristocrats switched to hunting with guns instead of hawks. "And so these guys were all of a sudden shooting instead of hawking," Carnie says. "And anything that took a bird away from them—that's one less bird they can shoot—was the enemy . . . So you're going from a highly revered, protected critter and you switch to a situation where you're in competition with man, so the game keepers started shooting hawks . . . killing them and destroying them. Anything with a hooked beak became fair game."

The Puritans in England compounded matters by trying to ban any form of sport on Sundays. According to Carnie, "The Puritans did not believe in people having fun and enjoying themselves, and they had a majority in Parliament, so they made a law that banned the practice of any sports on Sunday.

And James I, said, 'Wait a minute. I can not maintain a citizen's army to protect the British Isles. I'll have to depend on the ability of the British citizens to defend themselves, when called upon, and if they can't shoot with their bows and arrows on Sundays, we're up a creek.' This was about 1617. Charles the First, the son of James, tried to push it further in 1633.

"But a decade later, Cromwell's Parliament overrode the Kings' edict. No sports on Sundays. Due to that fact, a lot of people got rid of their hawks, and got rid of their falconers. After the monarchy was restored in 1663, modern firearms were becoming available, so instead of going back to the hawking thing, most took up the gun."

The demise of falconry in Western Europe coincided with the time when the Puritans and others sailed to America to gain independence. The religious philosophy of the Puritans eventually withered and lost favor, but the mentality of English hunters and farmers—that all hawks, falcons and eagles killed game birds, domestic animals and livestock and should be shot on sight—was directly exported to America.

Witness the slaughter at Hawk Mountain, Pennsylvania, that had occurred for decades up to the early 1930s. An October 1929 article in the *Pottsville Journal* carried the following headline: "SPORTSMEN SHOOT MIGRATING HAWKS. Pottsville Hunters Knock Down Pests from Point of Vantage in Blue Mountains. Kill 300 in single day."

"Thousands of huge hawks, redtails, marsh and goshawks, borne by a stiff northwest wind over a steep pinnacle in the Blue Mountains . . . are daily challenging hunters and sportsmen of Pottsville and vicinity," the article said. "Chilled by the early October winds, many thousand hawks are sweeping past the mountain pinnacle, inviting extermination, a challenge that has been accepted by local sportsmen and hunters who are shooting hundreds every favorable day. Impressed by the unusual opportunity to wipe out thousands of enemies to bird and game life in the state, a Pottsville sportsman today urged local hunters to cooperate in killing hawks. The migrating birds pass within a few feet of the ground at the mountain pinnacle, generally between the hours of 10 a.m. and 3 p.m., only when a

Hawk Mountain Sanctuary
Thousands of migrating hawks were slaughtered annually during the 1920s and 1930s at Hawk Hountain, in Pennsylvania. The participants claimed killing the hawks saved game birds.

stiff northwest wind is blowing. With ordinary shotguns, 300 hawks were killed last Friday."

At the time, no state or federal laws protected hawks, and the state of Pennsylvania had passed a law in 1929 placing a $5 bounty on goshawks, a bird considered rare in the Appalachian state. "Not one hunter in a hundred would know a goshawk if he should ever see one. A goshawk bounty is really an incentive to kill all hawks," author Maurice Broun wrote in *Hawks Aloft: The Story of Hawk Mountain* (Dodd Mead Co., 1948).

In the early 1930s, when the Craighead twin brothers were practicing falconry, setting up photography blinds and gathering material for their book, *Hawks in the Hand*, they, too, ran across sportsmen shooting hawks and falcons. A passage from their journal in September 1933, when the boys were seventeen described the scene they discovered at Cape May, N.J.

Thousands and thousands of hawks were migrating over an area about two hundred yards in width. The majority were

Sharpshins although later on the Cooper's and other species will be more numerous. We saw Sparrow Hawks, Pigeon Hawks, and Duck Hawks, all of which belong to the falcon family. The Pigeon Hawks were so tame that they sat in trees and on poles right above us and ate their prey . . . . Friday afternoon was quite windy. The Pigeon Hawks would bank on the wind and then coming tearing downwind on a long glide. I could hardly believe there were ever so many hawks . . . . On Saturday, we estimated as many as two-thousand hawks in the sky at once, mostly buteos (broad-winged, red-shouldered and red-tailed). They would come in long lines and circle above the Cape for 15 or 20 minutes and then move on as the never-ending stream came on.

We collected hawks that had been shot but only wounded by the hunters, who lined the roads in all directions, and shot the hawks as they came over the roads . . . . We thought that just about 90 percent of the Sharpshins that crossed the road on Saturday were killed . . . . While back at camp we would watch a hawk as he passed us and went toward the road. We would say, "There goes another one, watch them get him," and they usually did . . . . (The birds) were mostly from up north and had no fear of man, which made the slaughter even worse. John saw a man shoot a hawk out of a tree with a shotgun, never giving him a sporting chance. Shells were piled all over the road and hawks were piled all over the running boards of cars and scattered throughout the woods, for no one bothered getting a hawk that fell anywhere but in the road. When we think that these are not only hawks from New Jersey but from all over the Northeastern United States and Canada, it seems a crime that they should be so slaughtered.

Even the American national symbol, the bald eagle, was thought to be a pest in those days. The state of Alaska paid a $1 bounty for every dead bald eagle as an incentive to destroy eagles that scooped salmon out of streams, posing a threat to the commercial salmon-fishing industries. Native American tribes in the Pacific Northwest killed bald eagles to use and sell their feathers, too.

A 1939 pamphlet titled "The Two Eagles of North America," by Ellsworth D. Lumley for the Emergency Conservation

Committee in New York, displayed a chart summarizing state game laws regarding eagles. Only five of the forty-nine states protected bald eagles under state law in 1939. In Alaska, Lumley reported that in 1926, 41,812 bounties had been paid on bald eagles, and an estimated 70,000 bald eagles had been killed since the bounty program began ten years earlier.

Golden eagles were persecuted in the West because ranchers blamed them for carrying away lambs to the nest, and killing lambs and calves on the open range soon after birth. Hunters shot golden eagles because they believed they killed deer and antelope fawns, and the lambs of bighorn sheep. Up to the early 1960s, ranchers paid bounties of $25 or more to pilots-for-hire who shot defenseless golden eagles in the air. This was a common practice in Texas, Arizona, New Mexico, Wyoming and Montana. Strychnine bait stations put out by ranchers and the federal government to protect livestock from coyotes indiscriminately killed untold numbers of golden eagles and other birds of prey.

Perhaps the most persistent and laughable myth in those days was that eagles preyed on children and carried off babies. "It is doubtful if a search of history would reveal one true instance of an eagle stealing a baby," Lumley notes. Knowing that the largest bald and golden eagles weigh eight to twelve pounds, "It is doubtful whether any eagle could lift and fly away with a weight of eight pounds, except from the top of a slope or the edge of a cliff. Babies are not left unprotected in such places."

However, prevailing myths about birds of prey abounded, and American attitudes toward predators of all kinds were generally the same. During the same era, in the first half of the 1900s, state Fish and Game departments encouraged the killing of wolves, grizzly bears and mountain lions. In Idaho, for example, a $50 bounty was paid for dead cougars, as an incentive to protect deer and elk populations from predation. In 1954-56, the Idaho Fish and Game Commission paid $4,400 to bounty hunters for cougars.

Hence, there was scant little public understanding of the positive things that predators contributed to the natural world.

In essence, a zero tolerance policy toward predators prevailed in the United States at the time when Morley hit center stage.

Prior to Morley's crusade, early writings by members of the Audubon Society, the Craighead brothers and Aldo Leopold began an educational effort that presented alternative points of view about the value of birds of prey and predators in general.

T. Gilbert Pearson, for example, made a number of key points in a speech to the National Association of Audubon Societies in December 1929. Pearson noted that nearly all birds of prey were considered "vermin" because they killed game birds. "I would define vermin as any wild creature that kills something you want to kill," he said. "It is perfectly natural and altogether commendable that the game-keeper should desire to protect his birds, that the farmer should seek to guard his poultry, and that the fruit-grower should discourage predations on his cherry crop. . . . If a cherry tree is raided by a robin, the owner of that cherry tree has the recognized fundamental right to protect his fruit, but he does not have the right to start shooting all the robins of the neighborhood. The bird is of value to all agriculturalists and gardeners of the community because of the great numbers of insects and their larvae it destroys, and such gardeners and agriculturalists have property-rights that the cherry-raiser must respect.

"If a hawk catches a pheasant in a breeding enclosure, we may readily concur in the game-keeper's wish to dispose of that hawk, but there are many who would not agree with the idea that the game-keeper automatically has the right to make war on all hawks found within the boundaries of the county or state. Very few people breed pheasants, but many make their living by cultivating gardens, orchards or farms on which those hawks render valuable services as destroyers of rats, mice and various insects. In short, such people have property rights which are served by some species of hawks."

In the 1940s and 1950s, Morley and a number of other falconers across America gave public speeches about the need to stop shooting birds of prey. Falconers Hal Webster in Denver, and Walter Spofford, Al Nye and Robert Stabler in the northeastern United States frequently gave speeches to agricultural and hunting groups.

Morley was perfectly suited for the task. He knew about property rights and the perspective of ranchers and farmers, having grown up on a North Dakota farm. Being the Snow Survey supervisor of the Columbia River Basin, he already had an established relationship with farmers and ranchers. They trusted him because of his accurate water-supply forecasts. When he addressed farmers, ranchers or hunters, the audience sensed that he was *one of them.* As a falconer, he had visited hundreds of nests, and collected evidence of what the birds ate in various locations around the West. He had vast knowledge of all the different kinds of hawks, falcons, eagles and owls.

Nelson family collection

A previously "untrainable" saker falcon perches on Norm Nelson's fist. The bird was given to the Nelsons by Harold Webster, a famous Colorado falconer who was fed up with the saker.

Being a World War II veteran helped, too. He often played a subtle sympathy and patriot card by noting that he was a "bullet-holed" veteran with shrapnel still lodged in his back. No one would doubt his honor and duty to his country.

Knowing his subject well, Morley spoke with complete conviction. "Because he's completely convinced of what he's talking about, he says things in such a convincing way," says his former wife, Betty Ann McCarthy. "Morley's really good at that. He tells you what he believes, and by god, it's right."

When Morley hit the speaking circuit, he always took one of his falcons or eagles in tow. He spoke to school groups, 4H clubs, conservation groups, cattle ranchers, sheep ranchers, the Farm Bureau, and hunting groups—initially in Idaho, and as time went on, he was invited to speak to regional gatherings in the West and to national groups in Washington D.C.

Frequently, the birds that accompanied Morley to the

speeches made more of an impact than anything else. "The trusting, beautiful eyes, and manner of such a large bird like an eagle, does the same thing to humanity regardless of age, but it seems to affect children in a particularly strong and wonderful way," Morley says. "This is why the use of wildlife in education is so important. Wildlife becomes something beautiful, strong, intelligent and worth saving."

On one occasion, Morley was invited to speak to a group of cattle ranchers in the Castleford area in southcentral Idaho. From his Snow Survey work, Morley knew a rancher named Truman Clark, who was the supervisor of the local soil conservation district. Clark had told Morley that a number of ranchers were very active in shooting golden eagles in the South Hills because they believed that eagles were killing young calves. Other ranchers in the area did not think the eagles were to blame. Clark wanted Morley to come down and give the straight scoop.

"So in this meeting I gave a very impassioned speech concerning the eagles as I knew them," Morley remembers. "I talked about going to all the eagle aeries near their ranches, and I had gone over the cliffs and checked to see if there were remains of lamb in there, and of course, I found no remains of lamb but rather, I found as many as three or four black-tailed jackrabbits, magpies, rattlesnakes, bull snakes and ground squirrels.

"One of the men in the group said, 'Now Morley, we were out with our cows the other day and our calves. We were riding up on a high ridge and it was 3,000-4,000 feet down into the canyon, and an eagle came and attacked one of the calves. It just dove at that calf and obviously acted like it was trying to catch it. And the calf fell all apart at the seams, in the low cover of the rim, and it ran for the bottom of the canyon. So they had a heck of a time going down to the bottom of the canyon, and bringing the calf back to the rest of the herd back on top of the ridge.'

"And I asked the man, did the eagle ever bind to the calf? No, he said, it never did touch the calf but he just kept diving at it all the time. And I asked him, did it have a big white band in its tail? Oh absolutely, the man said. It had a great big white band in its tail and it was a very large eagle, obviously a female. Well, I said, that explains it to me. We see this with our trained birds, and we see it with the wild eagles that we observe. The eagle you saw was just attempting to learn how to stoop and hunt. She was diving as if she was going to kill the calf, but she pulled up just before striking it, knowing that, OK I could have taken the calf by the head or whatever, but it's too big and I'm going to let it go. She pulled up and rolled over, and detected that the wind angle was different, and she stooped down at the calf to practice its attacking and stooping capabilities and all of the control that is so important in catching a rabbit or other types of wild prey. And I said this is what that bird was doing. It was not attempting to kill your calf, but it was just learning how to make a proper attack.

"This made good sense to these men. So I went on and told them other things such as comparing the wildness of a horse to the wildness of an eagle to start with, and how you can see a gentle change in understanding and cooperative spirit with an eagle after they're comfortable with you. In other words, it's very much like training a wild horse.

"All of these points made sense to them, and I was very happy with my presentation. Then, I went over to show everyone how my eagle Otis had this kind of trust in me. Otis had been fooling around on the T-perch made of a two by four, and he went over to the edge of the wood block, and checked the strength of his talons. He came so close to the edge of the two by four that one of his talons got in between a split in the wood, and he broke off a piece of the wood, which truly opened everybody's eyes because it gave them an idea of the strength of the eagle's foot. The piece of wood fell down on the floor, and all of those men looked at it and thought, Holy Mackerel, he isn't going to pick up that eagle with those great feet after it just splintered that piece of wood is he?"

But Morley did. He reached behind the eagle's feet, with a super slow movement, and it stepped up on his bare wrist and forearm. "I talked to my eagle and said hello. He stood on my fist very carefully, and I let him fly back to the perch. After I was done, a sheepman came up to me, and he was one of those guys that chewed Beechnut tobacco. He spit out a wad of tobacco, and he came up to me and he said, 'By god, Morley, you wouldn't have had to said a word as long you brought that eagle, I'll never shoot another one.'"

"That really brought a beautiful philosophical thought to me. The success of what I had been doing in giving talks to all kinds of groups, that is, the authenticity of what I was talking about, was really dependent entirely upon the falcon or the hawk or the eagle that I brought with me on the stage. It was as if someone had hit me with a true thought that was very, very strong. That was a wonderful evening for me, and I'll never forget it."

In the meantime, as the knowledge of the "eagle man" began to spread, Morley received more and more wounded birds—all kinds of falcons, eagles, hawks and owls. Most of them had gunshot wounds and couldn't fly anymore, and nearly all of them were close to starving to death. Morley fed them fresh meat and egg yolks, and nursed them back to health, if possible. Some of the birds had to be destroyed. If their wings were broken, Morley would amputate the wings and give the birds away to zoos. Often times, he had more than twenty crippled birds to care for, in addition to all of his trained birds for falconry.

Morley did his best to keep up with all of the wounded birds, but it was never enough. If the birds were going to survive, he had to speak to more groups to stop this senseless slaughter. The wounded birds pushed him to try harder to reach a broader audience.

A new opportunity arose in the summer of 1958, when the National Wildlife Federation held a regional meeting in McCall, Idaho, a scenic alpine village about 120 miles north of Boise. A politically active group of hunters and anglers, the Federation had a strong interest in increasing the populations of wild game for hunting. Many of the men at the meeting still thought it was a macho thing to shoot a big hawk or an eagle. It was an influential group that would need the full treatment. Morley brought a peregrine falcon and a gyrfalcon to the speaking engagement.

"Before I started to speak, I took the hoods off the birds, and they looked out at the men and women in the audience with their terrifically powerful and intense black eyes. That got their attention. I told them that long ago, before there were any men shooting at birds of prey, there was much more game than there is now, and there were more peregrines then, and the birds played a strong role in taking the weak and the sick and the diseased in the evolution of the strong-flying ducks. I also talked about finding ground squirrels in prairie falcon aeries, and the variation of bird life from crows on down that you could find in a falcon aerie, and that most of the time, they had to kill something they could carry both for their own safety, and so they could bring it back to their young successfully. And then I showed them an educational film about the birds of prey."

Morley was ready to wrap up the talk. He looked around at the audience, and he couldn't see any signs of compassion. Deep inside, he felt angry that these men didn't seem to get the message. His mood suddenly changed.

"I just can't understand you," he said, his blue eyes staring like daggers at man after man seated before him. "I can't understand you men who call yourselves sportsmen and hunters killing these birds that are the greatest hunters of all, whose very lives depend upon their aerial skills. How can a man who knows the true joy of the chase and the hunt, a man who respects the clever elusiveness of wild game, how can that same man destroy one of these magnificent creatures for no other reason than to test his marksmanship? Is this life, this beauty so meaningless to you that it becomes a mere target?

"Gentlemen, these birds should be saluted as the most spectacular hunters on earth. To me, it's such a waste to see them shot for target practice. When society understands the true worth of these noble birds, they will be protected from indiscriminate shooting forever. The killing has got to stop. Thank you very much."

There was a short pause, and the men broke out in loud applause, a standing ovation. Morley's face flushed with surprise. In a matter of seconds, he was surrounded by federation members who wanted to see the falcons up close. Several legislators in the audience came up to him and pledged to work toward a new state law that would ban the shooting of all birds of prey.

“We’re going to button-hole every man in the Legislature and make good and certain that he knows the facts,” said Ted Wegener, the federation president. “Let’s get it done.”

“Well, that sounds terrific,” Morley said, enthusiastically shaking Wegener’s hand. “Let me know when we can get started.”

Morley just about jumped out of his skin when he read an article in *Sports Afield* that was highly critical of bald eagles and golden eagles in the November 1961 issue. The lead sentence of the article, penned by Westbrook Pegler, set the tone: “The Bald eagle is the national bird of the United States. It is no good.”

Never mind that the bald eagle had been the national symbol of America since 1782. Never mind that the bald eagle had been protected by federal law since 1940.

Pegler visited several contract hunters in western Texas, crop-dusting pilots who got paid by ranchers to kill as many golden eagles as possible to protect livestock. The article also featured an 1827 painting by John J. Audubon that depicted a “defenseless” lamb being attacked by a golden eagle.

Morley could feel his blood-pressure rising as he read the scathing article. After all the progress that had been made in educating the public about birds of prey, the article seemed to push the movement two steps backward.

In western Texas, Pegler wrote, “Eagling is a profession as well as a sport. John Casparis, of Alpine, in the Big Bend, estimates his total kill of eagles at about 12,000 in twenty years . . . . Mr. Casparis has a firm opinion of the bald eagle. This is it: “Nobody knows where any bald eagle ever did a damn bit of good. It has absolutely no friends among other birds, whereas hawks and crows and most other birds get along right well with lots of other kinds. And a plump old yard-hen is useful in life and on the table.”

“Mr. Casparis flies an Aeronica Chief with a motor of 125 hp, almost double the normal muscle. “I need that additional power to get into canyons and draws on the downwind side of a range,” he explained. “That wind is like a whirlpool below a falls . . . . But if an eagle is crippled and flies into such winds I’ll go in and scoop him out. Sometimes he will not be more than twenty feet above the ground. I have shot eagles at 20,000 feet. One day I got one at 17,000 with another man in the ship. I hunt for the Big Bend Eagle Club, and of course, we have a lot of people out for them because they were slaughtering our crop of antelope fawns . . . . I get tired of the sentimental claim that bald eagles don't kill sheep. They were killing 100-pound yearling lambs. To a rancher, lambs and kids are a lot more important than the imaginary virtures of the bald eagle."

Then Pegler joins up with another aerial gunner, Alvin Miller. "A day's kill of lambs and kids by from one to five eagles can be a very serious blow to a rancher," Pegler wrote. "Many registered lambs are worth $50 each. Miller has seen clusters of dead animals. He reports that an eagle in the midst of plenty will eat only the hearts and drink his fill of blood. One stockman told me he roped an eagle so heavy with food that it couldn't get off the ground. The man yanked the big bird's head off and lambs' hearts tumbled out."

Now, for Morley, this was too much. He angrily rolled up the magazine into his fist and threw it across the living room. He had to cool off. He just couldn't believe that a reputable magazine like *Sports Afield* had printed a story full of lies. He went up on the hill next to his house and fumed. He looked into the gray skies and saw a red-tailed hawk high in the sky, patrolling the Boise Foothills. It soared into the wind and circled, looking for rabbits and ground squirrels. An uncontrolled shiver reminded Morley that it was late November, and he'd forgotten to put on a jacket. The chilly air forced him back inside.

He picked up the magazine again, and looked at another picture accompanying the story, a photo of Alvin Miller holding his rifle across his chest, with a day's catch of three dead golden eagles nailed to the doors of his airplane hanger, their wings stretched wide. One of the birds had no body, only the wings were tacked to the door.

"That son of a bitch," Morley muttered under his breath.

Pegler closed the story by lamenting the unforgiving nature of the 1940 Eagle Protection Act, which only outlawed the killing of bald eagles and the possession of their parts. Golden eagles were exempt.

"The national law is unscrupulously fierce . . . . for it forbids possession of any part, even a feather, of any bald eagle or any nest or eggs thereof," he wrote. "The honorable U.S. Court may impose a term of six months in prison as well as a fine of $500 on violators. Still, all things considered, a hunter in a plane, miles and miles from any witness but his conscience and his partner, may find it very difficult not to bang down a bald eagle . . . . We are now a nation of guardhouse lawyers, and one constitutional defect glares forth—the Bald Eagle Act. No allowance is made for self-defense against a demon which can pierce with its mighty talons the barrel of a snake-proof boot or crack the back of a sheep with one blow of its beak. Must a man have witnesses to swear the eagle was the foul aggressor?"

Morley didn't waste any time firing off a letter of protest to Richard E. Berlin, president of the Hearst Corporation, the publisher of *Sports Afield*. Here it was, Thanksgiving weekend in 1961, and he had to deal with this lousy article before he could do anything with his family. Morley wrote the letter on the stationery of his emerging film-production company, Tundra Films.

"This letter is a protest against Westbrook Pegler's article, "The Eagle." As a president of a huge journalistic enterprise, you have tremendous influence over the general public's knowledge, and articles such as these do serious harm to a rare and misunderstood bird. I realize that the subject of eagles is difficult and controversial. This is true only because so little is known about them. The eagles have been the subject of emotions and imaginations since the decline of falconry in the Middle Ages. In the future, I hope that you will make space for some facts on eagles to give them the just and even break they have deserved for the last 1,000 years."

Morley went on to refute the lies in a point-by-point rebuttal with ten points in all. The letter went on for five pages, a very lengthy letter for the usually laconic fellow. First he countered the outright lies, vis a vis, an eagle cracking the back of a sheep with its beak: Eagles "kill with their feet and talons. They <u>never strike with their beak</u>. It is a fantastic statement of emotion to say that an eagle can break the back of

Nelson family collection
Morley sent this photo of son, Tyler, holding a golden eagle, to *Sports Afield* magazine, along with a column disputing an earlier negative *SA* article about the birds.

a sheep with a blow of its beak, and is utterly impossible," Morley wrote.

In reference to eagles being hostile to all but its own kin, Morley enclosed two pictures of his grinning eight-year-old son, Tyler, holding a golden eagle. "Both the golden eagle and bald eagle are friendly to people, dogs, cats and children once the steady hand of friendship has been extended to them, rather than the malled fist, shot and shell, which has been their lot since the invention of gunpowder," he wrote.

He refuted the information about eagle predation on antelope fawns, by citing specific studies on the subject, which concluded that eagle predation on antelope fawns was insignificant in Wyoming and Idaho.

To suggest that bald eagles kill sheep is preposterous, Morley said. "Bald eagles have never been accused of killing sheep where there was any evidence—they are essentially fish eaters."

As for the contention it is "silly" that bald eagles are the national symbol of the United States, Morley said, "If we are understanding enough to realize that humans are also predators, then the characteristics of the bald eagle become a national objective truly symbolized by the eagles' loyalty, strength, courage and nobility."

Morley countered every fib that Pegler uttered in the article, that bald eagles make funny noises, that they don't have any balance, and on and on. In answer to Pegler's final line "Must a man have witnesses to swear the eagle was the foul aggressor?" Morley said, "Yes, indeed we must, and we have fought several wars to maintain that point with the bald eagle as a symbol riding high above those living and dead who will always believe in it."

Morley closed by saying, "I have faith, Mr. Berlin, that you will make space for some factual articles on these birds. They are among our most inspiring forms of life, and the world will lose a noble example of the characteristics we cherish unless men like yourself take an objective position."

He copied his letter to the editor of the magazine, Pegler, the National Audubon Society, the National Wildlife Federation, Ducks Unlimited, selected members of Congress, John Craighead, Tom Cade, Edson Fichter, Pierre Pulling, the Air Force Academy and "others."

Morley's letter was published in the next issue of *Sports Afield*, along with a photo of Tyler holding a golden eagle, and a correction.

Sweet revenge came the following year, on October 24, 1962, when President John F. Kennedy signed House Joint Resolution 489 into law, adding golden eagles to the Eagle Protection Act of 1940. The act provided for penalties of a $500 fine or six months in jail, or both, for anyone who was found guilty of killing, possessing, selling, offering to sell, purchasing, bartering, transporting, exporting or importing, "at any time or in any manner, any bald eagle commonly known as the American eagle, or any golden eagle, alive or dead, or any part, nest or egg thereof."

The debate over the issue pitted East Coast congressmen against several members of the Texas congressional delegation, who argued that an exception in the act was necessary to allow the killing of golden eagles caught in the act of preying on livestock. Several other members of Congress from western states opposed any protection for golden eagles.

The National Audubon Society and the Fraternal Order of Eagles were the driving force behind the legislation, along with a number of golden eagle experts, including Walter Spofford, an eagle expert and professor at State University in Syracuse, N.Y. Two articles in favor of golden eagle protection had been published in the spring of 1962, shining the national spotlight on the issue in a timely manner.

John J. Stophlet wrote a fine article in *National Parks Magazine*, titled "A Victim of the Bounty Hunter Is the Golden Eagle, King of the Bird World." Stophlet focused on the Texas eagle slaughter. "The bounty hunters kill eagles for a profit, collecting twice—once on a bounty paid by ranchers, and again by selling the skins and feathers to the Indian artifacts industry for conversion into Indian headdresses for the tourist trade. The ranchers of west Texas have an organization known as the Big Bend Eagle Club, in which they hire hunters to shoot eagles. One hunter boasted he had killed about 12,000 in twenty years. How many of these might have been immature bald eagles? One hunter hired by the club averages about 1,000 eagles a year; and beginning in 1945, he slaughtered 8,300 in about seven years. How many thousands have been killed since then?

"The commercial slaughter of the golden eagle is strongly reminiscent of the wholesale destruction wrought against the American bison—or the passenger pigeon—by the market hunters of times past," Stophlet said. "No animal—from the polar bear in Alaska to the golden eagle of the Texas plains—can long withstand the concerted and deadly hunting from airplanes that is taking place today."

Still, despite strong support for the bill, co-sponsor Senator Ralph W. Yarborough of Texas and Senator John Tower of Texas negotiated an amendment to the bill, stating that, "on request of the Governor of any State, the Secretary of the Interior may authorize the taking of golden eagles for the purpose of seasonally protecting domesticated flocks and herds . . . ."

Audubon officials and the Secretary of the Interior agreed to the amendment, sensing that the bill would not pass without it. Finally, golden eagles were protected by federal law, meaning it would apply to all fifty states. In many cases, however, particularly in the West, the killing of eagles continued due to lack of enforcement and the frontier mentality.

In the spring of 1963, Morley participated in a Mutual of Omaha's *Wild Kingdom* TV program, called "Hunters of the Sky." Morley had gotten acquainted with Jim Fowler, associate producer of *Wild Kingdom,* through falconry circles since Fowler was a falconer himself. Plus, Morley was known in the movie and TV business, having served as a birds of prey consultant in seven Walt Disney pictures and numerous other programs.

The focus of "Hunters of the Sky" served an educational function for informing the public about the different kinds of raptors—hawks, eagles, falcons and owls—and the role that predators played in the wild. Fowler and host Marlin Perkins also performed a much-needed demonstration about how much weight various birds of prey could carry when airborne.

"Many birds of prey are known as blood-thirsty killers, but this just isn't so," Fowler said, holding a Harpy eagle. In a science lab with powder blue walls, Fowler placed a baby chick on a tan rectangular table in front of the Harpy eagle, a large bird with a funny-looking head that's native to South America. The eagle took a good look at the chick, but he didn't pounce on it and kill it with its talons. "He's curious but that's about all," Fowler said.

The eagle's talons are very powerful and strong, he noted. They are the bird's main instrument for attacking and killing prey. TV cameras zoomed into the eagle's talons, and Fowler places a lion's skull next to a talon, showing that the incisor teeth of a lion are smaller than the length of a razor-sharp talon. Then they measured the strength of the eagle's talons, and computed that the bird can grip its prey with the same strength as the "weight of ten men standing on the head of a nail."

Next, they discussed how much weight a bird can carry in

the air. "The books say raptors can carry twenty-three percent of their body weight," Fowler said. "We all know there have been many stories about golden eagles carrying away small children . . . Can they really carry the weight of a small baby?"

Now, there is a transition from the lab to the Snake River Canyon. Morley rappels down the canyon wall with Fowler and Perkins to visit a golden eagle nest. They band the young, and take a wild adult eagle to the top of the canyon to see if it can carry the weight of a human baby. Fowler ties a small gunny-sack of sand to a jess on one of the golden eagle's legs. The eagle jumps on top of a rock, and then it tries to fly away, repeatedly flapping its wings, but it can't leave the ground, due to the eight-pound bag of sand.

"Somewhere there may have been a super eagle with far greater strength, but certainly our eagle was grounded by his weight of about eight pounds," Fowler said.

Perkins closes the segment with the moral of the story.

"Sometimes we think of predators as being synonymous with cruelty, but nature doesn't work that way," he says. "There is a purpose. In cases where predators have been controlled or eliminated, contrary to what might be supposed, the other animals in those areas haven't prospered, but have fallen prey to uncontrolled diseases, which in some cases have wiped out whole populations. Oddly enough, the predators protect and improve the quality of the animal populations around them. The land predators, the sea predators and the hunters of the sky are all a necessary part of the Wild Kingdom."

After the *Wild Kingdom* segment ran on ABC-TV, Morley used the film clip in his talks to underscore the fact that golden eagles can not fly off the ground with eight pounds of weight, not only the typical size of a newborn baby, but also the average size of a newborn lamb. It was a perfect way to make the point—on film, in living color—with the additional impact of *Wild Kingdom's* seal of approval.

In the spring of 1970, Doctor John Lee was out in his back-yard of his rural ranch-style home in the farm country, near Meridian, Idaho. A veterinarian, Doctor Lee lived on top of a bench overlooking sugar beet, seed corn and alfalfa fields in the Treasure Valley, where red-tailed hawks and other broad-winged hawks liked to hunt and soar. His boys were out playing with some neighbor kids that day, and they ran across a red-tailed hawk whose right wing had been shot full of holes. They brought it home, and Doctor Lee's wife immediately thought to call Morley.

She reached him on the telephone, and he agreed to look at the bird. Mrs. Lee knew that her husband "detested" chickens and she wasn't sure whether he wanted anything to do with the injured bird.

"We're going to take the bird to Morley Nelson, do you want to go?" she asked her husband.

"Yes, I want to go."

"Morley was very famous by then, you know," Lee says. "We knew he was doing this kind of rehabilitation work with birds of prey, and we'd seen him many times on TV and in the newspapers. So I went along, and we showed up there at his house with all of these kids, my wife and I.

"Morley looked at the bird, and he said, oh boy, this is really bad. The humerus was shattered, it was a compound fracture, and the bone was exposed. We had wrapped up the bird to stabilize it, and protect the kids from its feet.

"Morley said, well, the wing is too badly damaged. I'll try to save the bird. But we'll have to amputate the wing. And I said, well, I'm a veterinarian, and I like to do bone work, and I don't know anything about the birds. But I think I can repair the wing. He had never met me, didn't know who I was or anything, and he grabbed me and gave me a great big bear hug. And he said, 'I've been looking for you for twenty years!'"

Morley had worked with several veterinarians in the Boise area since he had got into the business of trying to care for injured birds of prey at his Mews, but he had never met a vet who knew how to do sophisticated bone surgery with modern hardware. Lee was his man. In fact, Lee's decision to work on the red-tailed hawk sent him down a completely new road in his vet practice and his life. He turned into a committed conservationist, a disciple of Morley Nelson.

Lee's experience with the red-tailed hawk taught him quite a bit about many things. For example, when he surgically repaired the wing, he wondered how much anesthesia to use. "I was concerned about the anesthesia, so I used a small amount," he says. "I didn't want to hurt the bird by giving it too much. And I found out that birds of prey are very stoic. They do not freak out like a songbird, or parakeet, no matter how badly damaged they are, they just stand and take it.

Then Lee repaired the wing. "I drove the pin, from the break, out to the elbow, and then back up into the shoulder. I left one end of the pin sticking out of the wing so it could be removed after the bone healed. Then, with Morley's help, I learned how to put on a type of splint, where you use a leather thong, which goes over the top of the point of the wing, so it holds the distill feathers down tight to the radius and ulna, with one piece hanging above and below, you come around the humurus and tie it. The bird can move, and still extend the other wing for balance, and it can get around . . . .

"With Morley's wonderful help, I knew what to feed the bird, and how to feed it. I learned a tremendous amount from Morley about how to care for the birds. He was extremely beneficial with the general care and nutrition. He's an absolutely amazing person. From that point on, I worked very closely with Morley. If he got an injured bird, it automatically came to us. If it was a bird that just needed to be rehabbed back to health, with proper feeding and care, Morley did that."

What Doctor Lee didn't anticipate is how he fell in love with that red-tailed hawk, and many other birds later on.

"I kept this red-tailed hawk for six weeks, took the pin out, and with the help of Morley, we put jesses on the legs. With his instruction, we kept the jesses loose, but tight enough so the bird couldn't get away. I built a perch for him outside here on our lawn. By that time, he was doing beautifully. I fed him all of the mice that I could catch or steal from our cats, and fed him red meat without fat. As the days went on, I would hear him trying to exercise his wings. There was a big ring the lead was tied to, and he'd try to fly, and then he'd drop back down and rest. I just followed Morley's instructions on how falconers take care of their birds. I could see in less than a week that he was strong enough to take off on his own."

"By that time, though, he had captured my heart. I decided, and Morley encouraged it, that I would keep the bird for a while at least, and lecture, to schools and scout groups, and show people how beautiful the birds really were. We named him Tonka, after my boys' Tonka trucks. I began to fly him. First on a long line. And finally, I flew him free. I'd go to these different groups and fly him free and call him and have him come back to my fist. I kept him for about three months.

"Eventually, I thought, well, it's time to release him. It's not really just mutual affection as mutual respect. I began to feed him fatter and fatter, he felt less dependent on me for food, and I could see he'd come back to me a little more reluctantly. And that's what I wanted, since he was a wild bird and knew how to hunt. Although I must admit, I hated to let him go, because of the respect I'd developed for him and the birds of prey.

"Finally, I said today's the day to let him go. This was in early July. We had the whole neighborhood out to watch the release, but I don't think we had news cameras for this one. Pretty soon, I could see just a speck in the sky, and I called him and he came back to me. And I thought that was really neat. I fed him some more, and released him, and he flew down to a tree by the canal. And I called him and called him and he wouldn't come. I'd whistle. And hold up my fist with meat on it. And he stayed in that big willow tree, he'd ruffle his feathers and jump around. And he wouldn't come. And I thought, that's fine. When I went down under the tree, he looked at me. I could see those piercing eyes, you can see that stare for a long ways. They have a look about them. But he wouldn't come.

"So I climbed up on this straw stack, and the look on his face changed, like that's your place down there, not up here. This is my territory. It totally changed, like he didn't know me. And he flew off to a power pole and landed out there. About an hour later, I went to the pole and he flew off. I knew he was back to the wild. It was a great feeling to know that he was a healthy bird again, and he'd do fine."

This was the beginning of a twenty-five-year endeavor for Lee as a wild bird rehabilitator. "After that, I had so many birds," he says. "In that first year, I had 100 birds at various times. To start with, I saw mostly gunshot birds, a lot of red-

Echo Films

Morley and his golden eagle "Slim," pose for a poster shot to promote the film *Vertical Environment.*

Stephen Stuebner

The spot on Steen Nelson's South Dakota farm where his grandson Morley Nelson first saw a bird of prey in action.

Nelson family collection

Kuwaiti falcon handlers show off their birds.

Morley cradles a Bell and Howell movie camera he used to make his early films.

"Thor," one of Morley's favorite gyrfalcons, was stolen but later returned.

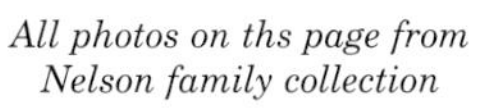

All photos on ths page from Nelson family collection

"Clyde" the golden eagle

A prairie falcon poses

Nelson family collection

Norm Nelson, left, Tyler Nelson, center, and Morley Nelson with "Amigo" a male peregrine falcon, shoot a standup for the movie *Peregrine* for the Peregrine Fund.

Nelson family collection

The golden eagle "Jessica" flies for Disney camera crews shooting "Ida The Offbeat Eagle. Morley's son, Tim, who died in an auto accident in college, is on the right.

Nelson family collection

Despite laws designed to protect birds of prey, irresponsible humans shoot thousands of raptors each year.

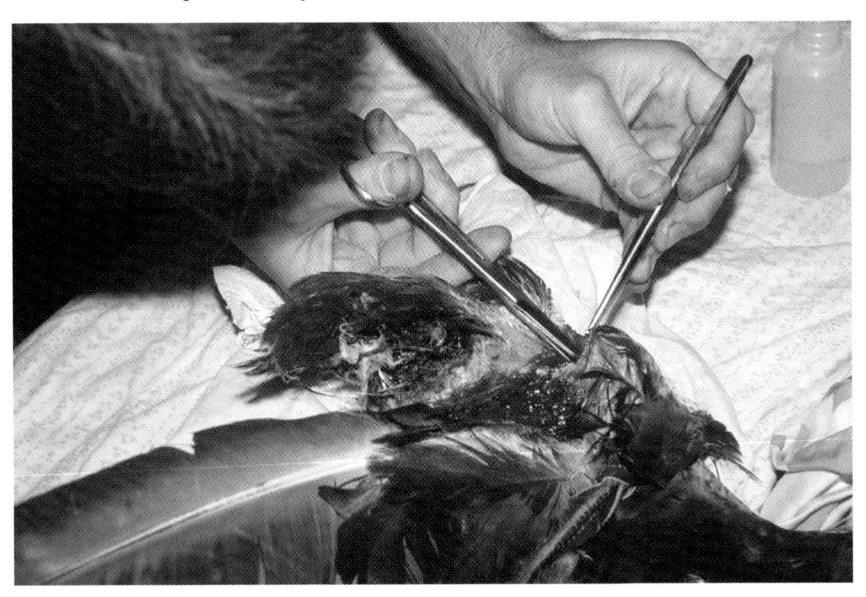

Nelson family collection

A veterinarian amputates the wing of an eagle suffering from a gunshot wound. Because it has lost its ability to fly, the bird will have to spend the rest of its life in captivity.

Morley prepares to film how an eagle lands on a power pole. Idaho Power set up the poles near Morley's home so he could conduct experiments to determine how to reduce raptor electrocutions.

Echo Films

Golden eagle chicks huddle on a power pole nesting platform above the Snake River Plain.

Dave Boehlke

Dave Boehlke

A bald eagle come in for a landing on a nesting platform atop an Idaho Power Company pole. The platforms improve the survival rate of chicks by protecting them from power lines and shading them from the sun.

The Peregrine Fund

Morley Nelson takes his turn at the podium at the dedication of the World Center for Birds of Prey. Morley's reputation and work with raptors was a factor in the Peregrine Fund's decision to build the center in Idaho.

John Tinker

Actor Paul Newman and daughter Nell in the Snake River Canyon. The Newmans became acquainted with Morley during the filming of The Eagle and the Hawk.

Sandy Ostertag

Singer John Denver and Morley Nelson on the rim of the Snake River Canyon. Denver wrote the musical score for the NBC special "The Eagle and the Hawk." Denver's visit to the canyon also provided the inspiration for his album, *Aerie*.

tails, a lot of Swainson's hawks, Amercian rough legs, very few ferruginous, and golden eagles. I began to get so much notoriety, the newspapers came out to take pictures of me releasing birds, and I continued to talk to many different groups about the birds of prey.

"Then I started to get more birds that were sick, and birds that were injured in other ways than being shot. I started to think we're starting to have an impact. I was getting birds from Weiser, Homedale, Mountain Home—towns that were sixty miles away or more.

"I'll never forget one time when the state police called and said we're bringing in an eagle from Mountain Home. The guy came in with his lights going and everything, and he pulled into the clinic, rolled his window down and said the eagle is in the back seat. There was nothing wrong with the bird. It was a juvenile eagle that got into trouble, like they often do. And of course, juveniles by the time they leave the nest are bigger than the adults, their muscles aren't toned down, they're kind of flabby and fat from their parents feeding them, and they're dumb! The bird was just out there in the median in the interstate. The cop was really afraid of it.

"If you're gentle and slow and deliberate, you can reach right under them and pick them up without a glove on. They fluff up and look as big as possible and watch you, but they don't strike at you. I've gotten an awful lot of golden eagle juveniles over the years. I just fed up the bird, and released it. It was fine."

Lee was almost overwhelmed with calls and visits from people with injured birds. "It was interesting, I began to get so much publicity, that people were bringing in all kinds of birds, not just birds of prey. They'd bring in pets—parakeets and cockatoos and parrots, and since I couldn't turn them away, and they'd say, oh, we saw you on TV. . . so I really had to scramble to become knowledgeable."

Having been raised on a dairy farm, Lee grew up with an attitude that red-tails were chicken hawks, and they were worthless to society. His veterinary practice provided new insight into the birds. Eventually, he even worked on chickens.

"After a while, I found out, as much as I detested chickens, I had chickens for patients that had real personalities. I was surprised. I worked on a lot of different kinds of birds and thoroughly enjoyed it."

It didn't take long for Lee to realize how much money the raptor rehabilitation work was going to cost him. He grew to enjoy the birds so much, that he didn't care, but it was expensive just to feed and care for the birds. "I had a mixed practice. The bird part of the practice was not large, I mean the paying birds. Initially, I kept track of my expenses. And Fish and Game came to me and said we'd like to help you with the expenses. I said, great, the eagle that I'm releasing today, the cost would be $75. And they said, whoa, we can't afford that. So I said, that's OK. Why don't you just give me what you'd like to give me. I sent them a bill, and they did help some, once.

"Fish and Game brought a lot of birds to me, so did the State Police. They all got a hold of me. I said, I don't want you saying to me, this bird is going to be too expensive. Bring the birds to me. Ah, in fact, I was keeping track of all of this, and when I got to $10,000 of primary expense, I quit keeping records. It scared me so bad. Here I was just starting out my practice, and we didn't have a lot of money. But I wanted to keep working on the birds, so I quit keeping track. This is something I wanted to do. If people brought a bird to me, and asked me how much it would cost, I would say there's no charge. If you want to donate something, that's great. I'm going to take care of it.

"If people said this is my pet bird, or I'm a falconer, then I'd say, well, I'm going to charge you. Anybody that had a private bird, I charged them just like it was any other animal. If it was a wild one, there was never any charge. Luckily, my partner went along with it, and as I got new partners in the business, they were gracious enough to accept the same terms. Doctor Dixon, who bought my practice, kept up the same tradition."

After he'd been working on birds of prey for more than a year, Lee had an opportunity to visit the Air Force Academy in Colorado Springs, Colorado, to compare notes with a veterinarian who worked on injured falcons. Since the AFA had adopted the falcon as its official mascot in 1960, they had developed a big falconry program. Lee was a graduate of Colorado State University, and his wife's parents lived in Fort Collins, Colo., so he drove down to Colorado Springs to see what he could learn. It was a fortuitous trip.

"I visited with Doctor James McIntyre at the Air Force Academy, and I got some ideas from him on nutrition. He used any of the strained meats from Gerber baby food products—lamb, beef, chicken, flat coke and egg yolk. In fact, he was the one who told me that you could take a badly injured bird that was close to death, and give them Coke, not diet, but either real Pepsi or Coca Cola, that's warm, and they would revive almost immediately. You don't even have to force their mouth open. You just put a few drops along the commasseure of the beak, and it would go right into their mouth, and they would revive. It was almost magical. And if you think about it, it makes sense. The drink is a combination of caffeine, sugar, and carbonated liquid."

Doctor MacIntyre taught Doctor Lee another trick to treat raptors that had somehow eaten poisoned meat.

"It was amazing, we had a number of sick birds come in, very weak, debilitated, like a Swainson's hawk or a red-tailed hawk or rough-legged hawk, they were very mild in temperament, beautiful, absolutely gorgeous birds, and they looked like they were going to die. I'd start them out on Coke, and then work into a different formula. You could always tell from their breath that they'd gotten into poisoned meat, whether it was roadkill or a poisoned gopher, or whatever. I'd smell their breath, and it was just rotten, really foul. Doctor MacIntyre used a special diet to treat that. You mix up a little powdered cloves with soft butter until it tastes really good. Then you stick it in the refigerator until it gets hard enough that you can roll it up in the shape of little balls, like peas. By giving them these balls of butter and cloves, it would sweeten up their breath, and it got the juices going in the lining of the crop, and all of a sudden, they'd regurgitate this big mass of rotten food. And within a week, they were strong enough they were ready to go back into the wild."

Hanging around with Morley, it was inevitable that Lee would get drafted to participate in a movie. Just a year after he started to work on the birds, Morley was co-starring in the made-for-TV film, *The Eagle and The Hawk,* with Nell Newman and Joanne Woodward. Most of the movie's plot focused on the

Snake River Canyon, but it also touched on the raptor-rehabilitation aspect of birds of prey conservation. In the movie, Morley and Nell visit Doctor Lee at his veterinary clinic to look at X-rays of a red-tailed hawk with a broken wing. Lee puts in a pin, and repairs the bird's wing. In the end of the film, Morley and Nell ride horseback into the Owyhee Mountains to release the red-tail.

Lee says he enjoyed working on the film. The vet clinic segment required only one take. "I just had a little tiny piece in it, looking at X-rays. Later on, I worked on a *National Geographic* special. Morley got me involved, but I actually never saw it on TV. I also was in a small role in the *Vertical Environment* with Lynn Redgrave. So we worked quite a bit trying to get support and exposure from the media to get the word out, and I think it made a big difference."

By the dawn of the counter-culture decade of the 1970s, complete with the advent of the first Earth Day celebration in April 1970, and a growing environmental ethic in the United States, national media coverage began to tilt decidedly in favor of endangered wildlife.

In 1971, during a visit by CBS-TV news crews in Boise, Morley teamed up with his good friend, first-term Governor Cecil Andrus, to illuminate the continuing problem of people shooting golden eagles and other birds of prey, even though most of them were now protected by federal law. Lack of enforcement was part of the problem, and the frontier attitude was still alive and well.

When the national TV crew came to town, Morley's mews was crammed full of injured birds, not to mention the increasing number of maimed birds that Doctor Lee had taken in himself. In the spring of 1971, the two calculated that they had received fifty-six wounded raptors, forty-nine injured by gunfire.

"I met Morley before I was elected governor in 1970," Andrus says. "I had always been an active duck and pheasant hunter, and Morley told me that he could always use the wings, heads and gizzards from the birds for his operation, so I used to go up there to his house and deliver the goods every now and then. Morley was an ad hoc self-appointed doctor for crippled birds of prey, wounded by indiscriminate hunters, and the last time I visited his place, he was more than full. We talked about trying to get on the national news to get some assistance in finding new homes for the birds, and of course, to talk about the fact that the indiscriminate slaughter ought to be brought to a halt."

Andrus arranged for Morley to bring Otis, a trained golden eagle, to his wood-paneled office in the Idaho Statehouse for the initial interview with CBS. A former logger from northern Idaho, Andrus was an extremely popular governor. Andrus was extremely effective in front of the camera, frequently unleashing humorous one-liners that left his foes groveling for a counterpunch.

During the interview in the governor's office, "I gave the TV crew all of my best lines," Andrus says.

The same day, Andrus fired off a letter to all of the governors in the West, asking the question, "What do you do with a wounded eagle?" He explained the situation to his fellow western governors, just as he did to CBS-TV, and urged the directors of zoos and wildlife sanctuaries to consider adopting a wounded eagle.

Morley took the TV crew up to the mews, where they filmed recently killed birds lying on the grass in front of his house, in addition to many of the injured birds. Morley's hospital included ten golden eagles, seven red-tailed hawks, seven sparrow hawks, four Swainson's hawks, two rough-legged hawks, two goshawks, a ferruginous hawk, a Harlan's hawk, a sharp-shinned hawk, three prairie falcons and 16 owls of various kinds.

Noting how many injured birds that he and Doctor Lee had cared for thus far that year, Morley said, "To understand the true significance of this situation, you must realize that these birds came from only one county in western Idaho. If you assume that every western county in the western United States has a similar situation of thoughtless shooting, and only one bird is turned in alive for every five or ten birds left dead in the field, multiply that figure by the number of counties in the

West, and you can see what a significant drain is taking place on these bird populations."

When the news story was broadcast on national news, CBS-TV, Walter Cronkite reported the slaughter, and Andrus barely got a couple sentences of air time, compared to about seven minutes for Morley. "That moth-eaten creature in a wool shirt and a green Army jacket stole the show," the governor says, laughing it off.

That evening and the following morning, Morley's phone rang off the hook. "We got a tremendous request from many zoos for the injured birds," he says.

But there was one small detail in Cronkite's report that caused even more concern. He said, "If Mr. Nelson can not find a home for these eagles, he will have to dispose of them." The office of Interior Secretary Rogers Morton received more than 900 phone calls that morning.

"The federal agent in charge of the Boise area came up to my house that morning terrifically upset because he had phone calls from the Interior Secretary's office wanting to know if I wasn't aware that golden eagles were protected by law, and when Walter Cronkite said I was going to dispose of them, people thought I was going to kill them," Morley says. "That, of course, was not true, and I told the agent that I didn't tell Cronkite I was going to kill them, but that I could try to call CBS and see if they'd give an update the next evening and clarify the issue. The federal agent said, well, you're never going to get anything like that on TV, and he was very negative about the whole thing. But I said I will call him and demand that this correction be made.

"And that night, Mr. Cronkite came on the news and said he was very happy to report that Mr. Nelson had found a home for all of the wounded eagles and that the Department of Interior would determine who would get all of these birds. There were various organizations that were set up to handle the injured birds, including Hawk Mountain Sanctuary in Pennsylvania and various zoos around the country. So the story had a happy ending, and we probably had better response than we might have otherwise since Mr. Cronkite inferred that the unwanted birds would have to be destroyed."

The headline in the December 24, 1971, issue of *High Country News* was unusually understated: "Eagle Shooters Charged."

Four Utah men had been charged with killing twelve golden eagles and one bald eagle from a helicopter over a Casper, Wyoming, ranch in December 1970, the article said. The arrests were just the beginning of a major federal case in which investigators were trying to apprehend numerous eagle-shooters for killing up to 800 eagles from the air. Helicopter pilot James Vogan of Murray, Utah, had flown the men to kill the birds over the Bolton Ranch in Casper, the article said. Vogan had testified about the shooting activity before the Senate Appropriations Subcommittee on Agriculture and Environment.

Vogan testified that he had kept track of at least 570 golden eagles that had been shot by aerial gunners. The rancher paid $50 for every coyote killed and $25 for every dead eagle, HCN reported, for a total of $15,000 in bounty fees. But Vogan hadn't been the only pilot who had flown the aerial gunners. The actual death toll turned out to be 838 bald and golden eagles, with 770 shot by aircraft gunners, twenty-eight killed by gunfire from the ground, twenty-five died from eating strychnine bait, and fifteen were electrocuted by powerlines.

The blatant eagle slaughter made national news, bringing even more attention to the problem of illegal eagle shootings, ranchers paying bounties and the lack of enforcement of the Eagle Protection Act. Many ranchers from Wyoming to Texas still considered eagle-predation to be a serious problem, and their actions were taking a heavy toll on the birds. At the same time, bald eagle and peregrine falcon populations were declining at rapid rates, primarily due to the heavy presence of DDT in the food chain, which caused the birds to lay thin egg shells. The young did not hatch or survive.

Golden eagles were not affected by DDT because of their diet, but of course, there were other factors at play that were causing populations to decline.

The main coyote-control agent widely used at the time was Compound 1080, a poison that was canine-specific. Federal

Animal Damage Control officers put out bait stations—horse meat laced with Compound 1080—on western rangelands to kill coyotes. Out of concern that the bait stations killed other critters, President Nixon was considering an executive order to ban Compound 1080.

A young third-generation sheep rancher from Kimberly, Idaho, Laird Noh, was the chairman of the National Wool Growers Animal Damage Control Committee at the time. Noh (pronounced "Nay"), along with practically every sheep and cattle rancher in the western United States, was very concerned about the loss of the use of Compound 1080 because ranchers needed a way to control coyotes. In contrast to strychnine bait stations, Compound 1080 did not harm golden eagles.

Knowing Morley from Snow Survey meetings, Noh sought him out to discuss how to respond to the impending proposal to ban Compound 1080.

"The question was, what in the world should we do?" Noh says. "If the government didn't give ranchers a way to control coyotes, they would take matters into their own hands. And the result of that could have been worse than the use of Compound 1080.

"The other issue that I remember discussing with Morley was a problem out on the Helle Ranch near Dillon, Montana. They range-lambed in May. It was open grassy country, and they lambed out in the open (no sheds for protection from weather or predators), and there was a major eagle depredation problem on those lambs. That U.S. Fish and Wildlife Service was live-trapping eagles to avoid the problem, but we needed a better solution than that.

"So I called Morley to get advice and counsel. He was very willing to meet with me and others to talk about the issue. One thing I remember is that Morley said Compound 1080 was the eagle's best friend because it didn't kill golden eagles. It was selective toward coyotes. The 1080 bait stations were all over the Snake River birds of prey area, for example, but the eagles thrived."

Morley went out to check out the situation at the Helle Ranch along with Robert Turner from the Audubon Society and the Fish and Wildlife Service. In that spring, 4,000 ewes were dropping lambs, and hundreds of golden eagles were migrating through the area from Mexico to Canada, and points in between. When Morley arrived at the scene, the FWS already had live-trapped 135 golden eagles and released them elsewhere. Up to that point, rancher Joe Helle had lost 400 lambs. He wasn't a happy camper.

"Some of the raptor experts we had talked to said you could shoot one or two of the golden eagles in the morning, and the rest would leave and wouldn't come back," Noh says. "Morley agreed that it might work, but he suggested trying to communicate with the eagles in their own language. That was an approach I had never heard of before. It was never tried, to my knowledge. The Audubon people said there was no way that they would accept anyone shooting eagles. Governor Andrus, who was secretary of the Interior at the time, said he wouldn't agree to that either. So Animal Damage Control officials had to continue live-trapping."

Photo courtesy Dr. John Lee

Doctor John Lee holds Tonka, a red-tailed hawk. Morley Nelson brought the injured bird to Lee for treatment, the beginning of a life-long association between the men.

Ultimately, Morley suggested that the golden eagle problem would be reduced when the jackrabbit population cycle went back up to a higher level, then there would be plenty of food for the eagles to eat. "And three years later, that's exactly what happened," Morley says. "And the rancher gave up on his attempt to get special permission to shoot eagles."

In 1982, the jackrabbit population was so high in the Great Basin that people went out and clubbed them to death with baseball bats. The epidemic made national news. There's no question that the eagles had plenty to eat that year.

When the Compound 1080 issue was still hot in the early 1970s, Noh invited Morley to speak at the national convention of the National Wool Growers Association in Denver. Noh remembers it was fun to hang around Morley for a couple days and watch him interact with his golden eagle.

"We went out to pick him up at Stapleton Airport, and we picked up the eagle, which arrived in the freight section," Noh says. "I remember we didn't have a very big rental car, and Morley sat in the front seat, with the eagle on his left hand. Every time you turned a corner, the eagle would spread its wings and fill the car. Morley, with his sense of humor, started to explain to us how birds of prey go to the bathroom. 'You'd better watch out when their tails go up, because they shoot it straight out,' he laughed. So we're all waiting for that to happen. But it didn't.

"When we got to the hotel, he tethered the eagle to a perch and he hid the eagle behind some shrubs next to the sidewalk while we checked in. We said, Morley what are you doing? He says, don't worry. People are scared to death of these things. If they get close, you never have to worry. I don't want to raise a big ruckus while I'm waiting around in the lobby to register."

Morley addressed about 800 members of the Wool Growers the following morning. "He had the eagle on a perch," Noh says. "The first thing that happened is the eagle raised its tail and gave a big squirt right on this red vallure wall cloth. That certainly broke the ice, that's for sure. Morley went over to the lecturn and started his talk. And the eagle wasn't hooded, and it got real interested in something and it took off! Flying! It came right for the next table below the head table, and they just panicked. Water glasses got tipped over, everyone dove under the table, and the eagle gracefully circled around the room and landed right on the lecturn next to Morley.

"His message was about eagles—what they'll do and what they won't do. A lot about their intelligence. And why people shouldn't shoot them. He also covered the powerline-electrocution issue, and Compound 1080."

Morley had more credibility than a typical environmentalist because of his agrarian background. "I was so impressed," Noh says. "I think it was because he came from agricultural roots. And his work in snow survey. He had lived on the land. I thought I knew a fair bit about eagles because I saw them all the time on my sheep range. Well, he knew every eagle nest on my sheep range, where it was, and the life history of it.

"He could explain how eagles are able to kill enormous-sized prey. There are eagles who learn how to kill coyotes, or how they'd kill adult antelopes, and how it works. He was able to confirm some of the things the wool growers knew or had personally experienced. He would recognize those things, and then he moved you beyond that to a broader understanding of the bird and how it works. And you also feel that he's honest and tells you things that you don't want to hear. That does bring trust and credibility.

"People understand that he's well-read, and he knows that eagles can kill little lambs. He explains why they're doing that—after all, a little sheep the size of a jackrabbit is a lot slower . . . . that's just common sense. But he was very hard on people who shot eagles indiscriminately. The idea that you ought to go out and try to reduce the eagle population across the board didn't make any sense to him. He was good at explaining that. And he came down hard on people who used thallium and strychnine bait stations. He could be pretty blunt.

"I don't mean to say that he overwhelmed the crowd and that we all went away as eagle lovers. But he opened some doors in the mind. He emphasized the power of the bird, and the skills of the bird. He made them a little more human-like, or bigger than a starling or a dumb animal. There was something quite remarkable about it.

"Some people I know won't change at all. But I'm sure there were many people who rethought the whole process. A lot of ranchers grew up shooting anything that moves since they were kids. He was blunt that there was no reason to do that . . . but he knew the birds so well and their habits, and he convinced people to believe it, and how the eagles fit into the web of life. I think some people went away thinking twice about it."

In the summer of 1973, a lean, dark-haired fellow in his late twenties named Patrick Redig drove out to Boise from Minneapolis to see if he could look up John Lee and Morley Nelson. Redig was a sophomore veterinary student at the University of Minnesota at the time. He had been a falconer since he was eight years old, and he had started working on injured birds and taxidermy when he was only thirteen.

"I'm not sure why, but from the time I was a young kid, I had a vision and a dream of being in a position to take care of injured birds," Redig says. "Anytime you live in a small community, and everyone knows you're interested in birds, when they find an injured bird, it tends to show up on your front doorstep."

Redig had seen the movie *The Eagle and the Hawk* in 1972. "I saw that, and it just struck me to the core," he says. As a child, Redig had been inspired to get into falconry by several things, including the Walt Disney feature, *Rusty and the Falcon,* featuring Morley's trained prairie falcons.

After seeing Doctor Lee and Morley in *The Eagle and the Hawk,* Redig thought he'd drive to Boise to see if he could meet the men and see what he could learn. He pulled into Meridian, and called Lee. "It was really a spur of the moment kind of thing, but he and his wife were very gracious, and they invited us over for a backyard barbecue that night."

Redig and his wife spent the evening at Lee's home, and even went out with the vet to help him pull a calf that night. The following day, Redig asked Lee if he could meet Morley. No problem, Lee said, and he arranged it.

"I sat down with Morley and Pat, and I think Tyler was there at the time, and we had some drinks, looked at his birds, chitchatted about things, and built a lifelong friendship from that point on," Redig says.

From discussions with Lee, and some correspondence after the trip, Redig learned a few crucial things from the Meridian vet. "There was definitely some mentoring going on," Redig says. "He helped me get going, there's no question about it. He answered a couple of key questions. Back in those days, the early 1970s, basic anesthesia was a huge stumbling block with birds, and he had a method using a couple of injectible drugs that worked better than anything that was published in the literature or that I had ever learned in school. And I would say in the overall scheme of things, that was one of the huge contributions that led to my early development in this field, and then after that, it kind of flowed from there."

Morley was pleased to hear about Redig's dream of starting a raptor rehabilitation clinic. He encouraged the young man to follow his dream.

"Morley's just such a charismatic individual who had such a big heart for the birds of prey," Redig says. "He was absolutely committed to wildlife conservation and rehabilitation, and he was so enthusiastic about promoting anybody else who was interested and encouraged them to jump in with both feet."

The same year that Redig completed vet school in 1974, he got a Ph.D. in avian physiology at the University of Minnesota. He formed a partnership with Doctor Gary Duke and opened The Raptor Center on campus in the College of Veterinary Medicine. Their goal was to develop a medical and educational facility to deal with injured birds of prey, provide public education about why birds of prey get injured, and foster veterinary education. In the center's first year, the vets treated 131 birds.

At the same time that The Raptor Center was getting cranked up for business, a number of other raptor-rehabilitation clinics were beginning to open around the nation, in California, Florida, New York and Colorado. Another pioneering facility was the Lindsay Wildlife Museum in Walnut Grove, California.

Doctor Redig says the centers were led by committed individuals who grew up during an era of great environmental awakening, coinciding with an emerging national ethic about protecting birds of prey.

"In the early 1970s, there was an explosion of environmental interest and awareness," Redig says. "The Endangered Species Act was passed, the Environmental Protection Agency was formed, organochlorines (pesticides and fungicides) were banned, and you had Rachel Carson's book, *Silent Spring,* which came out in the early 1960s. All of these things were building a new environmental consciousness in our nation. In the late '60s and early '70s, everyone my age was either in Vietnam fighting the war or protesting the war, and after that ended, they moved from that to environmental issues."

By 1980, The Raptor Center was treating about 350 birds a year, and it continued to grow. Redig eventually became the full-time director. He had realized his dream. He created the world's leading raptor-rehabilitation clinic, and developed public and veterinary education programs that are second to none. The center reaches approximately 300,000 people a year nationwide through public education programs and events.

"We're pretty good at what we do," Dr. Redig says, with modesty. "We're the largest raptor rehabilitation center, the longest standing of our kind, we attract students from all over the world, and we've been effective at attracting a public audience to show the world what we're doing for the birds of prey."

Morley's tireless efforts to prevent people from shooting birds of prey and to care for injured birds of prey planted many seeds in the raptor-conservation movement. One of those seeds came to fruition in the form of The Raptor Center. State laws would be changed following an important amendment in 1972 to the federal Migratory Bird Treaty Act, which added all migratory hawks, falcons and owls to a law that protected them from indiscriminate shooting. In places like Texas, Arizona, Wyoming and Idaho, the laws wouldn't mean anything if they weren't enforced.

But it was heart-warming to Morley to watch public attitudes change and evolve. All of his efforts were beginning to make a dent in public awareness.

Chapter Nine

Wonderful World of Color

1955-1965

On a chilly Sunday evening in late October 1958, the members of the Nelson family had their eyes transfixed on the Zenith color TV in the family room of their Boise Foothills home. A fire crackled in the fireplace as Betsy Ann served dinner on TV trays to Morley and all the kids, Norm, Tim, Susie and Tyler. At 7 p.m., Walt Disney's *Wonderful World of Color* series opened with Tinkerbell waving her magic wand, unleashing a splash of color before a nationwide audience on NBC-TV.

"Tonight, we're bringing you a very unusual story about a very unusual bird, the falcon—nature's own guided missile, and the most deadly predator on wings," Walt Disney says in a personal introduction. "Yet, the falcon is noble in spirit, intelligent and, as you'll see in our story, capable of deep and lasting loyalty."

The TV feature is called *Rusty and the Falcon.* It's the first time in American broadcast history that an hour-long movie will be devoted entirely to a story about a falcon, the fastest flier of all birds of prey.

"Hey, there you are Dad!" exclaims Tim Nelson.

Narrator Jerome Courtland introduces Morley Nelson as "one of the top falconers in the United States" who trained and worked with eleven prairie falcons to make the film. A short film clip about the art of falconry shows Morley, in his trademark flat-top crewcut, walking down a sagebrush-covered mountainside with a prairie falcon on his gloved left hand. He's decked out in his usual falconry attire, a green Army jacket, tan slacks, black Army boots, with a falconer's leather satchel slung over his shoulder.

Using film that Morley had shot in Boise with his own Bell and Howell 16 millimeter movie camera, viewers meet a sample of the Nelson's flock—all perched on circular wood blocks in the backyard—Blacky, the peregrine falcon, Sherpa, the prairie falcon, Tundra, the gyrfalcon, and Clyde, the golden eagle.

Viewers see the falconer's equipment, such as hoods, a heavy glove, bells, swivels, leather jesses, a leather leash and a lure. The film depicts Morley training a falcon to fly to his fist, and then to the lure, and then to wild game. He looks confident, smooth and gentle as he works with the birds on film, sometimes cracking a smile as he peers into the eyes of a noble falcon on his fist.

Now the program segues into *Rusty and the Falcon.* It's a story about a twelve-year-old boy named Rusty, a blond-haired kid with a big-toothed grin who learns how to hunt with a falcon. Rusty lives a Huck Finn-type of lifestyle in the old mining town of Park City, Utah. He's got his own nifty fort, an abandoned underground mining shaft, where he can shoot his bow and arrow, read books and mess around with his friends.

Inside the fort, Rusty hears the mournful distress call of a falcon outside. He runs out to see a brown-and-white prairie falcon jumping around the ground with a wounded wing. Rusty tries to pick it up, opening his arms wide to grab it, and since the bird can't fly, he succeeds and brings it home.

"And now, wild thoughts went tumbling through his head," the narrator says. "He'd catch the bird, nurse it back to health and then he'd train it and they'd go hunting together."

As the film's technical advisor, Morley had a lot of input into the script. In fact, many of the plot twists could be traced to real-life episodes in the Nelson household. Ultimately, the story makes the falcon a hero.

Once he catches his falcon, Rusty walks to the town library

to read more about falconry, with the bird on his fist, causing many unwary looks from townsfolk. From the bird's brown-and-white plummage and ink-black eyes, he discovers that he had found a prairie falcon, a bird that's common in the Great Basin of the West. Using the book as a guide, Rusty learns how to make a hood, jesses and a leash for his bird.

Rusty takes his falcon with him everywhere he goes—even to church on Sunday.

Morley: One of the falcons that we used in the movie was a wild-trapped prairie falcon that we caught along the Boise River just before the fall migration. She was a big beautiful prairie falcon, and we called her "Blond." As a wild bird, she had this tremendously powerful stare. So our producer and director Paul Kenworthy came up with the idea that Rusty would take the falcon to church, and stare down a penny-pincher.

Rusty sits in a pew next to a man who never gives more than a quarter when the offering plate comes by. When the plate gets passed to the man, he digs into his coin purse. He is about to drop a quarter into the offering plate, when he looks up, and sees the bird's dark eyes staring at him with shame. Flushed with embarrassment, the man digs deeper into his wallet and drops a $5 bill into the plate.

Time for the next hymn, and the organist begins to play when the falcon bates and falls off Rusty's gloved fist, hanging upside down, shrieking. The music stops, and everyone stares at Rusty and the ornery bird. The boy flips the bird back onto his glove, just as his dad grabs the back collar of his sport coat and escorts him out of the church.

Now Rusty realizes that it's time to find a place to fly his falcon in the hills above town, where he and his bird won't get into trouble. He's been working on training the bird, and it's ready to fly. Rusty holds the bird into the wind and releases it, with no leash attached. Cameras pan to the falcon's rapid wing beat as it climbs into the air. It banks and flies back toward Rusty, who ducks and plays hide and go seek with the bird. Rusty has bonded with the falcon. He's proven himself worthy of its trust.

"Rusty's heart began to beat again, with a wild joy that he'd never known before," the narrator says.

For this sequence, I swung the lure off-camera, causing the falcon to come down at high speed. We used "Apache" for this scene, a very, very bold and wonderful bird that would chase the lure with enthusiasm. We trained the bird to catch Rusty's hat by putting a piece of meat inside. So every time that Rusty threw his red camp into the air, the bird caught it no problem. We had Rusty stand on top of two rock pinnacles, with his legs spread, and I swung the lure on the windward side. You can't see the lure in the film, but the bird turns up into the sky, turns and comes down in tremendous speed, going right between Rusty's legs and out the other side. It was one of the happiest scenes in the movie before all of the troubles begin.

"Unbeknownst to Rusty, nature had prepared a startling climax to this aerial ballet," the narrator says.

Morley: I threw a ring-necked pheasant into the air, and the cameras followed my gyrfalcon, Tundra, flapping its wings and gaining altitude until it sees the pheasant and makes a high-speed angular dive directly at the pheasant. I've never seen Tundra or any other falcon fly directly at a duck or a pheasant before or since, but for this scene, she flew right at it, and hit it head-on, smacking the pheasant with the fist of her talons, right in the breast plate, sending it tumbling to the ground in a heap. If Tundra hadn't made that kill so precisely, she could have snapped her own neck in a head-on collision. Fortunately, her aim was perfect, and it made for a spectacular shot.

"Just the way it said in the books, Rusty had cast his hawk upon the wind and it brought him a game bird prized by all hunters," the narrator says.

Rusty was a real falconer now. He had the perfect thing to convince his dad that he should be able to keep the bird, a fresh pheasant for supper. The next scene opens with Rusty's family sitting at the dining room table, ready to begin eating supper. The pheasant roast lies in the center of the table. Rusty's falcon is perched on a block above the dining room table, with no hood on.

Morley: On a hot summer evening in Boise, we had some people over for a dinner party, and I put my peregrine falcon, Blacky, on her perch in the living room. Everyone sat down at the table and I began to carve a rare beef roast on the dining room table. When I made the first cut, exposing a rare juicy piece of meat, the big peregrine flew up to the ceiling and landed on top of the beef roast and sunk its talons into it. The guests shrieked and jumped up from their chairs.

"Damn it, Morley, get that bird out of here," Betty Ann yelled.

"I guess he's hungry," I said, laughing. I thought it was all pretty humorous, but Betty Ann didn't think it was funny at all. She didn't have an extra roast in the oven.

Before Rusty's family starts to eat, his dad raises a large pheasant drumstick in a toast to his son and his falcon. Bad move. The falcon leaps from its perch and lands on the man's wrist with its powerful talons. He shakes the bird off his hand and shoos it away from the table, sending the bird to land on top of a tall lamp in the living room. The lamp falls over, crashing on the floor. Then the bird flies to a small wooden cabinet full of teacups mounted on the wall, and knocks it to the ground, shattering the fine china.

Rusty's disheveled family is covered with barbecue sauce, and the house is a mess. Everyone is in shock.

"Well, the verdict was obvious, it didn't need to be said. The falcon had to go," the narrator says.

Dad and Rusty climb up to the top of a tailings pile above town and release the bird. He pauses to kiss the bird on its feathered breast before he lets it go.

Morley: If you give a bird a kiss like that, you'd better know him or you'd get bit. Rusty had a good rapport with that bird.

But of course, the bird is now imprinted on Rusty, and it won't leave him on command. The falcon flies back to Rusty and lands on his fist. Dad leaves Rusty on the hill with the admonition, "Don't come home with that bird under any circumstances."

So Rusty takes the falcon to live with him in his secret fort. But on the way to the mine, the bird gets in more trouble. It chases a dog. A town resident releases his pigeons, only to watch the falcon nail one of them almost instantly. The man runs for his shotgun and fires, knocking a couple of feathers from the falcon's tail. Just missed!

To cap off all of the mischief, the falcon flies down the sidewalk just six feet off the ground, scaring a new mother who's walking her baby in a carriage. The whole town descends on

Rusty's house and complains about the marauding falcon. But Rusty is gone now, sitting in the mine shaft with his falcon, eating a fresh-caught pheasant. He's ready to live on his own, and eat the quarry caught by his falcon. He'd rather live with the bird than his family.

Then it starts to rain heavily that evening. Water pours down into the drafty mine shaft, and raindrops pelt the falcon. Rusty needs a new, dry perch for the bird, so he reaches for a stick and begins to pound it into the side of the mine shaft. In seconds, Rusty unwittingly knocks out a support beam, and the shaft collapses on the boy. He's trapped under heavy timbers and debris.

When Rusty doesn't come home that night, his parents begin to get worried. The next day, his dad searches for him, to no avail. Then a neighbor mentions that Rusty's falcon is flying overhead in town. His portly father runs out to look, and runs after the bird. The pigeon-keeper fires another shot at the falcon, but Rusty's dad yells at him to stop, that the bird might be leading him to Rusty. When the bird reaches Rusty's fort in an old mine site above town, Rusty's dad yells for his son.

"Dad! I'm in here!"

He goes into the mine shaft, pulls the timbers off of Rusty's legs, and lifts the boy to safety outside.

The whole episode was a wake-up call for Rusty's dad. Now he wants to spend more time with his son. They go camping together, fishing together, and other outdoor outings together.

"When the falcon found Rusty, Rusty found his father," the narrator says.

And everyone lived happily ever after.

"This film made the prairie falcon a hero, boy, I'll tell you what," Morley says. "No one had ever looked at the prairie falcon as a hero. They were known as the bullet hawk, and no one thought they had any value whatsoever. This film changed that forever."

The next morning, Morley's phone rang off the hook with congratulatory calls from his falconry associates across the nation, friends in Boise and his mom and dad in North Dakota.

"Boy, you're going to be really famous now," his dad, Norris, said in a playful way.

"Yeah, well, I don't know about that," Morley said, chuckling. "But I'll tell you what. That film will do more to change public attitudes about the bullet hawk than anything I have done. No matter how many speeches I give, I could never reach millions of people across the nation on a single night."

By now, Morley was totally sold on the power of national movies and television as the perfect media for visual education. *Rusty and the Falcon* was the third film that Morley had worked on for Paul Kenworthy Productions and the Walt Disney Company. Millions of Americans watched the program premier that evening. Later, the film aired many times again as a re-run. Disney programs were broadcast in twenty-nine nations around the world, translated into foreign languages.

In the first two projects that Morley worked on, *The Vanishing Prairie* (1955) and *Perri* (1957), Morley's birds were used in short segments as part of a larger theme. Still, the early films were important because they gave him a "foot in the door" with Kenworthy and the Disney company, which would lead to many more film projects in the future. Each film project advanced his reputation as an expert falconer whose birds could do anything the producers wanted to see on film. In addition, Morley knew from experience that film-making requires multiple takes, different birds for particular tasks best suited to their abilities, and patience.

Morley got the first big break when Paul Kenworthy called him to inquire about whether he could help with a sequence in the *The Vanishing Prairie.*

"I needed a falcon that would strafe prairie dogs when they emerged from their den," Kenworthy said. "So I called Frank and John Craighead in Jackson Hole, Wyoming. They were famous for working with animals and birds of prey, in particular, at the time, and I knew them slightly. I can't remember if I talked to Frank or John, but one of them said, "Call Morley Nelson in Boise."

Morley did not know the Craigheads well, but he had met them in McCall, Idaho, in the late 1940s, when they were teaching a snow-survival course for the Navy in the lakeside mountain village. The Craigheads had gotten hooked on falconry as teenagers growing up in Pennsylvania. Their father and falconer Capt. Luff Meredith helped them get started, and soon, Frank and John were rappelling down cliffs to capture and photograph hawks, owls and falcons. Intelligent and well-educated, the Craigheads wrote *Hawks in the Hand: Adventures in Photography and Falconry* in 1939, when the boys were only 19 years old. The book contains many outstanding photographs of a wide variety of birds of prey. Their work also appeared in *National Geographic.*

At the time that Kenworthy called, the Craigheads didn't have any of their own birds on hand. They knew Morley had a number of trained birds for falconry.

Kenworthy called Morley, told him what he needed, and Morley didn't pause for long. "Yep, I can do it."

Morley drove to Tucson, Arizona, with a hooded prairie falcon perched on the top of the back seat, to meet up with Kenworthy. The film producer had struck it big with his first feature-length film, called *The Living Desert* in November 1953. While working on his master's degree at the University of California, Los Angeles, Kenworthy had shot most of the film on speculation, hoping that he could sell the raw footage to Disney.

"When Disney came out with feature films in the 1950s, it wasn't typical to see spiders and snakes. I knew I'd have a hard time selling a film on spiders and snakes in the desert," Kenworthy says. "But Disney had the guts to buy it, and they added music and a script and turned it into a seventy-minute feature film."

It was a big hit. *The Living Desert*, the first long feature that Disney produced as part of its ground-breaking True Life Adventure Series, won an Academy Award.

Now Kenworthy had strong potential as far as Disney was concerned. His next project was *The Vanishing Prairie,* a 70-minute feature film that told the story of how the pristine American Prairie was being lost to the plow and other development. The film contained outstanding footage of spectacular wildlife sequences including bighorn sheep rams slamming head-first into each other during the rut, a buffalo giving birth to a calf, mountain lions chasing down deer and more.

Kenworthy and Morley traveled to South Dakota to shoot the sequence of a prairie falcon diving down from high in the sky to kill a prairie dog standing above his den. Morley took a wild-trapped prairie falcon with him on the trip. He had caught the bird in the fall—so it was known as a "passage hawk." As an adult wild bird, the falcon already knew how to hunt in the wild. Morley just had to train the bird to accept the hood, and fly to the lure, which he did.

About halfway into the film, Kenworthy shot a nice sequence on the life of a prairie dog, both above and below ground, and the predators that tried to attack them. He filmed the sequences at a sanctuary for prairie dogs in South Dakota. Upon request, park officials trapped some wild prairie dogs for Morley, so he could stage a prairie dog standing next to the top of its den hole when it was time for his falcon to swoop down and nail it.

Kenworthy set up his camera directly in front of the path that the falcon would take so he could shoot the most dramatic angle with a telephoto lens.

The sequence arises in the film just after a funny segment in which prairie dogs drop to their front paws and stand up—as if they're doing calisthenics—in synch with the song, *Home on the Range*. Then the camera pans to a close up shot of Morley's prairie falcon, standing on a rock.

"When the falcon takes to the air, the concert is automatically over," the narrator says. "Now it's down the hatch everybody, this bird means business."

The prairie dogs scurry into the den, single file. The falcon flies off the rock into the sky, and waits to see an over-curious prairie dog return to the surface.

"On silent wing, the falcon hovers, waiting."

Kenworthy shoots the black shadow of the falcon, and then Morley releases a prairie dog by the top of a den, and the falcon comes swooping down at warp speed. Just as the falcon reaches down with its talons to smack the dog, it sneaks into the hole. The falcon comes down again, and misses another one, just barely.

On the third try, the falcon strikes a prairie dog, sending it flipping head over heels five times before it lands and scurries back into a hole. On this day, the prairie dogs outfox the predatory bird. That's all part of the story line.

"That bird was so sharp-eyed, it just knew what to do," Kenworthy said. "We got a spectacular shot."

Seeing the film again, so many years later, Morley gets a kick out of seeing the falcon stoop on the prairie dogs. "Whoa, that shakes you, doesn't it," he says when the bird zooms past the opening of a prairie dog den. "That's a heck of a shot. Kenworthy was one helluva motion picture man. When the birds strike at that speed, they kill those little prairie dogs like nothing flat."

The Vanishing Prairie won an Academy Award, too. Kenworthy had quite a track record now—he was two for two, with Disney's assistance, of course.

Disney's True Life Adventure Series proved to be a big success, both in terms of critical acclaim and in box office receipts. According to *Reel Nature: America's Romance with Wildlife on Film* (Harvard University Press, 1999), *The Living Desert* grossed $4 million in box-office sales. *The Vanishing Prairie,* Disney's second theatrical feature, brought similar results, much to the delight of the Walt Disney Company.

Walt Disney took a bit of a dare to invest in nature films because he did so well before it was in vogue. The first release, *Seal Island,* occurred in 1948, as a thirty-minute theatrical film, featuring original footage of the untamed wilderness in Alaska. In the late 1940s, there were no other nature films of comparable depth and quality available for the public to see. Mutual of Omaha's *Wild Kingdom* did not air until the late 1950s, and *National Geographic* television specials did not emerge until the early 1960s.

"People could see a low-budget nature film if they went to attend a lecture, but Disney turned them into a source of real entertainment," Roy Disney, Vice Chairman of the Board of The Walt Disney Company said in a personal interview. "We were the first ones to get into it at a really deep level—we were way ahead of everyone else."

Author Gregg Mitman of *Reel Nature* agrees, adding that Disney's True Life Adventure Series showed real-life natural scenes that aided the conservation movement and contained an important diversion from the death and destruction of war.

"In their search for pristine nature, naturalist-photographers and American conservationists found in Disney's True Life Adventures a place of renewal to offset the oppressive conformist trends of an affluent consumer society and a means to increase public appreciation for wilderness areas they sought to preserve," Mitman wrote. "To a wider public, Disney's nature—benevolent and pure—captured the emotional beauty of nature's grand design, eased the memories of the death and destruction of the previous decade, and affirmed the importance of America as one nation under God."

The nature films came at a propitious time, when the Disney company was struggling. "In 1948, Disney studios were really clamoring their way along. We were as close to bankruptcy in 1946 as you could imagine," Disney said. "The war had just ended, and we hadn't made a film since 1940. The most popular films we had at the time were reissues of *Snow White, Bambi* and *Pinnochio.*"

Due to an increasingly strong following, Disney pumped out dozens of nature films in the 1950s, and moved them to television as more Americans purchased TV sets and filled a demand for Sunday evening programming.

Disney's interest in the nature films had a three-pronged advantage: Americans enjoyed seeing rare footage of wildlife that they had never seen before in person; the films had a strong conservation message; and Disney made sure they were entertaining.

"We always understood the conservation message," Disney said. "Our position was intentional, right from the beginning. We were at the end of World War II, and we were masters of the universe. Protecting nature was not a high priority for our nation at the time. Most people accepted that. It took the Walt Disneys, and True Life Adventure films and Jacques Cousteaus to begin to change the attitude."

As Kenworthy recalls it, the entertainment value of a film was paramount, and the conservation aspect came along as a good byproduct of the film. "It was sort of an extra-plus," he said.

Walt Disney made sure that even though the films focused on

the life history of a particular fish or an animal, the script writers showed how they all had different personalities, often times making them seem human in intelligence. "Walt realized that personality was crucial," Roy Disney said. "And he made damn sure that the films were entertaining, using animation, music, sound and scripts to liven up the raw footage."

Kenworthy's next project with Morley was called *Perri*, the story of a pine squirrel living in the Rocky Mountains. The film was based on the book, *Perri*, by Felix Salten, a pseudonym for Austrian novelist Siegmund Salzmann, the author of *Bambi.*

Perri was filmed near Murray, Utah, south of Salt Lake City, near an old mine and smelter. Morley was involved to help Kenworthy film several sequences with a goshawk and an arctic white owl. Perri was a female pine squirrel. The film focused on the life habits of a pine squirrel over the four seasons, starting from birth in the spring. It depicted realistic chase scenes in which predators such as a weasel, pine marten and goshawk, nearly take Perri's life.

As things turned out, Roy Disney was part of the film crew, working as a gopher and a cameraman. Morley hadn't met Roy Disney, as yet, and he was several hundred feet above a cliff on an extended boom crane. He looked down and saw a young dark-haired man getting ready to film the sequence with a Bolex camera.

"I called down in a joking manner while I'm hanging from the tower and said, "Hey, wait a minute. You can't have just anybody shoot my beautiful goshawk with that cheap Bolex down there. You have got to have a better camera for the shot."

Kenworthy said, "Hey, Morley, take it easy. That's Roy Disney down there. He'll do just fine."

"Oh yeah? Well, I'll be darned. Hi Roy."

Disney waved back and took the ribbing like a champ. But he remembers the goshawk shot most distinctly. "We had a great big construction crane set up there above the cliff so the bird would come zooming down from above the trees to nail the flying squirrel, and Morley climbed that big crane—I don't know how high it was, but it was higher than hell—and he went all the way to the top, climbing the ladder one-handed while carrying the bird.

"Then a guy threw out the flying squirrel into the air and it started to fly for the trees, and I have a totally clear memory of looking through the view-finder and that bird came straight at the lens. I was only twenty-four years old, and it was a beautiful bird, it nailed the flying squirrel perfectly and flew off with it. It was a spectacular shot."

That sequence introduces the goshawk as yet another fierce predator. Then, Perri gets chased by a weasel and she decides to scale a dead fir tree to escape the weasel. But the goshawk sees her and comes after her, knocking her off the tree to the ground. A big fight erupts between the weasel and Perri at the foot of the tree, and the goshawk soars to the ground and pounces on the pile of fighting furbearers. Precious seconds pass, and the audience knows that one of them will not escape. Surprise! As luck would have it, Perri is the survivor.

From a philosophical and educational perspective, Morley really liked the fact that *Perri* focused on predator-prey relationships in an entertaining story about a pine squirrel.

"People and wildlife have exactly the same problem in this universe, in order to eat something, first you must kill it, whether it's a piece of asparagus which you break off from the plant or whether it is a cow or a deer," Morley says. "The raptors aren't any different from us. They must eat to carry on the maintenance of their life."

To associate human traits with wildlife is called anthropomorphism. It's a term that Morley uses frequently to help people understand the intelligence of birds of prey. Lending human traits to animals in the story line was typical for the Disney True Life Adventure Series as a way to boost entertainment and ease of comprehension.

The film "was a new adventure in a lot of ways for us at the time," Roy Disney says. "It was instructive on how to handle the animals better and still get natural and unnatural reactions. We were trying to show, this is what life is really like for them in the wild."

For Disney to help the public understand what it's like for wild creatures to survive in their natural environment was helpful, Morley says. "It starts a wonderful feeling that I felt as a boy herding cattle and driving horses with a plow and working with the chickens and the hogs and the dogs. It's very similar to what humanity must have felt when they watched the Walt Disney movie hour. They were reminded that we must all kill something to eat, but at the same time, they could feel a part of their environment and feel the beauty and grace of life."

To make *Rusty and the Falcon,* Morley and his sons, Norman and Tim, had to collect at least ten prairie falcons to ensure that they always had a hungry, trained bird that could perform for a particular shot. Even if a well-trained bird was hungry enough to fly to the lure, or catch Rusty's hat, a camera man could miss the shot. Then Morley had to use another cooperative and hungry bird to get the same shot.

To accumulate enough birds for the movie, Morley started with several prairie falcons from his mews that had been injured and rehabilitated. He also went to the Snake River Canyon and Malheur National Wildlife Refuge in eastern Oregon and retrieved several juvenile birds. When Morley went to the Snake River Canyon to fetch a prairie falcon, he went to a familiar nest site that was about 100 feet below the top of the cliff. But when Morley tried to fit into the nesting cavity in the rocks, the narrow opening was too small for him to get through. So he climbed back to the top of the canyon, and his son, Norman, offered to help.

"Let me go over the cliff, dad," he said.

"Well, I don't know, son, it's a steep cliff. Do you think you're up to it?"

"Yeah, I can do it, dad. There's no danger as long as I'm tied onto the rope correctly. You always say there's no danger. Let's give it a whirl."

In truth, Norm could feel a whole colony of butterflies fluttering in his stomach.

"I was scared. I was waiting for a rattlesnake to get me. My legs were quivering. But my dad's barking out orders, get your weight back, lean back into the rope, use your legs—and I just did it."

"That's a boy, lean out now, that's good."

Norm made his first rappel to the nesting cavity in the rock, climbed into the nest, and captured the biggest female of the five young birds in the nest.

"I got it, dad!"

They named the prairie falcon, Sherpa. It became one of the finest prairie falcons that Morley had ever flown.

To fill out the group of prairie falcons, fellow falconer Hal Webster sent him four juvenile birds that had been found starving to death in a nest in Colorado after their parents were shot. Those birds had to be treated with enheptine to treat a disease called the canker.

They ended up with eleven for the movie. Morley and his boys tried to man and train all of them before they went to the set in Utah, but the three rehabbed birds had not been trained at all. So Norm had to work on training the birds to accept the hood as they drove to Park City from Boise. Morley had all of the birds lined up on a large perch on top of the rear seat in his Chevrolet wood-paneled truck.

"These birds had never been trained to the hood, and we never had time to break them in," Morley says. "We got them hooded all right, but if anything touched them, they immediately flipped on their back and held their feet up and hissed. This caused a tremendous problem for the trained birds, because once one of the untrained birds flipped on its back, hissed and tried to grab the others, there would be a horrendous scene, and I'd have to pull the truck over and get the birds separated and calmed down."

While Morley and Norman drove the birds to Utah, it was hot, with temperatures hovering near 100 degrees. After several more episodes of the birds going nuts in the back of the truck, Morley began to lose patience. They still had eighty miles to go after driving for seven hours straight.

"I think we'd better stop for the night and rest. This is getting to be too big of a problem," Morley told Norm.

And his son, who was twelve years old at the time, said, "Come on, dad, we can make it. We're doing all right. I'll hood those birds again, and I'll hold a couple of them in the front seat with me. Let's carry on and see if we can't get there tonight."

Morley had to smile, realizing that his own son was showing more patience and greater understanding than he was.

When they arrived at the movie set, Kenworthy had a first-class mews set up for the prairie falcons. "Dad told them what size the perches had to be, how much distance to put in between the perches, and they did an excellent job putting it all together. It was really deluxe," Norm says.

Now, Morley and Norm had nothing more to worry about in terms of caring for the birds. They would be able to train the wild birds, fly and feed the others, and decide which birds would work the best for a particular scene.

Kenworthy was operating as an independent producer, so he had to find his own projects, produce and direct them, and then submit them to Disney for final editing, rewriting and finished production. He started with a seven-page script that he adapted from a book, *The White Falcon*, by Charlton Ogburn. He recruited a boy, Rudy Lee, to play Rusty, who was an experienced Hollywood actor. Kenworthy recruited the rest of the cast from the Little Theater in Seattle.

"I had met Morley, so I knew I had a guy who could deliver all of the scenes I needed with the falcons, and I thought I would plan a story around a boy and a falcon, and see where it ended up," Kenworthy says. "As it turned out, I improvised in the field, and we added the church scene and the library scene at Morley's suggestion."

Because the film focused entirely on prairie falcons, Morley was needed on the set for most of the summer of 1957. He had received a leave of absence without pay from his snow survey job, and he received a daily contract fee from Kenworthy. Betty Ann brought the rest of the kids down to live in a 13-foot trailer with Morley and Norm so the family could be together. She was five months pregnant with their fifth child. The Nelsons lived on the set so they could watch over and care for the falcons, and the kids had fun playing in the small and quiet mining town. Kenworthy and the rest of the crew commuted to the set every day from lodging quarters at the University of Utah in Salt Lake City.

"We stayed up in Park City, so we got to know these miners pretty well, and they were a lot of fun," Betty Ann says. "There weren't many people who lived in Park City in those days. It wasn't a high-brow tourist town like it is today. It was just a small town with a few folks. And you know, we could have bought any of those homes for $500 in those days. Can you imagine?"

Norm remembers all of the nifty Hollywood devices used on the set. "Rusty's hideout was a real cave, a mine shaft," he says. "It was cold and drab and smoky in there. But they had a lighting grid in the roof of the cave. And it was lined with fake wooden timbers and two-by-fours, and fake rock, so when the cave collapsed on Rusty, he wouldn't get hurt."

Morley had to teach the child actor, Rudy Lee, how to be very slow and deliberate when he picked up the falcons, or cast them off to fly. Eventually, the birds trusted Lee to the point where he could carry the birds without spooking them or causing them to bate. "Basically, the boy had to learn all of the techniques of falconry just to be able to carry the birds," Norm says. "Rudy was more of a city kid, but he took to working with the birds pretty well."

Kenworthy put a tremendous amount of effort into making a quality film. "It was unbelievable how much work they put into every shot," Norm says. "If the bird was the actor in the scene, it would take several takes and several birds to make it happen. Hell, it'd take half the morning just to get the shot of the bird jumping up on a branch."

Filming in the high mountain location also caused delays. "We had days and days of summer thunderstorms," Kenworthy says. "By the end of the summer, we were running out of money, with people sitting around in hotels while we couldn't work. Fortunately, Disney gave me some extra funds to get the project finished."

Each falcon had a particular personality that worked best for a given scene. They used a bird named Apache for many of the flying sequences because he was trained to fly to the lure—even through the boy's legs—and the falcon had a good rapport with the boy. They also had a very tame prairie falcon with a bad leg, named "Limpy," that they used for all of the scenes in which the falcon walks into Rusty's cave.

By the end of the summer, Betty Ann had to drive the kids

back to Boise so they could go back to school. She was seven months pregnant now. Morley stayed on at Park City to finish the film.

Nothing could have prepared either of them for what would happen next.

"I came home, put the kids into school, and it was very dry in the foothills after the long, hot summer," Betty Ann says. "Somehow, a fire got ignited in the hills right above our house. There was no auxiliary pump, and the BLM was up there looking at the fire, and the city was there, and we didn't have any water.

"So I went up and cut the birds loose. I thought the house and everybody else was going to burn up. And then my water broke, and I had to get to the hospital. I wasn't due to have the baby for another month, so it was quite a shock."

Betty Ann called a neighbor, Margaret Anderson, and asked her to drive her to the hospital. Anderson called Morley in Utah, and told him he'd better get home because this was pretty serious stuff.

"I had a placenta previa baby, that's when the placenta comes before the baby and cuts the air off. The baby was three-pounds and five ounces. I lost a lot of blood. And my regular doctor was gone. It was a bad scene," Betty Ann says.

Morley jumped into his car and gunned it for Boise. He got to St. Al's Hospital while the baby was still alive. But the infant was in intensive care, and the prospects were not good. It died five days after birth. The family arranged for a graveside service, and Morley drove back to Utah on the same afternoon.

"I told Betty Ann that I couldn't stay there with her because I couldn't hold up the movie crew, and I had eleven falcons down there that needed attention," Morley says. "It was a sad situation to lose the baby, and I knew that she wanted me to stay there but I couldn't. I was getting pulled in all directions. I was demoralized in an impossible situation."

"That was a very dark time for me," Betty Ann says.

For Norm and the kids, it was a scary time.

"I was at the house when mom came home. She was all white, and limp and depressed," Norm says. "She was in the living room, lying on the sofa. When she spoke, her words were garbled and slow. Most everyone else was too young to know what was going on, but it did register with me. I was quite scared about the whole thing. The whole situation was really strange and uncomfortable."

Betty Ann says that she never forgave Morley for leaving her alone with the kids at that difficult time. "The marriage was never the same," she says. "I needed more support than I got, but I just didn't demand it. It's something that most people would recognize as important . . . but Morley was too absorbed."

When Betty Ann got strong again, she had her tubes tied so she couldn't get pregnant again.

It was an eventful summer in 1957, a mix of opportunity, tragedy and sadness. Morley laments that his biggest break in the movie business came at the same time that his family suffered a personal loss. "It was a very difficult time for the family, but we had to keep going on," he says.

In the late 1950s, Morley met George Oliver Smith, a knowledgeable and experienced movie producer who had moved to Boise to escape the rat race in southern California. Morley had had his own second-hand Bell and Howell 16 millimeter movie camera that he'd used for a number of years, and he was interested in making a real movie about falconry.

He'd practiced making home movies on the birds and the kids, and he was quite serious about it. Betty Ann remembers him setting up all kinds of lights on Christmas morning to film the kids opening presents, and he'd take the camera on snow survey trips in the mountains.

Morley continued to see the need for more public education about falconry and birds of prey. His experience with Kenworthy and Disney had taught him a great deal. So he approached Smith about helping him make a movie to be called *Modern Falconry.*

"George said, 'Well, I'll help you produce it and show you how to edit and make prints of the film, if you'll do a film for me showing how to make snow caves and how to sleep in the snow,'" Morley says.

It sounded like a fair trade, so they agreed to work together on making a movie on snow survival, which Smith titled, *Winter World.*

A twenty-minute educational film, *Winter World* starred Morley Nelson. The movie was used to help train snow survey personnel in the Soil Conservation Service and U.S. Forest Service on how to survive in the snow. Loggers and others that worked in the mountains in the winter also purchased copies.

Winter World came across as a serious, educational film without the bells and whistles of a more well-financed commercial enterprise. A professional narrator reads the script, and it has somber background music. Morley does not speak in the film. The camera follows him on a winter-camping outing.

"Man must come to the great outdoors well-prepared," the narrator says in a deep baritone voice. "In the winter, the mountains must be approached with knowledge and respect. Nature is no place for ignorance or bravado."

Morley skis into the picture with a pair of large black goggles on his face, Head metal-edged skis on his feet, aluminum ski poles and climbing skins. He's wearing a light pullover waterproof shell with a hood, beige pants, gators and gloves. He climbs up a mountain, and peels some clothing on the way up to underscore the point that it's important to remove layers of clothing as you get hot during a climb.

The film shows Morley using the tip of his ski pole to dip his water bottle directly into the creek from a loft of six feet of snow to replenish his drinking water. Morley demonstrates how to make two different kinds of snow shelters—a lean-to with pine boughs, and a snow cave. Smith set up close-up shots of Morley snuggling into his sleeping bag inside the snow cave, and heating up some soup over candlelight, using his ski poles as a platform for his soup cup.

"Man has shelter if he has snow," the narrator says. "There is no danger of freezing."

The snow-cave sequence climaxes the next morning with Morley bursting forth from the snow bank. Of course, it would be too hum-drum to use the front door. The film wraps up with Morley skiing down the mountain in broad, giant slalom-type turns through the powder, just like a 10th Mountain Division ski trooper, against the backdrop of a musical crescendo.

"We used that for snow survey personnel, so they could learn how to sleep in the snow and survive in the snow," Morley says. "And then we'd go out in the field, and I'd drill them on the techniques."

In *Modern Falconry*, Morley wanted to produce an educational film about falconry and the conservation of birds of prey. "The art of falconry is the basis of the conservation of birds of prey in every nation and it always has been, since the beginning of time," he says. "I wanted the public to have a better understanding of birds of prey, what they do in falconry and in the wild, and what they're capable of doing. Falconry is one of the oldest past-times in the world. I thought that more people would want to protect birds of prey if they understood what falconry was all about."

The movie begins with a reference to the Air Force Academy's adoption of the falcon as its mascot, clearly identifying the Air Force's flying capability to a bird that has no equal in the speed of flight. "The stoop of the falcon is the fastest single action in nature," Morley says. "It's the perfect symbol for the Air Force."

Viewers are introduced to Morley's birds: Sherpa, the prairie falcon, Blacky, the peregrine, Tundra, the gyrfalcon, and Clyde, Morley's first golden eagle. As the birds are introduced, the film shows footage of their home habitat, including footage of the Oregon Coast, where peregrines nest in the cliffs, and the Mt. McKinley area in Alaska, where Tundra was captured from a nest as a juvenile, and the yawning Snake River Canyon, home to the largest concentration of prairie falcons in the world.

Morley stars in *Modern Falconry,* and a professional narrator reads the script. There is no dialogue, mainly because 16mm movie cameras in those days did not record any sound, just the motion pictures. The wind-up cameras were difficult to operate, with many manual adjustments necessary to control F-stops, focus, film speed, viewfinder distance, and more.

When Morley flew the birds for the movie, he had to train two young neighborhood boys how to operate the camera, or he had to coax George Oliver Smith to take the shots. "This was the first time that anyone did a film about falconry in America," he says.

Morley takes viewers through the steps of how to train a bird for falconry. He takes viewers out to the field, where Tundra learns to dive for the lure, and how to take pheasants, ducks and crows. When Morley flies the lure, he swings the lure in a counter-clockwise direction over his head, like a calf-roper. And when the falcons come diving from the sky to catch the lure, he times the presentation of the lure in a manner that allows the falcon to make a clean strike in the air, on Morley's left side. Sometimes he yanks the lure out of reach at the last moment, allowing the bird to set up for another attack. This was a great way to exercise falcons for hunting.

"Whoa, look at that bird take the lure in the air, isn't that fantastic?" Morley says. "A wild falcon would much rather take something in the air than something lying on the ground."

He hawks a few crows with Tundra, showing the public that birds of prey can be useful predators that keep the population of crows in check. Expert falconers know that catching crows with a falcon is one of the most sporting flights in the art of falconry.

Next, the film moves to a sequence in which Morley rappels down a cliff in the Snake River Canyon to band some prairie falcons. He keeps one of them for falconry. "Falconers always should leave at least one juvenile bird in the nest," the narrator says.

In the last bird-training sequence, Morley films two teen-age boys from his neighborhood, Ron Penninger and Bill Bailey, who helped train Clyde. Because female golden eagles are so big and heavy, and potentially very dangerous, Morley trained the bird to fly to a wooden T-perch. It's the same type of device that the Mongols used for carrying golden eagles by horseback.

Morley injects two humorous sequences into the film: He puts a lure on top of Tim Nelson's hat, showing how simple and fun it can be to teach a falcon to fly off with a hat, or catch one in the air; and he films a hilarious sequence of Blacky the peregrine taking a splash bath in the Nelson mews. It's funny to watch the bird douse itself in the water, and splash around like a playful toddler in the bathtub, and then shake its head and body and flap its wings to dry off. For a creature that seems so deadly serious all the time, it's a change of pace to see a falcon become so animated and playful.

Nelson family collection

A dozen eagles were needed for the Disney film *Ida the Offbeat Eagle*. Eight members of the cast, all trained by Morley and his sons, wait on their perches for their turns in front of the camera.

The film concludes with a nice silhouette of Morley's oldest boys, Norm and Tim, holding a bird on their fist, with the sun backlighting the whole scene.

After *Modern Falconry* was finished in 1960, Morley brought the film to many of his presentations to service clubs, wildlife conservation groups, hunting clubs, the Farm Bureau, Cattle Association, Woolgrowers and Falconry groups. It helped educate the public about falconry, and the intelligence and nobility of birds of prey.

"I took the film to every speech I gave, and people really responded to it in a positive way," he says.

In the summer of 1964, Morley and his oldest son, Norm, got an opportunity to work on yet another movie project for the Walt Disney Company, *Ida the Offbeat Eagle.* Charles L. Draper, an independent producer who did pictures for Disney, got a contract to film *Ida* on location in the Snake River Canyon, south of Boise.

The story focused on the relationship between a golden eagle and a crusty old Snake River rancher. Eventually, the eagle saves the rancher's life. It's a positive story, like *Rusty and the Falcon,* that depicts realistic scenes and dialogue on how ranchers viewed golden eagles at the time.

In the winter of 1963-64, Morley spent many hours on the phone talking with Draper about the script. When the snow began to melt and the onset of spring occurred, Morley and Norm began to visit a number of golden eagle aeries they knew about to collect and train at least ten golden eagles for the film.

At the time, Morley had only one golden eagle, Clyde. The bird had been turned over to him when it was found injured, and nearly starving to death. He nursed the bird back to health, and trained it with the help of the neighbor boys who appeared in *Modern Falconry,* Ron Penninger and Bill Bailey. Worried that Clyde might strike him or one of the boys, Morley had the eagle trainer wear a football helmet with eye goggles, and a heavy elephant-skin coat for protection. They learned that Clyde could be taught to fly to a T-perch, and kill wild game such as rabbits and rattlesnakes. It just took more time and patience than a falcon. Golden eagles were so wild and fierce in a natural setting that it took more time to tame them down to trust a human.

In his sophomore year in high school, Norm became passionately intrigued with a male golden eagle named Otis, that he found on the ground orphaned and starving below a nest in the Snake River Canyon. "I formed a stronger emotional bond with Otis than other eagles I've trained because I saved him," Norm says. "I mean he must have been really close to starving to death because he ate fourteen chicken heads in the car on the way home."

He manned the bird, taught it to accept the hood, and flew it to the T-perch. Norm and Otis developed a close personal relationship, and the eagle returned the favor by becoming a magnificent hunter and flier.

In the summer after Norm's sophomore year, he was hired to fly Otis for Mutual of Omaha's *Wild Kingdom.* He traveled to Arizona to work on a TV movie for Larry Lansburg, the award-winning producer of the Disney film, *Appaloosa Run.* Just like his dad, Norm had worked so closely with Otis that the bird completely trusted him and was totally loyal. He could make the bird do anything the movie producers needed to capture on film.

"For *Appaloosa Run*, I had to train Otis to catch a parachute," Norm says. "The idea was, these guys were flying some drugs into the country, and she flies over and grabs the drugs. That was the story line. She did it twice, and then she took off. I was with Jay Schiffler, one of the most famous dog trainers of all time. He trained collies for *Lassie*. When I lost Otis, I spent four days riding a horse in the desert, looking for Otis. Then I'd go back to the set, have dinner with Schiffler, and he'd tell me where to look and where to go, and where the wind was, what canyons to look in, and meanwhile, the whole movie production was shut down because the eagle is not there.

"Finally on the fifth day, I'm riding along, and I look up and I see two little dots in the air. I have no idea what they are, but they look like eagles or turkey vultures, so I jump down on the ground, put my T-perch up, and I start calling, "Yo yo yo," like crazy, and all of a sudden, one of them peels off and does like a 2,000-foot dive all the way down, and lands on my T-perch. It was Otis, but I didn't recognize him, the Arizona sun had bleached him out, and his eyes looked weird and kind of freaky. I grabbed his legs, and started to cry.

"I put his jesses on, made sure I had him, you know, and I'll never forget, I was underneath him, and I was putting on the jesses, and he rolled his head and looked up, and his girlfriend that had been feeding him for five days flew through the sun. I figured it was a spiritual moment, she was saying I'll see you later as she flew through the sun. I'll never forget that experience, and the absolute exultation that I felt getting the bird back."

By the time that *Ida the Offbeat Eagle* came along, Norm was well-prepared to help Morley catch and train twelve eagles—no small undertaking.

One thing that Morley and Norm learned was that young and hungry golden eagles could be downright dangerous. A captive eagle had to be hungry in order to get it to fly to the T-perch for food, so when it came time to shoot a scene for a movie, the same thing had to be true—the eagles had to be hungry so the trainers could get them to perform for a particular scene. But when the birds are that hungry, they can try to attack the eagle trainer, with little to no warning.

For example, when Norm was filming a scene of a golden eagle chasing a rabbit, he had made sure that the eagle was plenty hungry. "It was a muddy spring day, and she was hungrier than hell," Norm says. "We released the jackrabbit into the sagebrush, and the eagle went after it with a total single-minded purpose and killed it. But the cameramen said, "You've got to go down there and turn her around, because her back is facing the camera and we can't see the kill." So I went down there, and she had her hackles up, she was cowering over the food, and spreading her wings around the fresh kill, the whole nine yards. So I went behind her, and put my T-perch down to push her around, and she jumped up, did an immediate 180, and she was right there in front of me. And what did I do? I slipped in the mud, fell right on my back, and she came at me. I reached up to shield myself with the T-perch and rolled away, and then she came at me again. I put my glove up, and she hit the glove, and I threw her off, and then she turned around and went back to eat the rabbit. That was the closest call I've ever had. If I hadn't had the T-perch, she would have gotten me."

That situation illustrated how difficult it could be to position eagles for a movie scene. "No true falconer would ever try to turn his bird around while on a kill," Norm says. "It just shows you that to get a particular shot for a movie, you sometimes have to betray the rules you've learned from practicing falconry. You don't like it, you know the bird doesn't like it, but you have to do it to get the shot."

Another potential hazard is that golden eagles grip the T-perch, or a gloved hand with all of their might when they accept the hood. "I remember one time when dad picked up a golden eagle in the hawk house," Norm says, "and after he put the hood on it, it clenched its talons and put one of them through the leather glove and kept driving it into Morley's hand. The talon was putting incredible pressure on Morley's wedding ring, and he yelled, "Christ! She's going to take my finger off!" And

Nelson family collection

The camera crew and actor prepare to shoot a scene for *Ida the Offbeat Eagle.* The cabin was located near Swan Falls Dam in Idaho's Snake River Canyon.

fortunately, I was nearby, and I had to reach down and take the hood off of her, or Morley would have lost his finger."

After weeks of hunting for the right birds for the film, Morley and Norm rounded up a total of twelve golden eagles. Most of the birds were females, because they are larger and more impressive than the males. All of the birds had their own personalities. They had a nasty male named Cesar, which, Norm says, "was mean. I reached down to change his water dish, and he jumped off his perch and grabbed me, and took part of my finger and ripped it open. He was really aggressive."

They caught two female birds, Cleo and Jessica, from a pinnacle rock nest in the lower part of the Snake River Canyon, near Oregon. Both of those birds performed perfectly for the cameras in *Ida.* So did a bird named Buddy. "He was a little tiny high-performance guy with a short head, but we used him because he was so agile," Norm says. "He was really smart, and could do anything, and he was a really good flier."

The Idaho Department of Fish and Game gave Morley a

large female they named Betsy, which was used for a number of high-flying sequences.

Draper and his film crew arrived in Idaho in February to film wild golden eagles performing spectacular aerial maneuvers during the courtship phase in the Snake River Canyon. Morley told them to set up by a golden eagle nesting territory in Walter's Butte, and wait for the eagles to do their thing. More than 30 pairs of golden eagles nested in the Snake River Canyon at the time, so there was an abundance of eagles to follow.

The grandeur of the canyon provided an awe-inspiring backdrop for the eagle-flying sequences.

As things turned out, Morley served as the senior technical advisor for the film. Norm did most of the actual handling of the eagles with another eagle trainer from California, Gary Young. That way, Morley could continue to work on his day job as a snow surveyor during the week. He helped with the script and came out to the movie set on weekends. Norm was out of school during the summer, and he got paid $50 a day to work on the movie six days a week. He slept in a wall tent on the movie set, and he had to feed and care for the birds on a daily basis.

"It was an unbelievable amount of work," Norm says. "We would always be working on finding locations, improving the camp, dealing with all kinds of things that had nothing to do with shooting the film, taking care of generators, cars, food, all of that logistical stuff. And then they'd be adjusting the script, and talking to Hollywood about the film and making more adjustments. Draper just kept moving ahead. He was very patient, and he knew exactly what he was doing. "

The only break that Norm got was on Saturday nights and Sundays. After the shooting ended for the week on Saturday afternoon, he'd drive his grandfather's DeSoto back to Boise in 90 minutes, get cleaned up and go to a dance with his friends. Even if Draper wanted to work on Sunday, Norm would still go back to Boise for the dance.

"One morning I came back to the canyon in the DeSoto, and there was a big corner down there that was nothing but sand. I came into that corner too fast, and I rolled the DeSoto all the way over. So there I am lying on the ground, beside the car, and it's all the way over on its top, and Chuck Draper is coming back from the store, and here's his eagle trainer lying in the dust, knocked unconscious. He took me back to my tent until I woke up, and I was fine."

"The eagle is master of the sky," Walt Disney says in his personal introduction to *Ida the Offbeat Eagle*. "Its spectacular flight is unmatched by anything in the world of man or nature."

The film opens with a pair of golden eagles playfully soaring in the wind during the courtship phase. The eagles ride the wind gracefully with their wings stretched wide, climbing higher and higher. Then the wild male and female soar close together, rise up and then descend in a close-knit spiral, before splitting off for a moment to soar again.

Norm: That's the kind of shot that takes days and days of shooting and a lot of dedication to get.

"In the castle in the crags lives a tiny princess with a big appetite," the narrator says. A white downy eagle, only a few weeks old, stares into the camera from a large cliff-side stick nest, with its mouth open wide.

The scene cuts to an adult female golden eagle, making an 800-foot angular dive from the top of the canyon to the valley floor, where an unwitting Townsend ground squirrel gets snatched up in the blink of an eye. Mother bird returns to the nest to feed her baby eaglet.

Ten weeks later, a rugged-looking outdoorsy fella rappels down the cliff to visit the eagle nest. It's Morley, of course. "Hi there," he says as he arrives at the side of the nest. "It's OK, I'm not going to hurt you."

He grabs the bird by the body, folds in her wings, and holds her while he puts a silver aluminum band from the U.S. Fish and Wildlife Service on her yellow leg. The bird nips at his hands. "Owe! Hey, cut that out!"

The camera zooms in on the terse inscription on the band. It says, "Ida," followed by a sequence of numbers. It seems like a fitting name for the female eagle.

Morley climbs back to the top of the canyon, and reaches up on the rim to pull himself on top. What he feels isn't rock, though. It's a rattlesnake. Morley can't see it, and he has to assume the worst, a rattlesnake. He reaches back and wings the snake over his head, presumably into the water.

Morley: I told them that no one in their right mind would ever reach up a second time to grab the snake, but that's what we had to do to create the next scene.

The snake lands in the eagle nest, much to the surprise of Ida, and a confrontation occurs. The snake lunges at Ida, who doesn't know how to fly and has never left the nest, much less ever been in a position of defending herself. She falls off the back of the nest and plunges fifty feet into Swan Falls Reservoir. She swims up to a pile of sticks and stands on top of them, as the slow-moving water advances downstream toward the dam, spillway and powerhouse.

"This is probably the first eagle ever that learned to swim before she could fly," the narrator says.

The next morning, the sun casts a yellow glow on a little cabin on a flat next to the Snake River. A small plume of smoke rises from a chimney. Ida is hungry, she's looking for breakfast, and she sees what appears to be an easy catch, a rooster perched on top of the cabin roof, doing its thing. "Cock-a-doodle-doo!"

Ida leaps off a basalt slab at the top of the canyon rim and makes a long, angular dive for the cabin roof top. She's zooming at 120 mph, lowers her feet to nail the rooster, but she misses, and knocks over the chimney pipe.

"You basket buzzard, trying to kill my Rufus, huh?"

It's Uncle Billy, a small-time rancher who raises chickens and goats on his flat next to the Snake River. He runs inside to grab his rifle, and raises the barrel to his eye. And then he sees the band on Ida's leg. The band means the bird is protected by the U.S. government. "I guess I've got to do my American duty," he says.

Morley: Draper didn't have a professional actor to play Uncle Billy, so we figured it would be best to find a genuine country hick kind of guy to play the role. And I found him in Melba, Idaho. He looked the part, talked like a real country kind of guy, and was a real natural in front of the camera.

Uncle Billy throws a gunny sack over Ida to fix her wing,

which had been wounded by the collision with the chimney pipe. He wraps it up in tape, and keeps her inside.

Several days later, Uncle Billy is up on top of the cabin, fixing the roof, when Ida sneaks out the door and flies to the top of the chicken house. Her wing is healed, and she's going to be fine. But Uncle Billy doesn't want her to eat his chickens. "Get out of here you gold-plated government buzzard," he yells, throwing a piece of firewood at Ida. She flies up and catches the piece of wood in the air, and drops it at Uncle Billy's feet.

Morley: We put a piece of meat on the firewood, and taught Cleo to catch the wood, no problem.

Uncle Billy is flabbergasted. Now what kind of eagle knows how to catch a piece of wood? He throws it at her again, and she catches it, circles above the cabin, and then drops it on his bald head. He's ticked off at the eagle's tactics, but he's intrigued and impressed.

The next day, Uncle Billy decides to see just how smart Ida is. He brings out three red tin cans to engage Ida in a shell game. He puts a piece of meat under one of the tins, and shuffles them around in the dirt. Ida flies over and promptly picks up a tin with her beak and eats the meat underneath.

"Let's see you do that again, you old coot," he says.

He puts another piece of meat under a tin, and shuffles them around. Ida looks them over carefully, and flies directly to a tin, knocks it over with her foot, and eats the chunk of meat.

"OK, you were lucky."

Uncle Billy shuffles the tins, takes his straw hat off, and sneaks a piece of meat under the hat.

Ida stands on the fence post and thinks long and hard about this one. She turns her head to the right, swivels her head to the left, and turns her head upsidedown. Uncle Billy mimics her, and turns his head this way and that. "Stop your peekin' now."

Then she flies to the hat, lifts it off the ground with her foot and eats the meat.

"I don't believe it! Nobody, nobody can be that good without cheating. Even if you could, it wouldn't be fitting for a bird to know more than a man," Uncle Billy says.

Morley: That scene showed how intelligent the eagles really are. Uncle Billy couldn't move those tins around fast enough to fool Cleo. She always knew where the meat was. And I could imagine how the bird must have stolen the hearts of so many Americans when they watched the eagle tilt her head around this way and that way. That lent so much to understanding the depth of relationships that can occur between people and eagles. If enough time is taken to bring out these relationships, then a greater understanding can occur about the importance of conservation, and humanity can recognize that preserving eagles must be part of the equation.

The next morning, Uncle Billy goes to town in an old pickup to get some supplies. He leaves Ida inside the house. Soon after Uncle Billy drives away in a cloud of dust and blue smoke, a cougar pays a visit to the ranch, causing quite a flurry amongst the chickens and goats. Ida hears the commotion, flies up to look out the window and gets excited. And then she flies up to the top of the window, opens it and chases the cougar away.

Norm: We used Cesar for that scene because he's so aggressive. We put the piece of meat on the top of the revolving window, and he flew up to that no problem and tilted the window open, and tore out of there to get the cougar.

Morley: When we first talked about writing a scene with a cougar, I suggested that we could have the cougar try to steal some food from an eagle, and then have its mate come down and attack the cougar. But when we set up that scene with Cleo in an enclosure, the trained cougar came over and walked within about three feet of the eagle, and made a couple of swipes at Cleo. And I thought, I can't see the eagle backing off from a situation like this. Sure enough, she opened her wings, lowered her head, and her hackles came up and she lunged at the lion. The lion turned and jumped the enclosure fence like it wasn't even there at all and ran away. It took the cougar trainers two days to get that lion back, it was so scared.

When Uncle Billy returns home, he notices a broken window in the cabin and Ida is no where to be seen. It's the fall now, and it's time for her to find a mate. Draper switches to footage of wild eagles for the courtship scenes. He got those shots in the canyon due to the high density of golden eagles present—more than 35 pairs—in a relatively confined area, and Morley told them where to go to get the shots.

While Ida is off bonding with her new mate, one of Uncle Billy's goats decides to stray from the ranch and climb a cliff to the top of the Snake River Canyon. "Just because it was there," the narrator says.

Uncle Billy realizes that "Nipper" the goat is gone, and he can hear the animal's distress call in the cliffs above the cabin. She's stuck, no doubt. Uncle Billy grabs a rope and tries to scale the cliff to reach the goat. But he steps on a loose rock, falls and lands on his back.

"Uncle Billy was hanging on the edge of eternity," the narrator says, but the crusty rancher wasn't hurt.

Morley: They used me as a double to get that shot of the fall. It was a tricky situation because I had to fall, but I couldn't fall very far or I'd fall over the cliff. So I had to do kind of a controlled roll onto my back, and they had some fake rocks to make for a softer landing than the real thing.

Uncle Billy is in a real pinch now. He takes off one of his boots, ties it to the rope, and flings it toward the top of the cliff, hoping it will lodge in the rocks. He makes several attempts and the shoe doesn't get caught at all. Then he hears the familiar call of a golden eagle above him. He figures it must be Ida. Remembering the game of catch with a piece of firewood, he tosses up his shoe at the eagle. She catches it on the second try, and flies toward the top of the canyon. Uncle Billy holds onto the rope, and says, "Let go, Ida!"

Norm: My brother Tim was there to help with that shot. He'd throw up the boot with a piece of meat in it, and Cleo would catch it. It was just like putting a piece of meat on a hat or a piece of wood or anything else, when the eagle associates the object with food, she will go after it. But then it's not as easy getting the eagle to let go of the shoe. She still wants the meat.

Finally, Ida lets go of the shoe, and Uncle Billy yanks the rope, and the boot catches between two rocks. He pulls hard on the rope, makes sure it will hold, and then he climbs to the top of the canyon. He sets up a lasso, ropes Rufus and lifts her to the top. Ida saves the day.

"In Uncle Billy's world all was well," the narrator says. "He had his piece of ground, his patch of sky. And to top it all, his own private spectacular—it was staged by nature daily, the aerial ballet of the golden eagle."

The film closes with Draper's footage of wild golden eagles during courtship flight, soaring and diving together in the clear blue sky. Ida and her mate land on the edge of the rocks and nuzzle their heads together.

"And so in the end, our queen of the sky would fulfill her royal destiny."

Ida the Offbeat Eagle premiered on Sunday, January 10, 1965, as part of Walt Disney's Wonderful World of Color series. The Nielsen ratings for the show were very positive, Roy Disney says. "Ida got a huge rating. Forty-five percent of the potential audience were watching that show. It also happened to air on my birthday."

Morley was more than ecstatic when he saw the results of that hot summer of filming in 1964. "What's so fantastic in this film is she saves the old man's life on the cliff. Uncle Billy owed his life to Ida, and he wouldn't forget it. So everybody said after they saw this film, "How can you shoot an eagle? Why would you shoot something that's so intelligent and noble and beautiful?" I stopped it all with this film. It was one of the best things that I've ever done."

Norm recalls that it was a great moment for the family to watch *Ida* for the first time. "We sat down and watched it in our family room like we always did, and it was really exciting to watch that night because it was our movie. When the movie came on, and Walt Disney talked about the power and the majesty of golden eagles, that always turned me on, because to me, golden eagles were like god. I had such a close relationship with eagles, and I had worked with so many of them day in and day out. I knew them inside and out. So when he said the movie was about golden eagles, I got all choked up.

"I also remember that the family was just going berserk when they played the shell game. That was great. And everyone went nuts when Cleo caught the shoe. That was really cool. Seeing dad in the movie was really neat. I was glad it was all over. I got really burned out living down there in the canyon the whole summer."

Morley's and Norm's work on the Disney films served as a launching pad for both of them. Morley continued to work on more films as an eagle trainer, such as *My Side of the Mountain*, for Paramount Pictures in 1967. He made more of his own educational films about birds of prey, and starred in made-for-TV films about birds of prey, such as *The Eagle and the Hawk*. By the close of the 1960s, Morley had participated in twelve major films, including seven pictures for Disney. By the end of his career, he would have more than thirty-three film credits to his name.

Norm's experience of working alongside producer Chuck Draper in *Ida* actually increased his interest in learning how to shoot movies as a professional photographer. His dad already had taught him how to use a Bolex 16mm movie camera when he was in high school, and he had helped his dad make the movie, *Nature's Birds of Prey*. But he was intrigued with the idea of becoming a professional film-maker as a career. He was fascinated by the $2,500 Arriflex S movie cameras that Draper and his assistant, Frank Zuniga, used for filming *Ida*.

Norm asked Draper if he might have a shot as a professional photographer. "He said it's a really difficult field to get into, but once you get good at it, you can be successful."

Norm got a bachelor's degree at the University of Idaho, and he wrote Zuniga after he graduated. "I told him I was looking for a job on a movie project, and I wanted to be on the camera end of things, not the bird end," he says, "and he got me a job as a camera assistant on the Disney movie *Varda the Peregrine*. That experience gave me the ultimate foundation for becoming a film-maker."

Norm went on to start his own film-production company, Echo Films, in 1971. His younger brother, Tyler, joined him in the business in 1975. Morley had not only groomed two sons who learned how to be excellent falconers, they also became professional film-makers who were totally committed to the cause of saving birds of prey. Over time, Norm and Tyler worked together to make every conceivable film about birds of prey that helped raptors win over the hearts and minds of the American public.

They had only just begun.

Chapter Ten

Swapping stories with sheiks

1965

Deep in the shifting sands of Kuwait in the Persian Gulf, two white convertible Chevy Impalas with red trim race across the pancake-flat desert landscape at 60 mph with their tops down. The tan camel's hair robe and gold-embroidered turban headdress of Kuwait Sheik Abdulaziz Alsaood Alsabah flutters horizontally in the breeze, as he sits on top of the back seats next to Morley and his falconer. Decked out in traditional dress, Morley's white robes and black-and-white speckled turban fly behind his back. He scans the terrain for any sign of hubara, a long-necked game bird, the targeted quarry for the day.

"There's a bird over there," Morley says, pointing out to the left side of the speeding car. The speedy hubara runs even with the car.

"You've got good eyes," Sheik Abdulaziz says, flashing a smile to Morley.

The sheik's falconer instantly casts off a saker falcon from the cast-like leather guantlet on his left arm. The big, powerful brown-and-white falcon—nearly as large as a gyrfalcon—takes off like a rocket and rips across the desert about thirty feet off the ground, pursuing the hubara in a thrilling tail-chase. In a matter of seconds, the saker grabs the hubara's red-capped head with its sharp talons and wrestles the bird to the ground, causing them to roll in a heap multiple times in the sand. When the birds come to a rest, the proud saker grips the lifeless white-feathered body of the hubara with its talons.

They've got one in the bag, and ten more to go for supper.

It's November 1965, prime time for hubara hunting in the Middle Eastern nation of Kuwait. Morley and Betty Ann are the honored guests of Sheik Abdulaziz for a three-week escapade of hunting and sight-seeing in Kuwait and Saudi Arabia.

"It was like rubbing Aladdin's lamp and finding ourselves transported into a chapter of Arabian nights," Betty Ann says.

For Morley, it was a treat beyond comprehension to be invited to a distant land where the art of falconry had been practiced for more than 2,000 years. He would observe the Arabs' falconry methods in hunting wild game, and see the performance of their peregrines and sakers.

"This invitation was most interesting to me because the sheik came from a family of rulers who had been very famous in falconry circles since the beginning of time," Morley says. "They have written books about falconry, and I had read some of them, translated into English from Arabic, and used some of their falconry techniques."

The trip was very significant from a cultural perspective, too, regarding the fact that Morley and Betty Ann would join the rulers of Saudi Arabia on a hunt in the desert. Betty Ann would be the first woman in the world to engage in a falconry hunting excursion in Saudi Arabia.

"To be accepted by them was a fantastic thing," Betty Ann says. "It never occurred to them that a woman like me would know how to fly eagles."

Morley and Betty Ann were invited to Kuwait by Sheik Abdulaziz Alsaood Alsabah, after one of his brothers saw Morley's first film production, *Modern Falconry,* at the annual meeting of the British Falconers Club, "The Falconer's Feast," in 1960. They were impressed by Morley's ability to get his peregrine falcon and gyrfalcon to "wait on" high in the sky, and they were dazzled by his use of the light lure, when the birds zoomed down to nail the lure at more than 100 mph. In the Middle East, falconers always dropped the lure to the ground when a falcon approached. Morley developed his own technique in the field, thinking the birds preferred to take the lure in the air, mimicking the action of a wild bird killing its prey in the air. The Arabs also were fascinated by the falcons' big stoops on wild game such as pheasants and ducks. *Modern Falconry* showed numerous spectacular shots of prey getting killed in the air, feathers flying.

The journey to Kuwait began with a rapid-fire exchange of letters between Sheik Abdulaziz and Morley in the fall of 1960. The sheik's first letter arrived on Sept. 15, 1960:

Dear Sir,

Mr. M. Woodford, Secretary British Falconers Club, has been kind enough to give your reference to my younger brother, Yousuf Alsaood Alsabah who had visited Mr. Woodford recently and had seen with him a film on the subject of falcons, and we have been given to understand that this film has been supplied through your source.

Since we are very much interested in Falconry, I am taking the liberty of addressing you this communication and am pleased to inform you that we belong to the Ruling Family of Kuwait, Sabah Family, and we own quite a verity of falcons for hunting purposes. And to increase our knowledge in this line, we are interested to own a similar film.

We shall therefore be obliged if you could let us have details of this subject film such as the MM size, length, and display duration of the film including also the C&F Kuwait price delivered by airfreight. On receipt of this information, we shall provide you with the necessary funds.

Thanking you for the cooperation.
Yours sincerely,
Abdulaziz Alsaood Alsabah

Morley promptly replied to Sheik Abdulaziz, and told him the cost would be $239.50 (U.S.), including shipping. Upon receipt of the funds, he would be happy to send the film right away.

Sheik Abdulaziz didn't waste any time. He sent a bank order for $375 from Hong Kong.

"I couldn't understand whether the thing had any value or not," Morley says, "but when I went to the bank and asked for my $375, the president of the bank just laughed and said, "You bet your life that order is as good as anything can be in this whole world. So I quickly realized that money was not a problem for them."

Even before he received a print of *Modern Falconry,* on Oct. 24, 1960, Sheik Abdulaziz sent another letter to Morley, extending an official invitation to visit Kuwait.

Dear Mr. Nelson,

I would like to extend my sincerest thanks to you for receipt of your kind and very interesting letter dated 17th instant and appreciate very much the useful information contained therein and now look forward to the arrival of the film very shortly and I am sure we are going to benefit a lot from this film and increase our knowledge of the birds.

Now I would like to introduce myself to you and am pleased to inform you that I belong to the Ruling family of Kuwait known as Sabah and have pleasure to have three younger brothers named Nasir Alsaood Alsabah, Yousuf Alsaood Alsabah and Faisal Alsaood Alsabah and we all jointly operate and own The Sabah Trading & Contracting Company Ltd., which enjoys worldwide business connections and represents some of well-known American companies such as Kelvinator International Corporation, the General Tyres (sic), Montague gas ranges, White Horse brand motor oils, etc. etc.

We all the brothers have got a very active hobby in training falcons and hunting and as you know this is a very very old hobby in our part of the world and we have received and learnt it from our forefathers and our cousin brothers to enjoy this hobby and the hunting.

I personally I always like to increase my experienced knowledge on the falcons and with this view in end have contacted many friends in the foreign countries and have also read many books on this art and one of these books is "Art of Falconry" written by Casey A. Wood and F. Marjorie Fyfe and had got it from the States, and backed by the same idea I had taken the liberty of contacting you.

Now that I have taken the liberty of self introduction, I would like to extend you a cordial invitation to visit us in Kuwait and stay as our private guest and I am sure your visit will provide beneficial to both the sides.

I am pleased to inform you that the hunting season for "Hubaras" (Bastards) generally starts here by the end of October and in other words, it has already started and will continue up to end of March and on the whole the kinds of "Hubaras" have reduced in the last few years due to the appearance of many hunters. The Falcons and peregrines are generally caught in the current month when they come towards this side after the expiry of summer, and they remain in these parts until the end of March and then they go back to North West, but some species remain here for breeding in North of Arab Island and Turkey and this species does not migrate.

Now I have the pleasure in sending herewith some of the pictures though I am sorry that they are not very clear and fine and hope to send you some more decent ones in the future....

I have in my employment 30 men who catch the birds for me and up to now I have received about 90 falcons caught by my staff from various centres....

For the time being, these are the news from us and I trust you will enjoy them.

With kindest regards from all us all, the four brothers.

Yours sincerely,
Abdulaziz Alsaood Alsabah

Sheik Abdulaziz sent twelve photos to Morley, mainly shots of the sheik and his birds, a picture of a hubara, and falcons with sealed eyes (a long-time practice used by many falconers in that part of the world to assist in the process of manning and hooding a bird).

Two days later, Sheik Abdulaziz sent Morley another letter, indicating that he and his brothers had received *Modern Falconry* and watched it immediately. "We have a had a screening of the film and have been immensely pleased with it and increase our experience in the light of its subject and have sensed a great amount of happiness that your method of training the birds nears very much the method we follow in our country with the exception of minor differences," he wrote.

The sheik concluded the letter by asking if Morley could send him a white gyrfalcon from Greenland. The request underscored the fact that gyrfalcons are highly prized by Arab sheiks. Morley wrote back, indicating that it would not be easy to get him a gyrfalcon, but that he might be able to secure one in the next year or so.

In the meantime, Morley wrote the sheik that he wouldn't be able to visit him in 1960, but possibly in the next year or so. He was anxious to go, but he was just too busy with his work at the time. In 1961, he made a fruitless trip to Greenland with Harold Webster in hopes of capturing several gyrfalcons, but the local government would not allow it. As things turned out, Morley got consumed by three film projects in the next couple years, two segments for Mutual of Omaha's *Wild Kingdom,* and *Ida the Offbeat Eagle.*

Finally, in the spring of 1965, Morley wrote a letter to Sheik Abdulaziz and informed him that he'd like to visit:

Dear Mr. Alsabah,

It has been a long time since we exchanged letters but we have thought of you and your country often. There were films on Kuwait that came over television and several feature stories with pictures in our national newspapers. We enjoyed the films and the newspaper feature story on the progress being made in Kuwait.

The reason for my long silence is covered in the pictures that are enclosed. My family and I have worked in three motion picture films on eagles and falcons. We had 11 eagles and several falcons flying at the same time. The film was made by Walt Disney Productions and took a year and a half to finish. However, we are very happy with the finished film entitled "Ida The Off-beat Eagle." It presents the eagle in the wild and a story concerned a goat herder. The eagles are made heroes in the film rather than problems and one saves the life of the goat herder....

In one of your letters, you invited us to come to Kuwait for a visit. We were very grateful and pleased by the invitation and now are in hopes that we can visit your country. If possible, we would like to film Falconry as you practice it in Kuwait. Also to cover life in your country as well, but the major work would be on Falconry.

If your invitation is still open my wife and I would like to arrive in Kuwait sometime in October for about two weeks. I believe that we could bring the Walt Disney Film on Eagles as well as the new one that I have produced which might be of interest to you. These would give you some idea of what we have been able to do in Falconry.

We hope that you and your Family have been well and look forward to renewing our correspondence. If for any reason a trip to Kuwait would not be expedient at this time, we will understand and hope to make it at some future date.

Sincerely,
Morley Nelson

Sheik Abdulaziz wrote back to Morley, indicating that, indeed, the invitation was still open, and he and his brothers looked forward to greeting him and Betty Ann in Kuwait. He told Morley to arrange for a roundtrip plane ticket, and they would pay for it. The sheik asked Morley to bring some of his birds to Kuwait, but it didn't work out. Morley was putting the final touches on a new film of his own, *Nature's Birds of Prey,* which he promised to bring along and show to the Alsabah brothers.

"My wife and I are very anxious to get to Kuwait to see you, your country, and the Falconry that you follow. If we can go to the other countries that you fly your falcons in, that would be great, too," Morley wrote in an August 20, 1965 letter. "We do not want to change your plans for the season and will do whatever is convenient for you. We hope to arrive in Kuwait in the first week of November and will keep you informed of our progress more often from now until we leave. Best personal regards."

The trip was on. As always, Morley had big hopes of selling a movie to Disney or another film company about the art of falconry in the Arab nations. He would take plenty of film, and his Bell and Howell 16mm camera to record the whole affair.

Sheik Abdulaziz and his brothers all were on hand when Morley and Betty Ann arrived in Kuwait City in early November 1965. They had flown to Kuwait via Ireland, England, Italy, Spain and Greece. It was like a hero's welcome. The sheik escorted the Nelsons to a big white Cadillac, and they proceeded to the six palaces of the Alsabah brothers.

Nelson family collection
Morley "goes native" during the 1965 Kuwait trip.

"Their homes are huge stucco structures of three floors, right down on the sea, with a covered swimming pool, a boat house with three motor boats and a cabin cruiser, and tennis courts," Betty Ann says.

The Nelsons walked through a tiled courtyard in which the sheik's twenty-five falconers sat amid an indoor garden of flowers, holding their peregrines and sakers. The couple was escorted into a reception room for a formal Arabian dinner. Morley's new film, *Nature's Birds of Prey* would be shown that evening. It was an educational film that Morley produced with George Oliver Smith to use in his talks to school kids, hunters, farmers and ranchers about the different types of falcons, hawks and eagles, where they lived, how they hunted and what they ate. The film also featured a fair amount of footage about the art of falconry, and in one segment, one of Morley's red-tailed hawks kills a bull snake after a short struggle.

"They didn't understand the English sound track, but they understood what they were seeing," Betty Ann says. "They got so excited they hit the floor with their hands and shouted."

"A bull snake looks like an Arabian cobra," Morley says. "And when they saw that hawk nail the bull snake, they just went crazy. They were crying "hajt!" the word for snake. And I'm thinking to myself, what's the big deal, it's just a bull snake, and any red-tail is going to kill a damned bull snake with no problem whatsoever. But they took on the mood of every action element of the film. They were really into it. By the time the screening was over, the Arabs thought I was one notch from Allah."

At lunchtime, Betty Ann went to Sheika Bedria's palace for lunch. Sheika Bedria was the sister of Sheik Abdulaziz. Morley went off with the sheik and his falcon trainers to discuss the birds. This was one of the major traditions of falconry in Arabia. Falconers would sit in a circle with their birds placed on fancy embroidered perches in the center, and they talked about their favorite falcons.

Walking into the sheika's place "was like an unbelievable dream," Betty Ann says. "The first floor had white marble floors, with solid glass around the edges on the walls, overlooking an enormous courtyard with fountain and white Italian marble statues of deer and lions."

"Then we went into a gold elevator with crystal seats to be taken to the living quarters on the second floor. All of the floors were made of Italian marble. The rooms and the furniture were enormous—couches were thirty feet long, green marble cocktail tables were fifteen feet long, and the living room seemed to be as large as a city block. There was a sixty-foot expanse of circular floor-to-ceiling windows that looked out at the Persian Gulf. The library had teak walls, and a glass case containing gold and jeweled swords, gifts from other rulers and the Pope to the sheika's late husband."

Morley was curious about Kuwait's economy and natural resources.

"Kuwait is an arid land bordering the Persian Gulf, without a single river or lake. Only the size of Connecticutt, it's dwarfed by its neighbors, Iraq, Iran and Saudi Arabia. But tremendous oil reserves have changed an undeveloped nation to a modern state," Morley says. "For centuries, the desert nomads counted their wealth in sheep and camels. Bedouin tribes still roam the desert in search of vegetation for their flocks. Camels are seldom seen in Kuwait now, but they number in the thousands in other parts of the Arabic world. They are the source of food, milk, transportation and clothing to the nomad.

"After the discovery of oil in the post-World War II era, American, British and other government and business interests provided geologists, engineers, and technicians to help construct pipelines to carry crude oil to the tankers docked in the gulf, ready to carry their precious cargo to all parts of the world. New parks, hospitals and schools—financed through Kuwait's oil wealth—have created a modern city in less than twenty years. Education and hospitalization are free."

Without any live water inland in Kuwait, the nation was one of the first on earth to develop a salt-water conversion system for clean drinking water. "Salt water from the Persian Gulf is piped into a distillation plant, filtered, then heated by natural gas to boil off pure water," Morley says. "The water is condensed efficiently and stored in great tanks throughout the city. It's the largest distillation plant in the world. Soon, water will be piped directly from the plant to homes, but now water trucks deliver the water."

Next to religion, falconry is the oldest surviving aspect of the culture in Kuwait. The training methods used by the ruling family and its professional falconers have not changed in centuries, and they use the same types of birds for hunting—peregrines and sakers. Morley considered the saker to be a close cousin of the prairie falcon, in terms of its size, coloration and performance. But the saker is slightly larger than a prairie falcon or a peregrine, and just smaller than a gyrfalcon.

In contrast to North American falconers, the Arabs trap all of their birds in the fall, just prior to the hunting season, giving them about three weeks to carry and man the birds for the hunting season. Morley had trapped a few passage falcons, but most typically, he preferred to take a juvenile (eyass) bird from the nest and train it to fly like a wild bird. For the Arabs, the advantage of capturing falcons in the fall was that the birds were accomplished hunters, who knew how to catch prey on their own in the wild. The challenge presented during training was to tame down a wild bird in a short amount of time.

"The sheik's falconers were full-time trainers, so they carried the birds twenty-four hours a day," Morley says. "Over here in

Nelson family collection

A Kuwaiti falconer watches his bird knock a hubara out of the air. The bird handlers release their falcons after the hunting season and catch and train new ones the following year.

the United States, most of the falconers have to work for a living, so they can only spend time manning the birds in the evening."

Morley didn't like sealing the eyes of a bird, or using Dutch hoods. But the Arabs had done both for thousands of years; they weren't about to change now. The Dutch hoods fit tightly around the bird's beak, sometimes pinching the skin or the eyes. Morley preferred Indian-style hoods because they were custom-trimmed to fit a falcon's head, and around its beak without pinching any tissue. "Those Dutch hoods aren't worth a damn," Morley says.

The next foray in falconry took the Nelsons to a desert hunting camp in Saudi Arabia, where they were the guests of the Emir of the Northern Provinces. Knowing she would be the first woman to participate in a falconry hunt in Saudi Arabia, Betty Ann wondered, what should she wear so as not to offend the emir? She chose a tan blouse and skirt.

The immensity of the Saudi Arabian desert was apparent,

seemingly extending into eternity. The soil had a dull yellow color, a combination of sand and clay. The emir picked them up in a Mercedes Benz.

"There was no road," she says. "We just drove across the desert in two Mercedes Benz sedans into the black-tented territory of the Bedouin shepherds. The hunting camp included several large tents, a trailer, a generator for electricity and water tanks. Two lines of the emir's bodyguards and servants formed an aisle into his large tent. It looked like a movie set."

Much like the scenes that Morley experienced in Kuwait, the emir had an elaborate white tent, lined in bright colored silk. The servants laid down a ground cover of large blue Chinese rugs. Falcons were placed on perches in the center of the tent, and the people sat around the edges.

Before the hunt began, the group had a light breakfast. They loaded up the falcons and shotguns in large custom-made Ford four-wheel-drive vehicles, and headed out to look for quarry. The emir's truck had a green-velvet throne mounted to the roof. "We roared out of camp at about ninety mph," Betty Ann says.

They chased hubara, casting off the falcons for the kill, and if the falcon missed, they shot the birds with shotguns. They always harvested enough game for lunch and dinner.

"We had the hearts and livers prepared like shish kabob over coals every day for lunch, and the breasts of the birds were boiled for dinner, along with camel hump, dates and other fruit, and sweetened camel's milk," Betty Ann says.

As the honored guest, Betty Ann received the honor of eating a sheep's eyeball. "I just washed it down and swallowed it whole," she says.

During one of the hunts in Saudi Arabia, Betty Ann stayed behind to travel to Beirut, Lebanon, to experience the city. She donned traditional robes for the trip.

"I put on an ahba, a heavy long black silk gown that all Arab women wear in the streets, and black headdress that covered everything except my eyes. And I walked around the city getting acquainted," she says. "But I didn't fool anyone in my disguise. I thought I might get better bargains in the market, but they pegged me for an American right away. They said American women flounce when they walk. Arabian women are

Nelson family collection

Morley and Betty Ann relax with their Kuwaiti hosts.

so used to carrying weights in their heads; they have wonderful posture."

Morley, meanwhile, unwittingly got into serious trouble. While hunting close to the border of Iraq and Iran, a landowner got uneasy when he saw Morley's movie camera, and he didn't want anyone hunting on his land. He summoned the Iranian Army Police.

"They were saying, they are driving on our land and scaring our sheep and they're probably smuggling," and our whole party was arrested," Morley says. "Our men spoke Arabic and they spoke Persian. Nobody could understand anybody. One soldier cocked his gun and was ready to shoot our heads off. If he had, he might have started a war between Kuwait and Iran. The Iranian Army confiscated my camera, which looked suspicious to them, then we got shuttled in and out of three different jails in three old walled cities."

Finally, ten hours after their capture, Morley and his hunting party were released, and letters of apology were distributed.

The near-miss capped an exciting and enriching three weeks of adventure for the Nelsons in the Middle East. There were many press reports in Saudi Arabia and Kuwait about the rare visit from an American falconer and his dashing wife, including an article in the *Weekly News Review,* published by the U.S. Information Service in the Middle East and Africa.

Morley returned home to a letter from the Westinghouse Company, inviting him to appear on a Sunday night hour-long program on national TV called *Twentieth Century Falconer*. The program featured Morley's prowess as a falconer, with clips from his Middle East trip as well as his home-produced movies, *Modern Falconry* and *Nature's Birds of Prey.*

Betty Ann was so inspired by her experience in the Middle East that she put together a lecture-presentation on the whole adventure, with photos of the sheik's magnificent castle and Morley's candid films from the trip. Her first presentation occurred at the Middle East Center of the University of Utah in Salt Lake City.

Betty Ann's presentation was a big hit. An amateur actress, she had performed in a number of plays at the Boise Community Theater, and thus, she was a confident and entertaining speaker. Plus, she had quite a story to tell about her journey to the Middle East. Few Americans had ever been there. She was invited to speak at a number of locations in the West, including Portland, Ore., and San Francisco, where she appeared on a TV show with an eagle on her fist. She also spoke to many service clubs in Boise.

For once, the spotlight was on Betty Ann instead of just Morley. She enjoyed the recognition.

Chapter Eleven

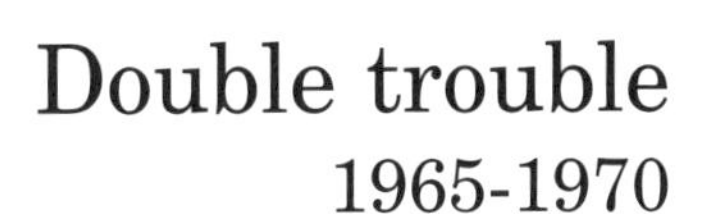

Double trouble

1965-1970

One falcon, one wife;
two falcons, no wife
— old falconry proverb

A few weeks after Morley and Betty Ann returned home from their trip to the Middle East, Betty Ann got a call from a friend in San Francisco, Nancy Clark Reynolds, inviting her to the premier of the film *Doctor Zhivago* and to meet California GOP Governor Ronald Reagan. Reynold's husband, Frank, worked as a public relations manager for Jack McCarthy, the Republican minority leader of the California Senate.

"Sure, I'll go," Betty Ann said. "That sounds like fun."

Betty Ann packed some of her nice clothes for the trip, and boarded the plane for San Francisco. She and Reynolds went to the *Doctor Zhivago* premier, and then they attended a press conference at which Jack McCarthy was speaking.

"When Jack got through, he came up the aisle and he leaned down and asked me if I'd like to go to lunch," she says. "I said sure."

Betty Ann was very impressed. McCarthy was a handsome fellow, a big man about town, being a Republican Party leader.

But McCarthy was just as impressed with Betty Ann. She was attractive, smart and funny, and she had a spirit for adventure, quite obviously, considering she just had been the first woman to participate in a falconry hunt in Saudi Arabia since the dawn of time.

When the two went back to McCarthy's office in the state Capitol, his secretary whispered to Betty Ann, "Do you know what he just told me? He said, Did you see the redhead in the gray flannel suit? I'm going to marry her."

Betty Ann blushed. Both she and Jack were married. She had enjoyed their lunch together, and didn't mind flirting with him, but she didn't realize that he already had serious intentions. By the time her California trip was over, the seeds of an affair had been sown. McCarthy had discovered a woman who had qualities that he had never experienced with his own wife, and he wasn't going to let go.

For Betty Ann, it was nice to receive that kind of attention from a handsome, wealthy and powerful man. She had grown tired of all the exposure that Morley received in the media, and him stealing the limelight at social parties to talk about his movies. She was bothered by the fact that Morley was extremely tight with money.

Perhaps most of all, she had grown tired of how Morley's life revolved around his passion for falconry and birds of prey. He didn't hang out with guys to play poker. He wouldn't go to a ball game with his buddies, and he wouldn't even attend his own children's baseball or football games. When he got home from work, he'd pull on his falconry jacket and his leather glove, and go directly to the hawk house to fly his birds. A variety of people stopped over at 4 p.m. to fly birds with Morley, and then he'd bring them into the house after they were done. Betty Ann never had much time with Morley.

Jack McCarthy's overtures came at an opportune time. Betty Ann was beginning to view life in a broader perspective. She yearned for more freedom and independence, an emerging feeling that many women experienced in the 1960s. Betty Ann had come of age at a time when women were expected to find a man, get married and have children. There wasn't ever any thought

Nelson family collection

Betty Ann and Morley Nelson in happier times. Betty Ann says her husband's preoccupation with birds of prey and the celebrity status he achieved was a major factor in the breakup of their marriage.

or consideration about what opportunities she might have as an individual.

"For the first time, women were saying, what's my life about?" Betty Ann says.

The year before Betty Ann and Morley went to the Middle East, she traveled to Burlington, Iowa, to visit her mom, who had been stricken with cancer. She was gone for several weeks. A neighbor told Morley that he ought to go to Iowa, too, to help Betty Ann through a difficult time. So he did.

But soon after he arrived, he was interviewed by the

Burlington newspaper and talked about his crusade to save birds of prey, the Disney movies he had done, and more.

"I remember thinking, did you come back to Burlington to talk about yourself, or did you come back to support me? Because I have no feeling of that," Betty Ann says.

The whole episode reminded Betty Ann of the time when Morley left her and the family to finish *Rusty and the Falcon* in Park City right after their fifth baby died three days after birth. "I had felt so alone during that period," she says. "I wasn't feeling well at all, and the four kids were crying, and I was crying, and it was a big mess. Morley recognized how you made babies, but none of the rest of it. Before my mother died, I remember she told me, 'I think you'll be divorced someday.' Because Morley did not recognize the singular life that he led in the family."

If Jack McCarthy had never come along, Betty Ann says she doubts that she would have considered leaving Morley. If her fifth child had lived, she couldn't have left. But Jack did come along, and she would leave Morley, nearly four years later.

The affair started while Betty Ann was traveling around the West to universities giving her presentation on the Middle East adventure. McCarthy sent a private plane to some of the cities where she talked, and they would rendezvous. Later on, they found other ways to get together.

"It went on for at least three years," Betty Ann says.

None of the kids had any idea that she was having an affair, but eventually Morley found out. He was mad. He called McCarthy to tell him to stay away in no uncertain terms, but it didn't work.

In the fall of 1968, Betty Ann told Morley that she wanted a divorce. They were sitting by the swimming pool when she spilled the news. She doesn't remember exactly how she put it, but Morley was "furious," she says. "He'd known about the affair, but he thought I'd get over it. And of course I didn't."

Many years later, Morley says he regrets that he didn't make more efforts to give Betty Ann the feeling that he loved her and appreciated everything she did in raising the kids and taking care of the household.

"Betsy Ann is a real hero because she did a great job raising our kids," he says. "I had to be gone a lot, and she had to take care of them. She did a marvelous job."

Betty Ann got jealous sometimes, too, Morley says.

"She didn't like the fact that all of those young girls wanted to come up to our house and learn about falconry. She thought I was taking an interest in those young girls, and they were very sexy young ladies. I could see how she might feel that way. That's where the problems came up in our misunderstandings between Betty Ann and me. It was partially money, and partially the interest I had started by going on national television and doing all of these things with birds of prey, and all of the girls that would call up and say, I want to come up and help you with your birds."

Most of the time, it was a wide variety of teen-age boys and Morley's friends who came up to fly birds. "We had falconers coming up here all of the time, from all over the world, and a lot of the people I knew in the Ski Troopers," he says. "My door always has been open to anyone who's interested in the birds. So I had too many people coming in here all the time, and Betty Ann complained that there wasn't enough privacy in the house."

Although Morley was upset about Betty Ann's affair, he knew that he had not been loyal to her throughout their marriage. In Italy, near the end of World War II, when Morley grew so numb to the massive numbers of soldiers being killed around him, he succumbed to the temptation of sleeping with an Italian woman.

During several days of leave, Morley went to Rome, and met a college-educated woman who was training to be a physician. They went to dinner, met her family, and spent several days together. "I had openly discussed with her that I was married, that I loved my wife and I hoped to go back to the United States to be with her after the war, and that she was about to have a child," Morley says. "She knew all of this, and yet we enjoyed some intimate moments at a time when every soldier, including myself, wondered if he would live to see the next day."

In the fall of 1968, Betty Ann moved her things to a rental place on the Oregon Coast. She married Jack McCarthy the following spring, in April 1969.

"I'm the bird that got away," she says.

Betty Ann had waited until most of her kids were out of the house. Only her youngest son, Tyler, was still in high school when she left. Norm had graduated from the University of Idaho, and he was an officer in the Army, stationed at Fort Lee in Petersburg, Virginia. Tim was a junior at the University of Idaho. Suzie was a freshman at Arizona State University.

Norm recalls that when he heard about the impending divorce, he was surprised—to say the least. "It was as if a car had hit the family," he says. "I didn't have the slightest concept that my parents were going to get divorced. I'd been in Canada and Alaska filming, and then I went to Georgia and Virginia in the Army, and all of that stuff happened in Virginia. It was all a total shock to me."

Suzie Nelson says she didn't see it coming, but she knew her mother felt lonely.

"I had no idea, it was kept very quiet," she says. "I never saw my parents fight in the thirty years they were married. My dad is so Norwegian. I mean he was quiet, and it took a lot to make him angry. He just didn't pay enough attention to her."

Both Betty Ann and Morley did their best to raise their four children in a warm and supportive home environment. Family always had been important to Morley. He enjoyed taking the kids on all kinds of outings whenever possible, whether it was to fly birds, check out nest sites or go skiing and snowmobiling in the winter. Betty Ann took care of the household, washed the clothes, cooked the meals, cleaned the house, and took care of the children's day-to-day needs.

The Nelsons never had a lot of money, but still, their kids had a lot of fun and opportunities growing up, even though the four youngsters shared the same bedroom in the Nelson's two-bedroom home on Ada Street. Betty Ann and Morley took the kids swimming in the summer, and they'd go to drive-in movies.

"Every night we went swimming by the 16th Street bridge," Betty Ann says. "Everyone would be furious and fighting because it was 104, and we'd get in the car and go down to the river and go swimming, and then everyone was happy."

"Dad liked to take us to the 16th Street bridge because of the rapids below the diversion dam," Norm says. "Once we knew how to swim, he taught us how to swim in the current and the rapids. It was definitely a sink-or-swim type of deal, and we all did great."

When Morley had to go away on trips, Betty Ann would take the kids to McCall and rent a cabin near the lakeshore home of their friends, the Beyerles. "We'd hang out at the beach all day with their kids," Beav says. "Mom was really close to Dorothy Beyerle, so we hung out together and went water skiing together. It was fun."

For the most part, the Nelson children got along well. Norm and Tim fought each other all the time, being boys, and being only two years apart, and the older boys picked on Suzie and Tyler, when he got old enough to take some punishment. All of the kids learned how to train eagles, hawks and falcons, and how to rappel down a cliff to assist their dad in banding birds, filming birds, and checking out nests to see if they were actively occupied.

Beyond Morley's personal clutch of birds, he kept a good stock of pigeons on hand for the birds to chase and to eat. In the course of feeding and handling pigeons, Suzie bonded with the birds.

"Dad had some junky city pigeons that he picked up from somebody, and then he got some nice Belgian homing pigeons, the kind they used in World War II to fly messages to the Allies," Suzie says. "I thought the pigeons were cool. I raised and raced Belgian homing pigeons, and I became the president of the Boise Valley Pigeon Fanciers Club when I was fourteen. I bred the birds and sent notes around. I took them to the Oregon Coast and they'd beat me home."

Picture the boys and their dad flying falcons at pigeons in the draw next to their foothills home, and after they finished, they'd run into Suzie coming up the trail to check on her Belgian homing pigeons. Norm would tease her about one of the falcons taking down Fraidy Cat Clyde, her favorite pigeon.

"You did not!" Suzie would say, her face all twisted. She had short brown hair, styled in a pixie, and she wore blue glasses with butterflies on each corner. She was skinny and tall. Tim

Nelson family collection

Morley Nelson and son Tim

and Norm were comparatively short and skinny for their age, with flat-top crew cuts.

"Clyde was the dominant pigeon, and we'd make jokes about it getting killed or eaten, but she was really into it, and so girlish and innocent about it," Norm says. "We took advantage of her innocence. But the only pigeon that was off-limits was Fraidy Cat Clyde."

Actually, Morley did allow his peregrines to kill a few of Suzie's birds, and she and her dad had a big discussion about it. "When the falcons strike a pigeon at over 100 mph with their talons out, the bird is killed instantly," Suzie says. "I'd get all upset and start crying, and my dad would sit me down and give me a lifelong lesson on the food chain. He'd say, now, this peregrine falcon has to eat something to survive, and the pigeon is part of its diet, and it's nature's imperative for an animal to kill or be killed, and all of that made perfect sense. Dad is so honest and strong, he's like a mountain. I understood what he was saying, but I still didn't like it."

Morley's second son, Tim, was a bit like his dad, considering he learned to handle hawks and falcons with finesse and skill, and he liked science and engineering in school. He built model cars, airplanes and buildings. His first engineering project was to build a go-cart with an old lawnmower engine.

"Tim was always building contraptions of ingenious kinds, and the highlight contraption was the go-cart," Norm says. "It was built out of plywood and two by fours. He constructed a pulley on a post and ran a cable up to the pulley to make a steering wheel. Morley helped him build the go-cart, and put a clutch on it. He drove it around the neighborhood with a leather flier's cap on, the kind with the ear flaps and glasses."

Norm admits that he harbored some animosity toward Tim growing up, considering that his younger brother was as good or better than he was at a lot of things.

"Yeah, we had an ongoing thing we had to deal with," Norm says. "I don't know why, but we had about four fights, one of which I still carry with me. I have a scar on my finger. I tried to hit him, I had a ring on, and he ran away, and I hit a door, and the ring cut my finger. In the instances that we had those fights, for the life of me I can't remember one fight as to why it started, but he had a temper and I had a temper, and I tried to be domineering, and in some of the cases, I started the fights, and I got spanked. There was more than one fight where my dad couldn't figure out who started it, so he whipped us both.

"He got really mad at us. At that time, it was typical for a parent to spank their child. Morley used his belt. The worst part was later on, when you'd be lying in bed and dad would come down, and say, OK, now we gotta have a talk. And that's when you had butterflies going throughout your whole body because he had this low masculine, Norwegian tone that was totally intimidating. He would make you feel this big (two inches tall). And then he would never come to the "I love you" kind of thing, but it would come to the emotional connection between dad and son. He was saying, 'Hitting your brother is the worst thing you can do in this family. We all have to work together.'"

In high school, Norm concentrated on training Otis, his golden eagle, and other golden eagles for *Ida the Offbeat Eagle* and flying a variety of goshawks. Tim helped train eagles, but he preferred to fly goshawks.

"Tim was a hawk guy," Norm says. "He and I liked the personality of hawks, accipiters, better than falcons. Goshawks were like god to us. They're about twice as aggressive as falcons as far as their mental presence. They're dangerous to be around, so they're hard to train. They don't do any of the spectacular aerials like the falcons; they don't even try. They just move in for the kill."

On weekends in the fall, Norm and Tim went pheasant and waterfowl hunting with their dad. Norm remembers their best day ever when he was in ninth grade and Tim was in seventh grade, and they went goose hunting.

"We didn't have any goose decoys or anything, so my dad says, we're going to do it World War II-style, we're going to sneak 'em," Norm says. "We got up really early in the morning, floated down the Snake River in this boat, and we saw all of these geese on an island below Centennial Park. So we floated down the other side of the island, laid down in the boat, and let the boat drift down the far shore. Then we got out of the boat on the downstream end of the island. We walked for forty yards. Then we crawled 200-300 yards. I mean we crawled forever. Morley's got all of his Army outfit on, and we felt like we were in World War II, crawling three abreast.

While we're lying in the mud, we loaded the guns. Tim had a .20 gauge because he was really little. We crawled fifty yards through wet mud and tall grass, and now we can see the geese. We stopped and dad says, 'Ok, get ready.' Butterflies are just jumping all over in your stomach. He says, `Pick your bird,' and we stand up, run about four steps, stop and they started to fly. We knocked down four great big Canadian honkers. Somebody got a double, either Tim or my dad, definitely not me. Tim was a much better shot.

"So we got in the boat and we're all happy as a lark. Then we see a bunch of lesser Canadian geese, a smaller species, on the shoreline. So we pulled into the shore, snuck up on them, and all three of us got a bird. We came home with seven birds that day. There's a photograph of us on the lawn that day with the day's kill. That was one of the best hunting days we ever had."

In high school, Norm had a few friends that he hung out with, and he spent a lot of time working with and training golden eagles. Tim was more popular, he had more friends, and as a thin, sinewy guy with speed, he made the varsity cross-country and track team. He wore his Boise High School red letter jacket with pride.

Tim's size—5-foot-6, 117 pounds—didn't reduce his popularity. His bright personality, smarts, athleticism and interest in pranks made him fun to be around, according to his friends. "I just loved being around Tim, he was really tenacious, really spirited," says Rick Collier, who ran on the cross-country team with Tim. "He loved to live. He had a great passion for life. He had a great sense of humor, and he was someone who could raise hell and cause problems and be spontaneous about it."

Pat Beiter, Tim's coach and history teacher, recognized his talent when he saw the boy streak across a parking lot to fetch something for his mother out of the car during practice for an amateur play. Beiter asked Tim to try out for the cross-country team. Tim started running long distance, on the streets in town, golf courses and the foothills.

"We'd go run ten miles in the afternoon, and then when we got back to school, our coach made us run sprints against the wide receivers and running backs on the football team, and we'd always win," Collier says. "It was the most amazing thing for guys like us to win a race like that. We weren't seen as the paragons of athleticism."

In the off-season, Collier remembers going into the mountains to look for goshawks with Tim, and watching him fly golden eagles to the T-perch. "To this day, it's pretty rare for kids to have an experience like that," he says. "Tim was an expert bird handler. He was really intuitive, and really natural in the way he dealt with the birds, a lot like his dad."

Collier remembers Tim driving his red VW Beetle convertible, with the top up, on a Christmas Eve, making slalom turns on Boise's Harrison Boulevard. "We went up and down the boulevard several times, it was snowing really hard, and Tim laid down fresh tracks in the road each time we went around the boulevard," he says. "It was a riot."

Tim also hung out with Bill Evans, who dated his sister,

Suzie, as a freshman in high school. Evans was more of a trouble-maker, a kid with an attitude who had lost his parents before he reached high school.

"The rest of the family really didn't like me because of my reputation," Evans says. "But Betty Ann kind of saw through my tough-guy act, and she took me under her wing. She took me shopping for college clothes and stepped in as my mother. I was kind of like one of their stray birds."

Both Evans and Tim Nelson wanted to go to the University of Idaho. Morley and Betty Ann had saved money for Tim's college fund, but Evans was broke. Betty Ann helped Evans obtain financing to pay for his college education, through a number of businessmen who were UI alumni in Boise.

By the time Tim Nelson turned eighteen, he was ready to head off for college at UI. He was still just a wisp of a guy with a squeaky voice. It's not that he hadn't matured and gone through puberty, he just wasn't going to get any bigger.

He joined the Beta Theta Pi fraternity, and hung out with pledge brothers Don Farley, Mike Chaney and Tom Boreson. He remained good friends with Bill Evans. All of those guys were big fellas. Most of them played high school football in Boise.

Tim's brother, Norm, was a junior at UI, a Beta Theta as well. He remembers the hazing routine that Tim and his buddies had to withstand in their freshman year. In those days, it went beyond the drinking games. "We made them do push ups, cold showers at lunch, they'd have to get up at three in the morning, stand in the hallway and get paddled," Norm says. "We'd make them climb to the top of the stairs, take a drink of hot garlic water, and come downstairs and spit it into the fire, and do it again. Stuff like that."

Tim won a cross-country race for freshmen, too. "He just wiped out the entire field by at least half the length of the track," Norm says. "He could run forever."

Tim wasn't good enough to make the varsity track or cross-country team, however. So he concentrated on his field of interest, which was engineering, and hanging out with his friends.

"Tim was a very outgoing guy," says Don Farley, now an

attorney in Boise. "He was a lot of fun. He was somebody you could confide in, somebody who was obviously smart. You could probably count the number of bad days Tim had on one hand. He was always up, always had a smile on his face, and he was always quick with a joke and a laugh. And he had a personality in the sense of always being ready to play a trick on you if he could."

Before Christmas vacation, Mike Chaney and Tim went around to the sorority houses to pass out invitations for a Christmas dance at the Beta Theta House. Chaney, being a big hoss, carried Tim in a bag. "He wore little green leotards and a little red coat and a cap, and he'd pop out of the sack when Chaney showed up at the door," Farley says. "It was pretty funny because I know that Chaney dropped him a couple times."

The only area in which Tim didn't excell was dating girls. Evans says Tim enjoyed hanging out with the guys at the bars in Moscow, the college town, but he was really shy around women. "The University of Idaho was a drinking town," Evans says. "The drinking age was twenty, but all you had to do was to get a fake I.D., and getting into the bars was no problem. But while most of the guys were trying to meet women, Tim would be off in the corner, sipping a beer. I think a lot of the girls liked him, but due to his size, he didn't have very much confidence about trying to pick up a girl in a bar or asking a girl out on a date."

Tim Nelson visited Betty Ann in Oregon in January 1969, before she married Jack McCarthy, and told her that he supported her decision. "I'll tell you mom, I've got my life to lead, and I've got things I want to do. If I was married to someone who didn't love me, I wouldn't want to live with them," he said.

Betty Ann asked Tim to give her away to Jack McCarthy at the wedding in California, and he agreed to do it.

Farley remembers the trip to the Oregon Coast. "We were on semester break in January, and we went over there and stayed with Betty Ann. We had a great week over there, running around the beach, playing football in the sand, and going clamming. It was a great vacation."

In the fall of 1969, Tim was serious about finishing his engineering degree at the University of Idaho and moving on to the next chapter in life. He enrolled in the ROTC program, just like his older brother had done, knowing it was a virtual certainty that he'd be drafted upon graduation. The United States was heavily involved in the Vietnam War. Tim wanted to be a pilot in the Air Force.

When Tim began his senior year, he was house manager of the Beta Theta Fraternity, responsible for maintenance and upkeep. He also was a member of the Intercollegiate Knights, a service organization for students on campus.

As always, at this time of year, when the leaves began to peel off the trees in Moscow, Tim and his friends went to UI football games and would party hardy. On September 28, 1969, Tim and three of his buddies hopped into Tim's shiny red VW convertible to go watch the University of Idaho play Idaho State University in Pullman, Wash. Pullman was the home of Washington State University, which shared its football stadium with Idaho. The Kibbie Dome, a new stadium for UI athletics, was under construction.

Tim's pledge brothers, Don Farley, Mike Chaney and Tom Boreson, all rode with him to the game. The beer flowed freely on a sunny afternoon in the Palouse Hills. Farley says he doesn't recall how much they had to drink, but Evans says it was routine for the boys to get drunk at Vandal football games. "I didn't sit next to them, so I don't know how much they had," he says. "It could have been three beers, maybe six. But Tim was so little that three beers for him would be like six or seven for me."

Instead of driving the main highway back to Moscow, Tim decided to take a shortcut and drive on dirt farm roads to avoid bumper-to-bumper traffic. The top was down, and the music on the radio was cranking. "We were yelling and talking back and forth—it was just a happy-go-lucky trip back to campus," Farley says. "Chaney was sitting next to me in the back seat, and Boreson was sitting in the passenger seat."

Less than a mile from campus, Tim pulled onto a paved road

that winds by the Vandal Golf Course. He came zooming up to a big corner in the road, and all of a sudden, he lost control.

"I can remember looking up, and there was a wide turn before you get to the golf course, the upper part of the golf course, and he took the corner too wide, and the car came around the corner, and he cranked it back, and it flipped the whole car," Farley says.

Everyone was ejected from the car. Tim Nelson suffered severe head injuries. The passengers received scrapes and bruises but were OK. An ambulance took Tim to a hospital in Moscow, and then to Deaconess Hospital in Spokane, Wash., for possible surgery.

Mike Chaney called Morley in Boise to tell him the news. Suzie Nelson answered the phone.

"I was home cooking dinner, it was 5:30 p.m., and the phone rang, and my dad was up on the hill, flying his birds," Suzie says. "Mike sounded so awful that I knew instantly that something was big and bad. He said, "Is your dad there? I need to talk to your dad." I said he's flying his falcon up on the hill. And he goes, "Go get him. I need to talk to your dad now." And I said, is it Tim? He said, "Suzie, get your dad!"

She ran up on the hill and screamed to her father. "Dad, you've got to come down, something has happened to Tim."

Morley called the airlines and reserved a seat on the next flight to Spokane. Before he left, Morley grabbed Tyler by the shoulders and said, "Son, we've really got to pull together now as a family. We've got to be strong. I'll be back as soon as I can."

Tim Nelson had suffered severe brain damage. He died on Wednesday, October 1, 1969. He was twenty-one. His obituary ran in *The Idaho Statesman* on October 4. The eight-paragraph obit underscored Tim's achievements—house manager of his fraternity, a member of the Air Force ROTC pilot training program, an eagle handler and trainer for Walt Disney films, and an employee for the U.S. Fish and Wildlife Service, banding eagles in the Snake River Canyon.

The funeral was held Saturday, October 5, in Boise at St. Michael's Cathedral, an ivy-covered gray-stone building across from the state Capitol. Tim's friends Bill Evans, Don Farley, Mike Chaney, Bud Raymer, Taylor Gudmundsen and Randy

Smith were the official pallbearers, and every member of the Beta Theta was there for the service.

Morley remembers that it was a sad time. "It was a terrible loss. It was hard on Betsy Ann and me, and it was hard on my family, because my family loved Tim, too."

Being a sturdy Norwegian, Morley's way of coping was to remain quiet and strong. He drove the family in their Chevy wood-paneled station wagon to Cloverdale Cemetery in west Boise. Norm and Suzie cried and held each other in the backseat, next to a weeping Tyler. "My sister and I were crying on each other, and I was crying really hard, uncontrollably," Norm says. "It was really horrible. I felt a huge distance between me and Tim because I hadn't seen him in over a year. I had no idea what he looked like, or anything, and then he was gone. I was never going to be able to see him again. I'd never get to go hunting with him again."

Tim was buried in a plot next to his little brother, Eric Steen Nelson, who had died three days after birth.

The dynamics of the divorce and Tim's death shook up the Nelson family, and left an indelible mark on everyone. Instead of staying at Arizona State, Suzie felt a calling to return home after Tim died to take care of Morley, Norris and Tyler.

"They didn't have anyone to cook or clean, and Grandpa Nelson was burning up bacon every morning," she says. "So I moved back into the house, I went to Boise State (university) and worked full-time . . . . But I never got my degree. All of my girlfriends in college were mad at me. They said, you should be going to college and find a man and get married, you don't need to go back to Boise. But I felt compelled to do it. I didn't feel I had any other choice. Here was my dad, my grandfather and my baby brother living all alone in the house, and nobody was taking care of them."

Morley was heartsick about Tim's death, more so than the divorce, friends say. "The thing with Betty Ann was tough, but the thing with Tim was devastating," says Neil Sampson, a friend of Morley's who worked with him at the Soil Conservation Service. "That hit Morley really, really hard. But he's a tough old rascal, and he tried to lower his head and plow through it."

Morley says the experience of losing his brother before World War II, his mother to a stroke, and watching many people die in the war, including some of his own men, helped him cope.

Shay Hirsch

Pat Yandell

"Well, I had been through the war, and I had lost all kinds of men, and that was the same thing that I had to go through with Tim," Morley says. "You just have to go through it. You have to keep working at it, that's all you can do."

It helped immensely that Morley had met Pat Yandell, that he had a soul mate to comfort him during the ordeal. Pat was an attractive and outdoorsy gal who had raised three children after a divorce early in her marriage. She had a good job in the regional office of the Bureau of Reclamation, a federal agency charged with overseeing dams and reservoirs in the Pacific Northwest. Yandell also had managed her money wisely, and purchased a number of rental properties in Boise for extra income. By the time she met Morley, she had twenty-two rental units.

"If you opened her purse, there was always a hammer and a measuring tape—never any makeup or feminine stuff in my mother's handbag. It was always tools and things that were important to her," says Shay Yandell, Pat's eldest daughter.

Morley dated a number of women before he met Pat. Being a good-looking man who had made movies for Disney, practiced falconry with Arab sheiks, slept in the snow and traveled in Boise's highest social circles, he was a popular fellow.

"I'd get up to go to school in the morning, and I'd see him drive in from being out all night, and I'd say, Hey, where you been all night boy? And he'd just sort of chuckle and smile at me. That happened several times," Tyler says.

After several months, Morley started dating Pat. At first, she told Morley that he ought to wait a while to work through it. But he kept pestering her. Pat eventually relented, and they got pretty serious after a few dates. All of Morley's friends say that the two had strong sexual chemistry, and he liked the fact that she was game to go with him on outings in the Snake River Canyon, even rappel down the cliffs.

"My mother was gung-ho for anything," Shay says. "One reason her relationship worked with Morley was that they were true partners. She enjoyed what he was doing."

In May 1970, Morley and Pat married in Boise. It was a small ceremony, attended by a few friends, including Morley's pal from the Ski Troops, Forbes Mack, and a good friend of Pat's, Shirley Mix.

They honeymooned in Costa Rica.

"The two of them drove there in the Jeep Wagoneer," says Shay. "They drove around, fording rivers, looking for birds in the hills. They camped out the whole way. My mother spoke Spanish fluently. So that was a big help. This was a time when there were a lot of drug-runners down there. And they walked right into a drug bust. All of a sudden a policemen came up armored with guns, but pretty quickly they could tell that Morley and Pat had nothing to hide. They were just looking for falcons."

Morley was moving on with his life now with a new partner. Another important factor was that he had his feathered companions to help him through this difficult time. On a breezy evening in May, he took a favorite gyrfalcon, Tundra, out on the hill behind his house, and cast her off into the wind. She flew away with her powerful wings, and in a matter of seconds, she was flying 500 feet above Morley's house, even with a row of homes on top of the hills, hunting for a fresh treat.

Tundra kept climbing, and Morley watched the bird grow smaller, wondering where it was going to go. Then she saw a covey of quail flush from a blue spruce below. Tundra came zooming down and struck one of the quail before the bird even saw her coming. Morley walked over to Tundra, who stood erect and proud, yakking and screaming with delight, with the quail tucked under her foot.

Right in that split-second of time, when Morley watched the

gyr take the quail in a blur, his mind flashed back to that singular, awesome moment of truth, when he watched a peregrine take a duck when he was twelve, and flashed by hundreds of other moments when he'd seen his birds take quarry. It didn't seem to matter how many times he'd seen a falcon kill prey—each and every time it made him stop for a second and ponder what a magical and spectacular thing it was. During that moment, he'd forget about everything else, even the tragic events of the last year. It made him feel lucky to be alive.

"Step up, old girl," Morley said, reaching gently in front of her, offering a fresh piece of meat. "Gawd, you're one hell of a hunter."

Chapter Twelve

Snake River Birds of Prey Natural Area

1968-1980

Morley's movie projects in the Snake River Canyon, including Disney's *Ida the Offbeat Eagle,* and "Valley of the Eagles" on Mutual of Omaha's *Wild Kingdom,* showcased the canyon's outstanding environmental values to millions of Americans nationwide in the 1960s. From Morley's own research, he knew there were at least thirty pairs of golden eagles nesting in the canyon rims, and more than 100 pairs of prairie falcons, not to mention breeding pairs of red-tail hawks, northern harriers, ferruginous hawks, short- and long-eared owls and other raptors that flocked to the canyon every spring.

"What a magnificent empire these eagles rule," Marlin Perkins said in "Valley of the Eagles."

Yet, the Snake River Canyon did not possess any special management status to preserve the raptor populations. This gaping hole in land-management policy was well-known by several high-ranking people in the Boise District office of the Bureau of Land Management (BLM). Edward C. Booker, the Boise District manager, and William R. Meiners, chief of the division of resource management, were both good friends with Morley, and were supportive of crafting some type of national protection for the area. The key questions were, what kind of protection should be sought, and how much land should be set aside?

Booker and Meiners worked for a federal agency that was best known for overseeing mining and livestock grazing operations on the public domain. The BLM did not have any kind of reputation for protecting areas for environmental purposes, and it didn't have any congressional authority to balance development uses with environmental protection until 1976. So Booker and Meiners told Morley that they would need plenty of "hard" evidence to prove that the canyon deserved protection.

Morley anticipated the BLM would need empirical evidence about raptor populations in the canyon. In early 1968, when Maurice Hornocker was named the new research leader of the University of Idaho Cooperative Wildlife Research Unit, Morley gave him a call. Morley had met Hornocker in the mid-1960s when he gave a talk at UI about birds of prey at a Wildlife Socicty meeting. Hornocker was studying mountain lions in the Central Idaho Primitive Area. As two men who admired predators and understood them better than most, they hit it off.

"Morley called me right away, and told me, we've got this wonderful opportunity down here with the birds of prey," Hornocker says. "You should come down and take a look and get some students going on this. I got this stable of young guys down here who are just raring to go. And so I did."

Hornocker traveled to Boise and met with Morley, Booker and Meiners at the BLM office. They agreed that the number one priority was to document the number of golden eagle nesting territories in the area, see what they ate, and how many young were produced. One of Morley's protégés, Andy Ogden, had been working on a study for the U.S. Fish and Wildlife Service to determine the diet of golden eagles in southwest Idaho, to see if they could find any evidence of golden eagles eating domestic sheep. In the course of that work, Ogden had documented thirty-six nesting pairs of golden eagles in the Snake River Canyon in 1966 and 1967, one of the highest known concentrations of golden eagles in the United States at the time.

Hornocker recruited a young man named John Beecham Jr., a recent graduate from Texas Tech University, to conduct the first study of the nesting ecology of golden eagles in the Snake River Canyon, beginning in the spring of 1968. Hornocker raised the money for the study, and Beecham performed the research as part of earning a master's degree in wildlife management.

Andy Ogden met Morley when he was an eighth-grader at North Junior High School in Boise. He and his buddies, Steve Allard and Dave Rodenbaugh, wanted to learn how to fly hawks. Being from Boise, they knew that the best way to learn was to hang out with Morley, the legend. They weren't the first kids that Morley had taken under his wing, and they wouldn't be the last.

"After school, we'd go directly to his house, and we'd sit around and wait for him to get home from work," Ogden says. "We weren't allowed inside, so we just sat outside for an hour and a half and waited for Morley. We'd sit around and stir up ant piles, and poke them with a stick."

Never one to turn away interested kids, Morley put them to work. He showed them how to handle birds and how to work his Bolex camera, so they could take movies of him flying his birds.

Morley had a number of falcons and eagles on hand at the time. He had just finished work on *Ida,* and he needed to train a number of prairie falcons for his next movie project, *My Side of the Mountain*. "We had ten to twelve young prairie falcons, and he was training them all at once," Ogden says. "The remarkable thing was, we'd work for hours and hours with one bird to get one little accomplishment done in maybe a week, and Morley would grab a hold of a bird, and run his fingers up and down its breast and talk to it and say, 'hello there, big hawk, how're you doing?' and then he'd say, 'well, we better put that one back. We don't want to fly it today'.

"He knew all the birds by name, and we couldn't even tell them apart. He could tell just by the way the bird acted whether it was ready to fly that day, if it wasn't too fat, wasn't too lean, and it was ready to go. Then he'd go out and exercise that bird for a while, and then we'd go get another one, it was just like a production line of bringing these birds out, doing the kind of stuff that we would spend hours and hours to get our birds to do, he was doing in five minutes. It was just remarkable to see.

"When we were filming, he could tell you which way a bird was going to turn. Then he'd swing his lure, and tell you which way the bird would come at the lure, so we could get the camera turned around and focused at the right thing. Judging by how the wind was blowing or whatever, he could judge which way they were going to come down every time. And he was pretty much always right."

Ogden hung around the Nelson household long enough to see two horrific moments working with golden eagles. The first incident involved Morley's youngest son, Tyler, when he was in sixth grade. Norm was working with Jim Fowler and several photographers from *Wild Kingdom* in a netted enclosure behind the Nelson's foothills home. They were trying to get a close-up shot of a golden eagle, and they used Buddy, the male golden eagle that had performed so well in the shell game in *Ida*. Tyler was assisting Norm during the shoot. At one point, Norm gave Tyler some meat to hold onto while he untied Buddy's jesses, and without thinking, Tyler put the meat in the back pocket of his jeans. As always, Norm and Morley had made sure that Buddy was hungry for the shoot, so they could coax the bird to do whatever was needed with fresh meat. But when Buddy saw Tyler put the meat in his pocket, he flew directly at his butt, extending his razor-sharp talons. One foot latched onto Tyler's butt pocket and pierced the skin.

"I ran down the hill screaming with the eagle holding onto my butt," Tyler says.

"Oh, Jesus, that was nasty," Norm adds. "I ran down the hill with some fresh meat, and got the eagle to release Tyler."

Then poor Tyler got in trouble at school for not sitting up straight in his chair in class. He was too embarrassed to tell the teacher what had happened, so she sent him to the principal's office. But after the principal called Tyler's mom, he realized that the boy truly had severe, deep wounds in his butt. It would be a matter of weeks before Tyler could sit down comfortably once again.

Another time, Ogden worked with his friend, Steve, and Morley to get a shot of a golden eagle killing a bull snake. They used Cleopatra, a female golden eagle, who was equally hungry during the shoot. "Steve walked up the hill with the eagle on a T-perch, and we had a bull snake on the ground," Ogden says. "The plan was, we were going to film the eagle grabbing the bull snake. We wanted to film it several times, but we only had one snake, so Morley was going to step in front of the eagle at the last minute, and put a chicken neck on the T-perch, and save the snake. But it didn't work.

"She came in and she went after the snake, and all of sudden, Morley sticks the T-perch in front of the bird and she grabs the snake with one hand and goes right up over the top of the T-perch and hits Mr. Nelson right in the chest, knocks him over, and stands there with her talons right over the top of his face, and the snake in her other foot. She just stood there for a couple seconds and then she stepped off. She was a mean eagle. Gawd, if she'd clenched her foot, (he whistles) Morley would have had some big holes in his face. But fortunately, she just hopped off and finished killing the snake, ate it, and said, OK, we're done for today."

Later on, Morley took Ogden and his friends to the Snake River Canyon, to check on nest sites, and he taught the boys how to rappel down a cliff. Ogden and his buddies were amazed at how many birds they saw in the canyon. "All through high school, that was where we went," he says. "We climbed into nests just to find them and see if they were occupied."

Was it spooky? "Not really, because our overriding interest was to get hawks. Our hawks constantly would fly away from us. We didn't have any radio telemetry stuff or anything like that, so we went through quite a few hawks. We had to have a regular supply of them to stay in business."

After Ogden graduated from high school, Morley helped him get a job working for the Fish and Wildlife Service study on the diet of golden eagles. "I was dropping into the nests and seeing what they were eating," he says. "The funding agency was the gopher-chokers, the Animal Damage Control agency, and they were trying to show that eagles were eating livestock. So Morley was like, bring it on! I climbed into nests for two years, and we never found any evidence of livestock in there.

"But in the final report, (they) wrote that there was evidence of wool in the nests. It was a lie, you know, because (they) had never gotten . . . into a nest in the whole two years I worked on the study, and (they) lied in the report. So I called Hornocker and said, you know, this is a lie."

In the spring of 1968, Beecham launched his nesting ecology study, and hired Ogden, now a student at Boise State University, to work as his assistant. Ogden taught Beecham how to set up ropes and rock climbing equipment for rappelling into golden eagle nests. From the research he had done before, he knew where many of the nests were located. And of course, Morley knew where more nests were located as well.

Beecham's study area extended for about 150 miles along the Snake River Canyon, from Bliss to Marsing, Idaho. One of the first things that Beecham did was visit with Morley about nest locations. "We'd go up to his house, and get the maps out, and try to pinpoint the locations," Beecham says. "Morley had a tremendous amount of knowledge, but it was kind of frustrating because he never wrote anything down. All of his knowledge was in his head."

If Beecham had doubts about nest locations, Morley accompanied him into the field to help him identify locations. Many of the nests that Morley had discovered over the years had names that he and his older boys, Norm and Tim, concocted. The names included things like the "Otis Nest" near Marsing, where Morley and Norm had taken a male bird that became Norm's favorite eagle companion when he was in high school, the same bird that appeared on national news with Walter Conkite and in several movies.

"Morley always found the time to go out to the canyon," Hornocker says. "I'd call Morley from Moscow and say, I'll be down next Wednesday. He'd say, I'll be there. I'll take the day off. He'd take annual leave. And he'd lead me and the students to a particular eagle aerie, and take the time to instruct the student in how the nest should be approached, how the climb should be made. He always had the time, found the time, made the time.

"The combination of Morley's knowledge and experience was vitally important. The climbing skills. Anyone can learn those in a climbing school, but to go out on a crumbly canyon wall to get to an eagle aerie, it has to be approached the right way, to do it safely. There were some nests you couldn't go to because the cliffs were too high. And these young men didn't know these things, they couldn't know those things, but Morley could teach them."

Hornocker says he enjoyed going with Morley and his students on research trips because Morley was so full of stories and enthusiasm. "Oh, every trip out was an adventure," he says. "He'd tell us stories from World War II. And he had this uncanny ability to instill enthusiasm in anyone. It was a treat just to go out with him just for that. I'd find myself refreshed and energized by every trip."

Beecham's study was exhausting from a travel and time-investment perspective. He and Ogden had to cover an enormous patch of territory. "It was John and Andy going hard every day, from sun up to sun down," Beecham says. "We were constantly driving on dusty and nasty roads to get to the canyon, and then we'd have to work along the edge of the cliffs to find the nests. We were gone all the time. It can't believe my wife stayed with me. It's the hardest I've ever worked."

When Beecham's study was complete, he confirmed that the Snake River Canyon played host to a large concentration of golden eagles. He documented twenty-five nesting pairs of golden eagles in the study area in 1968, and 36 pairs in 1969. The density of eagles was similar to a high concentration of eagles found in comparatively rich habitat zones in Scotland, Montana and California.

Beecham's study also showed that golden eagle eggs in southwest Idaho were not suffering from egg-shell thinning. At the time of his study, the effects of organo-chlorine pesticides, such as DDT, were causing steep declines in the populations of peregrine falcons and bald eagles throughout North America. Pesticides caused peregrines and bald eagles to lay eggs with thin shells, ruining productivity. Beecham found that golden eagle eggs in the Snake River Canyon had an appropriate and normal thickness, .6 millimeters.

Golden eagles in the Snake River Canyon made a living from a wide variety of food sources, the most common being blacktail jackrabbits and desert cottontails—species that were not affected by pesticide use. They also ate ring-necked pheasants, pigeons, magpies, chukars, bull snakes and rattlesnakes. Beecham did not find any evidence of eagles feeding on livestock, confirming Ogden's research.

Beecham's study was written up in the form of a fifty-page master's thesis. It was approved on April 28, 1970, just in time for Meiners to prepare a proposed "protective withdrawal" for 26,255.5 acres of federal lands along thirty-three miles of the Snake River Canyon. Previously, Meiners and Booker had decided that they had the best chance of gaining federal protection for a rim-to-rim corridor through administrative action by the Secretary of the Interior. Morley agreed that it would be a good first step. Legislation would take longer, and there was no assurance of success. A bigger protection plan would come later.

The Bureau of Land Management laid the groundwork for the protective withdrawal with a two-part proposal developed under Meiners' direction. The BLM natural resources staff prepared a thirty-page document that described the resources at stake—wildlife, water, soils and vegetation. The report noted that the proposed withdrawal would not unduly affect livestock grazing or Idaho National Guard training activities adjacent to the Snake River Canyon. A proposed high dam on the Snake River at Guffey, which could have flooded the bottom lands in the canyon, was listed as one of the primary threats to raptors because it would inundate lands along the riparian zone of the river, a key habitat area where some raptors hunted for prey.

New development of agricultural lands on the top of the canyon posed another threat, because it eliminated the sagebrush-bunchgrass habitat on which jackrabbits and ground squirrels depended for survival. Those two species comprised the main prey for golden eagles and prairie falcons, the two spotlight raptor species in the canyon. Morley knew that future farmland development on the north side of the canyon was the biggest threat, but he also knew that more studies would need to be done to prove it.

The second document that Meiners prepared was called a "Pictorial Resume" of the Snake River Canyon. This was a very unusual report for a federal agency to prepare, because it showed that Meiners had an intimate personal knowledge of the canyon, from spending many years working alongside Morley. All of the photographs in the 42-page document were taken by Meiners, with the exception of one aerial photograph.

The Pictorial Resume opens with a photo of a desolate-looking scene in the bottomlands of the canyon. "Remote and desolate as the area seems, and in fact is, it still has been viewed by more people than possibly any other similar spot in the United States, or the world," Meiners writes. "The simple reason is that two nationally televised films have been made here, namely, Walt Disney's *Ida the Offbeat Eagle* and the *Wild Kingdom* series "Valley of the Eagles." Meiners notes that at least 130 million people had seen *Ida* in the United States alone, and more than fifty million had seen "Valley of the Eagles."

"Idaho has within its territory a unique geologic area, remote by nature, yet only thirty miles from Boise, the state capitol; desolate and forbidding in appearance, yet seen by 130 million people in the United States alone; a seeming worthless piece of land, yet praised by experts as priceless for what it is, namely, a wildlife sanctuary for raptors, specifically the golden eagle and prairie falcon," Meiners wrote.

The BLM's proposed protective withdrawal was approved by the BLM Boise District, State Director William L. Matthews and forwarded to the Secretary of Interior, Rogers Morton. The Bureau of Reclamation, U.S. Fish and Wildlife Service and Idaho Environmental Council all backed the proposal with letters of support.

"The Environmental Council votes an emphatic yes," said Lyle M. Stanford, chairman of the group's public lands committee.

The BLM advertised the proposal in the *Federal Register*, and received more than 200 letters in support and opposition. Proponents of a Snake River Birds of Prey Natural Area showed up for a public hearing on February 27, 1971, at the Boise Interagency Fire Center. Meiners opened the evening with color slides of the area. A majority of people testified in favor of the proposal, according to an article by Morley's long-time associate at *The Idaho Statesman,* Bob Lorimer.

One of the rare voices to speak in opposition was Homedale

farmer Rolf Geertson. "He felt 200 acres per bird was high and that he was unaware of any real evidence showing birds of prey benefit man," Lorimer wrote. "He recalled that he had hunted and fished in the area for twenty years and enjoyed the wildlife but felt more consideration should be given to humans."

Many Boise residents supported the proposal, as did adjacent landowners, south of Kuna, who had been assured by the BLM that their farms would not be affected. The hearing also attracted Robert Turner of Sacramento, representing the National Audubon Society, who spoke in favor of the proposal. "The raptors using the Snake River breeding area urgently need a large, undisturbed region to bring off their young," Turner was quoted as saying in the *Statesman*. "Indiscriminate shooting, illicit trafficking in falcons, and chlorinated hydrocarbons pesticides are taking a drastic toll nationwide of these species."

Due to the lack of opposition to the BLM's proposal, it sailed through. On August 24, 1971, Interior Secretary Rogers Morton made it official on a basalt rim overlooking the brand new Snake River Birds of Prey Natural Area on a warm, sunny day. The dedication ceremony attracted more than 100 people, including VIPs such as Idaho Governor Cecil Andrus, Senator Frank Church, D-Idaho, Senator Len Jordan, R-Idaho, U.S. Representative Orval Hansen, R-Idaho and Assistant Interior Secretary Nat Reed. Morley was there with a golden eagle on his fist, accompanied by his wife, Pat.

A front-page photo showed Mrs. Anne McCance, a daughter of Rogers Morton, holding one of Morley's golden eagles for the "hero shot," as he calls it. Leave it to Morley to script the perfect photo for the occasion.

Morton singled out the late Edward Booker, Meiners and Morley for making the protective withdrawal a reality. Birds of prey in the Snake River Canyon, he said, "deserve a sanctuary from the pressure of man." He vowed to look for more areas managed by the BLM that could be preserved to benefit endangered species. But the Snake River Birds of Prey Natural Area, as the area was now officially called, set the tone—it was the first such preserve created among more than 264 million acres in BLM ownership nationwide.

After all of the official remarks had been made, Andrus walked up to Morton and said, "Ok, now, what about the pantry?"

"I beg your pardon, what do you mean, the pantry?" a surprised Morton replied.

"Well, you've set aside a nice place for them to soar and a nice place for their nests, but what about their food supply?"

Andrus winked at Morton and proceeded to tell him what he had learned from Morley—that most of the raptors' food sources lived on top of the north rim of the canyon, ground squirrels, snakes, rabbits and everything else. "We need to make sure that their food supply is non-interruptable," Andrus told Morton.

Little did he know that later on, the first-term Idaho governor would have the option of protecting the "pantry."

From his own personal observations over twenty-plus years, Morley understood that a unique set of circumstances coalesced to create such an outstanding home for raptors in the Snake River Canyon. Steep cliffs lining the river provided a bounty of caves, ledges and holes in which raptors could build nests. Westerly prevailing winds provided a nifty way for birds of prey to soar to their hunting grounds on the north rim without expending any energy. After the birds caught their prey, they rode the wind drafts back to the nest with ease. Fine-grained loess soils on the north rim provided perfect burrowing conditions for ground squirrels, and the foundation for sagebrush-steppe vegetation to grow there. Sagebrush, rabbit brush and other shrubs provided cover for rabbits. The Snake River provided a critical water source in an otherwise hot and hostile climate, particularly in the summer months.

When Morley added all of those elements together, he saw the Snake River Canyon as an *ecosystem*, a term of art that no one used at the time. As time went on, Morley called it the *Vertical Environment*.

Now it was time for the BLM to take the research effort to the next level, to build the case for protecting the dining area for raptors as well as the bedrooms. Plus, there was so much more to learn about the canyon: What other species of raptors lived there? What species of prey were most important? What did the prey species need to survive over the long haul?

In 1970, Andy Ogden started to work on a master's thesis under Hornocker at the University of Idaho, focusing on the nesting density and reproductive success of prairie falcons in the Snake River Canyon. Up to that point, no studies had been done to pinpoint exactly how many prairie falcons lived in the canyon. Ogden proved there were more than 100 nesting pairs of prairie falcons, the most that had been found in a single geographic location anywhere in North America.

The research work on prairie falcons was hazardous, Ogden says, due to the variable kinds of rock he and his helpers ran across in the canyon. "It was nasty, nasty climbing," he says. "It's really bad rock, bad basalt. I don't know how many times I've fallen off of cliffs, how many times I've been knocked out by falling rocks and stuff because that basalt doesn't hold up very good. By the time we got done with my study, we knew which cliffs you could trust, and the cliffs where you had to be really careful. The cinder ash cliffs, the ones with rocks stuffed in them, they just fell apart at the drop of a hat."

Ogden's method to identify prairie falcon aeries was to scan a cliff at the base of the canyon, see where the birds flew into a hole, mark it visually, and then try to rappel into the hole from the top. For climbing, Ogden used the same ropes and equipment that Morley used in World War II..

"We learned how to climb based on what Mr. Nelson taught us," he says. "The gold line rope, it looks like a regular twisted rope, (nylon) and it stretched like nobody's business. It stretched forever. That was what Mr. Nelson used so that's what I used. Then I saw this braided climbing rope and it didn't stretch, and I thought it was like a dream! It wasn't until I was going to graduate school that I had enough money to buy some real climbing ropes."

Due to the hazardous rock, he fell frequently. "Oh, I fell all the time," he says. "Several times, I fell thirty or forty feet. After a while, you'd get good at falling. The way it happened usually was that you're standing on the top, unroped, and you're looking for nests, and you'd have a chunk of the cliff break off, and you'd go down with a bunch of rocks and stuff. After a while, I got to be a pretty good rock rodeoer, because I could ride them down, and slide away, and jump off to the side."

Ogden never broke any bones falling off the cliffs, but his partner, Dave Rodenbaugh broke his ankle once.

"We did lots of stuff to try to get to the nests," Ogden says. "I put a twenty-foot extension ladder in a twelve-foot aluminum boat and drove it from Swan Falls Dam all the way to Jackass Butte to get to a prairie falcon nest. They thought we couldn't get them, but we *got* them. It was an ash cliff—you couldn't put a piton in the cliff. So I drove a stake in the top and laid a rope in a rack in the rock, so the next year, all I had to do was grab the rope and climb up. So the next year I went up there, and I noticed a pack rat had nearly chewed the rope in two."

Beyond falling, a second significant factor climbing into prairie falcon nests was the danger of adult birds flying down to protect their young. "There were only two birds that were dangerous, the Sun Goddess, and the Sun God," Ogden says. "We were scared to climb into their nest because the Sun Goddess was out to try to hurt you, and she would. She dropped Dave colder than a wedge.

"Most of the falcons flew around and bitched at you. But the Sun Goddess never made any sound. She'd go real high, and get the sun behind her back and then she'd turn and come down. You had no idea where she was or when she was coming. One time, honest to god, I'm rappelling down the side of the cliff, and Dave yells, she's coming, and my only defense is to hug the cliff. So before I can even move, she went between me and the cliff with her feet out, and she's trying to kill me. That pretty much shook me up, you know. The next time she stooped, Dave yelled, she's in the sun, I'm hugging the cliff, and all of a sudden, the safety rope started coming down, I'm yelling "Up-slack! Up-slack!" and nothing happened, so I rappelled to the bottom of the cliff, climbed back to the top, and Dave had a tear that big on the back of his head. She'd come in and hit him, and he was just starting to come to when I got there, probably fifteeen to twenty minutes after he'd gotten whacked. He was moaning and groaning."

While Ogden worked on his research, Morley kept tabs on his findings. "I can remember Morley being so insistent about reporting what I was finding that basically once a week, I'd get word through my parents that he was looking for me, and I've got to go find him and report in," he says. "He always wanted to be kept up to speed on our progress."

Beyond documenting the largest single prairie falcon population in North America, Ogden discovered that prairie falcons ate ground squirrels like bears eat berries. Ninety-five percent of their diet came from Piute ground squirrels, the little brown mammals that lived on the north rim of the canyon. Falcons patrolled the north rim, waiting for a ground squirrel to stick his head above the portal of a burrowing hole, and they'd dive for the earth at high-speed.

Ogden's findings would lead to more research questions about ground squirrels, such as, what did they need to survive? What variables affected their populations?

On November 26, 1971, the premier of *The Eagle and the Hawk* was broadcast on NBC-TV, showcasing the virtues of the Snake River Birds of Prey Natural Area to a nationwide audience just two months after it had received national protection. The film provided continued momentum in Morley's political effort to expand the boundary of the birds of prey area, and it trained the spotlight on the festering problem of people shooting hawks at will. The film also marked a significant shift: the use of celebrities to impart the conservation message.

After working with Nell Newman, Joanne Woodward and John Denver in *The Eagle and the Hawk,* Morley saw the added value and impact that big-name celebrities could provide in his crusade to save birds of prey. In any new film projects that came up, he worked with producers to bring celebrities into the mix to increase the number of viewers that watched the films, and deepen the impact of the message.

Morley met the Newmans in Burbank, Calif., while he was working on some studio scenes on *My Side of the Mountain* with Producer Bob Radnitz at Paramount Studios. Radnitz was a friend of Paul Newman's, and he knew that his daughter, Nell, had a strong interest in birds of prey. She had a trained kestrel that lived at the Newman home in Burbank. Newman was working on a movie in another studio nearby, so Radnitz took

Morley over to meet him. Newman was obliged, and invited Morley and his wife, Pat, over to their house for dinner so they could meet Nell and talk about falconry. Radnitz and his wife came for dinner, and so did Newman's pal, actor Steve McQueen.

While the adults sat down to sip some cocktails, Nell and Radnitz' daughter, who was ten years old, went outside to look at Nell's kestrel. Nell let the Radnitz girl hold the kestrel on her gloved fist, but she forgot to hold onto the leash, and the kestrel flew to the top of a house nearby. The bird was loose!

Nell came running inside. "My kestrel flew away! Dad, can you help me get it back?"

"Sure," Newman said.

Morley jumped out of his chair to help.

"When you lose a bird with its leash on, not only is it the ultimate sin, but you don't usually get them back, and if the bird flies to a tree, you can permanently lose them because they'll hang themselves," Nell says.

"That's true, it's the most dangerous thing that can happen to a falcon," Morley says. "Because when it attempts to land the leash flips around the wire or the branch and the bird hangs upside down until someone can come to get it or else it dies. The only recourse anyone has when a falcon is lost in this way is to climb up and call the bird down. "

The whole scene turned into a humorous chase, with Morley and Paul Newman running around Burbank as the kestrel flew from one rooftop to the next.

"Paul and I went to the next door neighbor where the bird landed on the roof of a very big house," Morley says. "I asked permission to climb the house to see if I could take this bird down. My request wasn't well received. They had considerable questions about me climbing to the top of their house. Then the bird flew off again to another house a block or more away downhill from where the Newman's live. This time I asked Paul if he would ask permission for me to go up there. I knew that he would get a better reception than me because he is so well known and well-liked by the public at large. This proved to be more than true. I went up with some help from Paul onto the edge of the house and again, before I ever got close, the bird flew off again, landed on another house but lower down, dragging the leash."

Just as Morley and Newman reached the next house where the kestrel landed, it flew across a six-lane highway and landed on a powerline.

"Well, Paul and I attempted to cross the highway and that created a crowd of people when they saw Paul walking across the highway," Morley says. "Many of the cars pulled off of the road down into the ditch and stopped, and other people came around from all over and were impressed by the fact that Paul Newman was walking around on the highway. I was not adverse to climbing the power pole if I could reach up to the ladder on the pole where there were stakes driven so that you could climb it. Paul helped me get up there. The only chance that I had was to climb up to the pole underneath the line, grab the leash and when the bird flew, pull her down.

"However, the bird was relatively happy to see me when I reached the top of the pole, and I reached up and grabbed the leash. The bird did not try to fly, which meant it wanted to come back into captivity. As always, I used super slow movements to put the bird at ease. Then I pulled the leash, and pulled the bird backwards below the powerline. She tried to fly, and I brought her under control and onto my fist.

"Of course, everybody clapped, and when I looked down I couldn't believe that there were over 100 people standing there watching us. All I can say is we were lucky to have the opportunity to get Nell's kestrel back. Someone offered to give us a ride back to the Newman's house, and we went back to join the party."

"When I saw that they actually got the kestrel back, that was amazing," Nell says. "It made me very enamored with Morley."

During the dinner conversation, Morley asked Paul and Joanne if they would be interested in working on a movie with birds of prey. They thought about that for a moment, and said, sure, we'd be glad to help, and what about Nell, she might want to help, too?

Morley had been in communication with Robert Riger, a well-known producer and director with Tomorrow Productions, Inc., in New York, who was interested in movie script ideas for an hour-long TV special to be sponsored by General Electric and the National Audubon Society. When Morley told Riger that he had had dinner with the Newmans, and that they were interested in helping out on a film project, Riger hastily crafted a script outline and presented it to General Electric. It was approved.

"This seemed like a wonderful opportunity to me," Morley says. "To have people like Joanne Woodward and Nell Newman participate in the film would help create interest in the entertainment value of the film, and add significantly to the number of people who would want to see it."

Riger brought John Denver into the project, and he suggested that he ought to spend a few days with Morley in the Snake River Canyon to gain inspiration for the original musical score that he'd have to write for the film. Denver followed through. He went to the canyon with Morley, rappelling into golden eagle aeries to see young eaglets only weeks old, and watching the dynamics of eagles, falcons and hawks soaring and flying in the thermal drafts of the canyon. As things turned out, he not only wrote the title song for the movie and several others, he ended up with enough material for an entire album, called "Aerie."

The Eagle and the Hawk displayed spectacular footage of golden eagles in the prime time of courtship flight, all original footage shot in the Snake River Canyon, footage that most Americans had never seen in their life. "I'll tell you what, boy, that's an impressive sight to see a pair of golden eagles flying together like that during the courtship phase," Morley says. "They soar into the air, riding the wind currents, and they come together at the absolute peak of the ascent, and then they dive at great speed toward the bottom of the canyon. It takes my breath away every time I see it."

The film showed many spectacular scenes—a male red-tailed hawk handing off a ground squirrel to the mother hawk in mid-air, two hawks defending their nesting territory from an intruding golden eagle, resulting in dramatic acrobatic maneuvers, close-up scenes at a golden eagle nest with young downy birds, and a juvenile golden eagle's first attempt at flight. Several scenes showed Morley with injured birds of prey that had been turned in after being shot. He noted that seventy injured birds had been turned into him or Doctor John Lee in one month.

Looking back, Morley says the film carried an important message about the need for people to stop shooting birds of prey for mere kicks. "It is my hope that sportsmen in general will feel the message here," he says. "Unless you're going to eat something, don't shoot it."

Overall, the movie educated Americans about some of the most important things that Morley had been working on for 30 years—the majesty of the Snake River Canyon, the problems of people shooting birds of prey, the challenges of raptor-rehabilitation, and in the end, the successful release of an injured red-tailed hawk.

"The film really grabs onto you," Morley told Bob Lorimer of the *Statesman*. "People cry when they see it. After seeing it in New York, I walked around for two hours to get settled down. It's beautiful—sincere. To me, it was the culmination of 30 years of jumping up and down. I was absolutely flabbergasted."

There were a number of national articles about the film before it aired, including a national wire story produced by United Press International. Joanne Woodward granted an interview to reporter Jack Gaver of UPI and touted the upcoming film. "I know there are millions of people in this country who are just as concerned about ecology and wildlife as I am," she said. "Everyone does what they can in behalf of their favorite cause, whether it's donating money or actually going out and stimulating interest."

Woodward said her daughter was quite eager to see the movie for the first time. "Nell is very excited about the special," she said. "And not just because she has a big role in it and is making her television debut. I guess she's inherited Paul's and my love for nature, and she's just as concerned about the welfare of our wildlife as we are."

Morley had a strong impression on Nell, too, Woodward said. "Morley was a fantastic host. He literally took Nell by the hand and took her to all the far-out areas where she could see first-hand how these great birds live . . . She saw the birth of a young eagle and followed its growth until the moment it took its first solo flight. I was there, too, for that last event, and it was one of the most exhilarating moments of my life.

"Nell is still excited about her trip to Idaho. It is an experience that she will never forget. And she's already invited all of her friends to our house to watch the special on Friday night."

In 1970, Mike Kochert, a graduate of Purdue University, came along to join the budding Snake River research team. Kochert picked up where Beecham left off on golden eagle research. He completed a master's thesis under Hornocker that was titled, "Population status and chemical contamination in golden eagles in southwestern Idaho."

"This was more than a population survey," Kochert says. "I was looking at the ecological effects on eagles. There was a lot of concern about chemical contamination at the time, looking at the myriad of organo-chlorine pesticides and how they affected eagles, and there was concern in southern Idaho about heavy metals, primarily mercury, because farmers were using a lot of alkyl mercury, a fungicide for wheat."

Like everyone else who conducted research in the Snake River Canyon, Kochert had to meet Morley.

"Boy, that was an interesting place," he says of Morley's foothills home and the mews. "No matter when you stopped by Morley's, there was always a group of people over there, a lot of neophyte raptor enthusiasts. I don't mean that in a derogatory sense, but a lot of them were just kids, like Andy Ogden and Pat Benson, to name a couple. And then, of course, Morley had his own herd of boys. You could go over there, you felt that you could drop over anytime, and you were welcome. No hassles. And it was neat to go up there and see that menagerie of birds."

Kochert finished his master's thesis in 1972, and he was ready to begin looking for a job. At the same time, the BLM was getting criticized for failing to do anything following the creation of the Snake River Birds of Prey Natural Area. Inside the Interior Department, the Fish and Wildlife Service was talking about trying to add the Snake River area to the Deer Flat National Wildlife Refuge, a series of islands valuable for waterfowl on the Snake. BLM officials realized that they had better do something. There's nothing like a little interagency turf battle to cause action.

"There was a lot of criticism being heaved at the BLM for the fact that they hadn't done anything," Kochert says. "I mean, they'd had the new area for six months, and they hadn't done squat. They hadn't even hired a biologist."

BLM officials in Washington D.C. snapped to attention and sent financial resources to the Boise District to begin a new research effort. "It wasn't more than a matter of days, and they had a biologist—that was me," Kochert says. "I had just finished my research, I hadn't received my degree yet, and I got a call from the BLM saying, do you want a job? They were dead serious. They did not want to give this area up.

"I thought they were crazy. They wanted to pay me three times as much as I'd ever made in my life to do what I did for basically nothing in graduate school, hah, and they were going to give me a vehicle and all of this other stuff, it was like throwing me into a candy store."

Kochert took the job. The following year, Kochert received a memo from the national BLM office, asking him to detail the research and management needs for the Snake River Birds of Prey Natural Area. Kochert wanted to expand the prairie falcon and golden eagle research, he needed a detailed inventory of all raptor populations that lived in the natural area year-round, and he identified many other projects that would be helpful. In 1974, he received the funding. The research began in earnest in 1975, and it continued until 1978.

During the research phase, the BLM placed a moratorium on any further desert land entries, i.e., conversion of public land, desert habitat in its natural state, to irrigation farms on the north rim of the canyon, until research was completed. Plus, a new, research boundary was established for the Snake River Canyon, a much broader area consisting of 484,600 acres along an 81-mile section of the river.

Kochert laid out a visionary research plan that established the foundation for saving the most important areas in the Snake River Canyon for birds of prey. The goal of the research program was to determine the minimum amount of area necessary to maintain the existing nesting population of birds of prey, and to preserve a complex ecosystem in which raptors played a pivotal role. Research projects were divided up into 13 different studies, conducted by BLM scientists, or independent scientists from a variety of colleges and universities.

They put radio transmitters on Piute ground squirrels and tracked their movements. They put transmitters on prairie falcons to see how far they ranged outside of the Snake River Canyon to find prey. They monitored badgers in the canyon to see how they co-existed with falcons, since both of them ate primarily ground squirrels. They examined why prey species were more numerous in some areas than others. They studied lesser-known species such as long-eared owls to see what they ate, and how they fit into the ecosystem.

"Each study was designed to provide a key piece of insight into the complex, interlocking jigsaw puzzle of interactions that comprise the birds of prey ecosystem," natural resources expert Ed Chaney wrote in a fully illustrated summary report on the research effort.

The outcome was stunning. More than 600 pairs of raptors, representing fifteen species, bred and raised their young in the canyon each year. It was the densest-known population of nesting raptors in North America, if not the world. More than 200 pairs of prairie falcons nested in the expanded study area, representing possibly five percent of the species' entire population.

In essence, the area was even more valuable than Morley had thought. Now it was all documented for the next political hurdle.

In January 1977, President Jimmy Carter appointed Idaho Governor Cecil Andrus Secretary of the Interior, a fortuitous move that would benefit Morley's interest in expanding the boundaries of the Snake River Birds of Prey Natural Area. Andrus was squarely behind Morley's effort, and now he was in a position to have a major impact.

The first way that Andrus helped in his new post was that he controlled the purse strings of the Interior Department. Kochert needed money and manpower to work on the research program. Andrus made sure that he got it.

"It was easy to do," Andrus says. "We had discretionary

money in the secretary's office, and I spent it in Idaho. I made certain they had what they needed. I wanted to see that area protected."

"We had some real cooperation all the way up the line," Kochert says. "Andrus was extremely interested in the whole thing, and it all goes back to the Andrus-Morley link. Morley was extremely effective behind the scenes in terms of influencing the congressional types and the political people. Beyond everything else, that has been his most effective contribution. Morley kept tabs on the research project, talked to the politicians and kept them up to speed, and sold them on the project."

As the research progressed, and prairie falcons were observed hunting in a broader swath on the north side of the canyon than previously thought, Andrus extended the moratorium on agricultural development in the fall of 1977 to 539,000 acres of land. Now the total study area boundary encompassed 833,000 acres.

Agricultural interests such as the Idaho Farm Bureau did not take the boundary expansion lightly. This was a time when the Sagebrush Rebellion—a movement by Wise Use advocates to transfer ownership of the public domain to the states or private interests—was gaining ground in the West and on a national level. Agricultural groups had dominated Idaho political interests for years, representing the largest economic sector in the state. They opposed giving up any potential farm land for any purpose. Real estate agents from Mountain Home, Idaho, a small town with an Air Force base on the southeastern edge of the birds of prey area, hoped to make big bucks by converting desert land to irrigated farms.

"Mountain Home residents were very upset about the BOPA expansion," says former Senator James McClure, R-Idaho. "They were in favor of more agricultural development, and they were quite loud about it."

How irrigation water could be brought to the high desert was the biggest issue that would determine the fate of the lands south of Mountain Home, however, not the expansion of the birds of prey area per se. Either water had to be pumped from the Snake River to the top of the canyon rim, some 300-500 feet above, or it would have to be imported from somewhere else.

The rising price of electricity, and a depressed market for farm commodities made the high-lift water option a poor economic endeavor without government subsidy.

However, McClure was interested in backing a water-diversion project from the Payette River to the Boise River, and eventually to the Mountain Home desert. It was called the Southwest Idaho Development Project. "When I ran for Congress in 1966, the election turned on who could be most in favor of that," McClure says. "There was no dissension. It was just motherhood and next to God."

McClure was the former executive director of the Idaho Reclamation Association, a group that favored converting desert lands into irrigated farmland, and so he was able to convince voters that he, more than anyone else, would have the knowledge, expertise and political acumen to get the job done. More than ten years later, however, the Southwest Idaho Development project was still in the proposal stage, and McClure had not been able to win federal funding for the project.

Still, with the possibility that the project could come through, and a strong belief in Manifest Destiny, a number of agricultural groups lined up to oppose any expansion of the birds of prey area. People like Jack Streeter, a real estate agent from Mountain Home, wrote scores of letters and correspondence attacking the proposal.

"I opposed tying up the desert for those damn birds," Streeter says.

In the meantime, a firm proposal to expand the bird of prey area was offered by the Bureau of Land Management in June 1979. The agency crafted an environmental impact statement, as required by federal law, that proposed expanding the birds of prey area to approximately 482,000 acres. Andrus took the initiative to introduce legislation adopting the BLM's proposal. It was unusual for a secretary of Interior to introduce a bill in Congress, but Andrus knew that he didn't have much time, and he wanted to resolve the issue before he left office at the end of Carter's term in 1980, assuming that he might not get re-elected.

Andrus' bill embraced the concept that Kochert laid out in

his final research report in 1979. "A boundary for the proposed Birds of Prey National Conservation Area should be based on the biological needs of the raptors and their prey," Kochert said. "The eastern and western boundaries should encompass the unusually high raptor nesting densities along the Snake River Canyon. The northern boundary should include the actual and estimated prairie falcon home ranges, and the southern boundary should be based on the distances flown from the river by red-tailed hawks and golden eagles. The present opportunity to protect and manage an area on an ecosystem basis is unique in the history of natural resource conservation."

In the meantime, to diffuse critics of the expansion proposal, Morley and Andrus met separately with ranchers who grazed cattle or sheep in the area to try to convince them to support the birds of prey bill. The proposal called for allowing all existing uses to continue, including livestock grazing and national guard training.

"I talked specifically about the fact with sheep and cattlemen that their grazing rights, recreational rights, and hunting rights wouldn't change, it would all be the same when this new area was established," Morley says. "One of the most important meetings we had was with the sheep and cattle ranchers in Murphy, the county seat of Owyhee County. I knew many of the men at the meeting. I had been over the cliffs on many of their individual ranches, and they knew by the detail that I talked about that I had spent many, many years studying the golden eagle's relationship to the environment. And of course, I had an eagle with me at the meeting. The people who were there could look that eagle in the eye from three paces away or less, and that eagle made more of an impression than I ever could.

"After two hours of discussion, Dick Bass, the president of the Idaho Cattlemen's Association, came out and said, "Well, I think we all know that Morley is no sixty-second environmentalist, and I think we ought to go along with this proposal." And I let out a big sigh of relief. That caused the fence-sitters to come over to our side, and I knew that they'd support us in the public hearings."

Andrus, meanwhile, worked his own private connections, including a long-time sheep rancher named Phil Soulen. His family had grazed sheep in the birds of prey area in the winter for decades. "I said, Phil, if we do this, I'm looking at maintaining all of the existing uses out there," Andrus says. "The only thing that will destroy the snakes, the Piute ground squirrels, and the rabbits is the plow, a harrow and a disc out there cultivating the land. He wanted to make sure that he was not going to be affected. He took me at my word that I'd keep him whole, and he gave me his word that he'd support it, and he did."

To leave himself an extra option, if needed, Andrus drew the boundaries of the proposed expansion around an area south of Mountain Home that could be sacrificed for future farm development, if needed. "I purposely left one block of good agricultural cropland in the proposal," Andrus says. "I knew it was an area that we could compromise, and we could cut it out and we wouldn't lose anything, but it would give the opposition a chance to say, we changed Andrus' plan."

Beyond personal politicking, Morley did his part to bring national attention to the Snake River Birds of Prey Natural Area by accommodating national media requests for interviews. After Morton set aside the area in the fall 1971, and the *Eagle and The Hawk* appeared in November 1971, a number of national media outlets took notice, and one story led to another.

Life magazine ran a five-page full-color spread in its September 22, 1972, edition, featuring Morley rappelling down a cliff to save an abandoned eagle chick whose nest had tumbled off the cliff. Morley packed the bird to another golden eagle aerie, where it was readily accepted. Big photos of eagles and owls by photographer George Silk were displayed. "These birds are an inspiration," Morley siad in a brief article accompanying the photos. "They have loyalty and strength and integrity. They have been emblems of nobility and valor since the time of Caesar. When you know them, you realize why."

The National Wildlife Federation featured a story about the new Snake River Birds of Prey Natural Area in *Ranger Rick* magazine, a publication targeted at America's youth. A two-page article in the center-spread of the magazine showed Morley standing on the north rim of the canyon holding a prairie falcon. Idaho's own Ted Trueblood, a long-time conservation writer for *Field and Stream*, penned the story for the magazine.

The Federation ran a more in-depth story on the research program in the Snake River Birds of Prey Natural Area in the April-May issue of *National Wildlife* in 1976.

Not to be outdone, *Audubon* magazine devoted a thirty-five-page spread to the comprehensive research effort and political debate in the Snake River Birds of Prey Natural Area in an article by Ann Zwinger in July 1977. Morley wowed her by showing off his eclectic collection of birds in the mews, and flying a prairie falcon to a lure.

"I feel like a kid in a candy shop seeing birds at close range that I've only seen in pictures before, or seen from a distance," she wrote. "A magnificent Bateleur eagle with red nares and feet sits on a log among some trees. An American rough-legged hawk was brought here with a damaged wing . . . A handsome kestrel springs from his perch at our approach . . . Several prairie falcons are here, tethered to their boards by leashes, each bird with its own haunting dignity. Golden eagles in large breeding cages shriek or whistle at the presence of strangers."

But was when Morley flew a female prairie, named Shaikha, that Zwinger felt the thrill that Morley experiences when he flies his falcons to the lure. "Morley casts her off. She springs upward in a quick elegant sweep, rising easily, tilting into a lovely helix. In flight, she is a smooth, trim bird, a delight to the eye. . . . She lifts again. And again, in free yet controlled arcs that make the air sing. Bird and man communicate, reach out, separate, join, no sound except the burring hum of the lure, the soft whistle of the wing. That small sleek shape transfers to me such joy, tracing untrammeled parabolas against a fading afternoon sky, up and over and around and back, so swiftly that the lines of all her loops seem remain written in the air even as she scribes a new one. And then, by a split-second adjustment, Morley allows her to capture the lure.

"Her flight has not lasted long, and yet to me, it seemed an hour. In the intensity of concentration, time has expanded, elongated. And in that expansion of time I realize that all I have seen and learned here is encapsulated in the flight of this exquisite bird."

Quite obviously, Morley had worked his magic like a charm. Zwinger's prose would evoke similar feelings in *Audubon*'s 200,000-plus readers.

Public hearings on the expansion of the Snake River Birds of Prey Natural Area were held in the fall of 1979. In August, the Bureau of Land Management hosted a hearing in Boise, and in Washington D.C. The Idaho Land Board held its own public hearing in October. The issue of expanding the Snake River Birds of Prey Natural Area got people stirred up, drawing more than 200 people to each event.

In Boise, testimony was largely black and white—people were either for it or against it. Jack Streeter and other Mountain Home residents cast doubt on the need to set aside so much federal land for the raptors. "The ground should be put under cultivation as the need arises," Streeter said, according to an *Idaho Statesman* article on August 3, 1979. The land is the "rightful inheritance of the children of southern Idaho." He accused the BLM of "playing farmer against rancher in order to build a land dynasty."

Lining up in favor of the proposal were Bob Turner of the Audubon Society, Bill Meiners, who had retired from the BLM and represented the Idaho Wildlife Federation, and Phil Soulen, the sheep rancher. Morley stayed in the background at that event.

Turner pointed out that the canyon contained "the world's largest nesting population of raptors," and it was critical to protect the lands above the north rim of the canyon for "the birds' dining table."

Meiners urged Idaho's congressional delegation to act on Andrus' legislation, calling the expansion plan "a must." He cast doubt on the value of converting federal lands within the birds of prey study area to farmland, saying it would "further depress an already depressed agricultural economy." Further, Meiners said, it would cost $420 per acre in electrical subsidies to develop dry farm land into irrigated ground.

"Hold the phone! Who is doing what to whom and what for?" the *Statesman* quoted Meiners.

State Representative Dan Kelly, a Mountain Home

Republican, criticized the BLM for putting the interest of raptors above that of people. "We must also look to people, to whom the good Lord gave dominion over other life forms," Kelly said.

Oscar Field, president of the Idaho Farm Bureau, said he couldn't understand how the raptors needed so much territory to survive. "If the hen that lays my breakfast egg can survive on two square feet, why does it take 2,000 acres for a hawk?" he said.

In Washington D.C., a number of Mountain Home residents, including Representative Kelly, testified about their same concerns, but several new faces appeared to support the birds of prey area. "There is no other place in this country, perhaps on Earth, where one can see this many kinds of birds in breeding season as you can along this 80-mile stretch of the Snake River," said Bill Shands of the Conservation Foundation. Shands' comments were carried in a Gannett News Service story in *The Idaho Statesman* on August 10, 1979.

Christopher Earle of the National Audubon Society attacked the economics of converting public land to farmland, clearly a strategy that proponents were using to discredit the farm land issue. At 1977 farm level prices, a new 320-acre farm would suffer an estimated net loss of $29,000 to $33,000, Earle said. To pump water uphill from the Snake River to new farm land would require 500 million kilowatts of power, he said.

Hence, the issue boiled down to a contest between a globally unique refuge for birds of prey, and the wishes of Mountain Home residents to have the option of potentially buying federal land for farm land.

In Morley's mind, the contest was no contest. In a press conference on October 23, 1979, in Boise, Morley told the *Statesman* the debate "should be involving the birds themselves and their prey base," not the potential farmland issue. "If you plow the area up . . . it's gone forever."

Republican members of the Idaho Land Board, however, had such a strong alliance with the Idaho Farm Bureau that they could not support the birds of prey expansion. Idaho Governor John Evans, a Democrat and chairman of the Land Board at the time, stood squarely behind Andrus and Morley. But he was in the minority. The other four Republican members of the board cast doubt on the plan, and tried to undermine the credibility of the BLM research report. The Land Board owned 44,000 acres of state land inside the birds of prey study area boundary. They were concerned that if livestock grazing were curtailed, it could affect the board's ability to maximize income on state lands.

A week prior to the Land Board's own public hearing on the issue, Idaho Attorney General David LeRoy asserted that Andrus should delay moving ahead with legislation because of a purported "bias" in the BLM research report. LeRoy wrote a seventy-nine-page critique, claiming the BLM's report was "preordained" and that federal research scientists were in favor of expanding the area before they started research under Kochert in the mid-1970s. LeRoy even made a slur aimed at Andrus by saying the birds of prey expansion proposal was the former governor's "pet project."

Evans countered by saying, "There is no doubt in my mind that the secretary is as dedicated as I am to establishing a Birds of Prey Natural Area along the Snake River."

The truth of the matter was that Republican members of the Land Board were trying to stall the expansion proposal in hopes that the plan could be derailed until Andrus left office, presuming that Carter would not be re-elected. The board really stretched the limit of credibility when Republicans found a California oil company that claimed the birds of prey area could have vast reserves of oil and gas under the north rim of the canyon.

Great Basin Oil Company officials of Los Angeles told the Land Board that they owned 7,647 acres of oil and gas leases on state land within the birds of prey area expansion boundary, and they wanted to see what kind of potential the area had before Congress or the Interior Department approved an expansion plan. Curiously, no environmental groups were invited to the special Land Board meeting where the oil and gas issue surfaced.

At the Land Board's public hearing on October 23, 1979, Russell Hayward of Great Basin Oil Company estimated that the state could earn $238 million from oil and gas leases in the birds of prey area, based on estimates the area could yield sixty-three million barrels of oil or ninety billion cubic feet of natural gas, according to an article in the *Statesman*. But Hayward's estimates were whoppers—the company never drilled any production wells in the Snake River area.

State Senator Walt Yarbrough, R-Grand View, wondered "What's going to happen when they (the BLM) see one cow step on a ground squirrel?"

But the real surprise of the evening came when former Oregon Governor Tom McCall, a Democrat, rose to the podium. McCall represented the interests of The Nature Conservancy, a private non-profit organization that had raised $500,000 to buy five private inholdings totalling 900 acres along the Snake River inside the birds of prey area. He noted that 1,369 people from 50 states and nine foreign countries had contributed to the birds of prey campaign. Morley had flown to Saudi Arabia to help raise funds for the purchase, and Andrus had brought actor Robert Redford to the canyon for a special boat tour of the area, led by Morley.

"This is safely one (opportunity) that comes along every billion years," McCall told the Land Board. "If you muff it, there is no rerun. The assets of the proposal seem to outweigh its liabilities. Preservation is reversible; exploitation is not."

Clearly, the Land Board had tried to raise enough doubts about the project to scare the Idaho congressional delegation from acting on Andrus' bill. But if the board allowed two sides to testify, the debate still came down to a world-class area for birds of prey versus a few potential acres of farmland in a year when crop prices were depressed from oversupply. Still, the Land Board did not have any real say over the birds of prey expansion. The proposal involved mostly federal land, and therefore, it was a federal decision, not a state decision.

As the issue progressed during election-year 1980, a year in which Republican Congressman Steve Symms, R-Idaho, was running against four-term Senator Frank Church, D-Idaho, the Idaho congressional delegation did not act on Andrus' bill. Realistically, Church was the only member of the delegation who was likely to pursue it, and he was overloaded with other issues, including a 2.3-million-acre wilderness proposal for Central Idaho, the Panama Canal, re-election and more. Symms and Idaho GOP Representative George Hansen would never support a conservation-related measure if the Farm Bureau opposed it, and neither would McClure.

Andrus pleaded with McClure to amend the bill to create more agricultural land for Mountain Home interests. "I said, McClure, take that acreage out and make it your bill," Andrus says. "But he was just stalling it, thinking I would go away, so we did what we had to do."

In November, California GOP Governor Ronald Reagan soundly defeated President Carter. Andrus could see the handwriting on the wall. Reagan's conservative brand of politics included a strong anti-environmental bent, and that meant he would install a new Secretary of Interior closely aligned with his beliefs. Andrus would be gone in a month.

But all along, if necessary, he knew he had the option to use a little-known executive privilege provided in a 1976 law called the Federal Land Management and Policy Act, a privilege that allowed secretaries of Interior to withdraw valuable public lands for environmental purposes for twenty years. The week before Thanksgiving 1980, Andrus invoked executive privilege and set aside 482,640 acres of federal land for an expanded Snake River Birds of Prey Natural Area.

McClure and Symms went berserk. Their stall tactics had failed.

Andrus' executive order is "an act of extreme arrogance and abuse of discretionary power," McClure said in the *Statesman* on November 25. Symms called it "excessive, arrogant and a slap in the face to the Idaho congressional delegation."

Under the law, Congress had ninety days to overturn Andrus' order, something McClure pledged to do immediately. As it turned out, however, he apparently couldn't muster the votes to get it done. Congress never took action at all.

Andrus, meanwhile, felt extremely pleased about expanding the birds of prey area. "It is one of my most satisfying achievements," he says. "I think Will Rogers probably said it best when he talked about the land, and he said, they just ain't making it no more." That's a thought that I've always kept in my mind. Once you destroy it, it's pretty hard to get it back, whether you're talking about a river or a piece of land or anything else."

The same day that Andrus took action on the birds of prey

Nelson family collection

Morley's long-time friendship with Cecil Andrus paid off when Andrus, as one of his last acts as United States Secretery of the Interior, used an executive order to add nearly 500,000 acres to the Snake River Birds of Prey Natural Area. Andrus, who also served as governor of Idaho, said his respect for Morley played a role in his decision.

area, the leaders of Sagebrush Rebellion efforts in Idaho threatened to file a lawsuit, claiming the BLM's environmental impact statement had numerous flaws and would be thrown out by a judge. "We have determined that a vigorous pursuit of that legal action has a strong possibility of totally negating the secretary's hasty, ill-conceived and unneeded action," said Vern Ravenscroft, a frequent candidate for governor and president of Sagebrush Rebellion Inc.

The lawsuit did not prevail, however, because the EIS was solid. The BLM research report contained detailed scientific work that justified the expansion of the boundary. Kochert et al. had done an outstanding job. Their work stood up to the test.

Morley, meanwhile, was jumping for joy. Andrus had called

him on the day he signed the withdrawal, and told him the good news. "Morley, we got 'er done," he said.

So the next morning, when Pat and Morley saw the headline in the *Statesman* "Andrus acts to expand birds area," they both were so pleased. Pat whooped for joy and gave him a big hug.

"Hey, Morley, look at this!" she said, holding the front page of the paper.

"How about that!" Morley said, beaming.

"I was really happy to see that Andrus went with me, because he was such a powerful man," Morley says. "He believed in what I was talking about because we had gone over the whole program in great length, for over twenty years, and he never wavered an inch—he went along with me the whole way."

What a remarkable string of events. Morley had personally discovered the heavy concentration of nesting birds of prey in the Snake River Canyon in 1948, long before anyone else knew the area was valuable for anything. He personally advertised the area to a national audience by working on *Ida* for Disney, *Wild Kingdom* specials, *The Eagle and the Hawk* and dozens of national newspaper and magazine articles. He initiated the first reseach projects in the area with the assistance of Maurice Hornocker at the University of Idaho. And he was so impressed with the way the research effort expanded under BLM research team leader Mike Kochert. All the building blocks had fallen into place to create a perfect sanctuary for birds of prey in the Snake River Canyon.

Maybe it was serendipity, maybe it was meant to be. But clearly, if it hadn't been for Morley Nelson, the Snake River Birds of Prey National Conservation Area wouldn't exist today.

"There's no doubt that Morley was the key player in bringing about the official Snake River Birds of Prey Area," Hornocker says.

"Yep, if there hadn't been a Morley Nelson, there wouldn't be a birds of prey area out there in the canyon, or the area I withdrew," Andrus says. "Morley Nelson stimulated within myself, and lots of other people, the thinking to do things right the first time. If you don't do it right the first time, when you're talking about land and water resources, you never get a second chance."

Chapter Thirteen

Raptors and powerlines
1972-1980

In the spring of 1972, a Boise TV anchorman declares on the evening news: "Thirty-three eagles were found dead yesterday under a powerline in Twin Falls. All of them apparently were electrocuted by high-voltage powerlines." TV cameras showed footage of an Idaho Fish and Game officer holding up a dead eagle, spreading the underside of its enormous gold and white-flecked wings.

The report made national news.

It wasn't the first discovery of dead eagles under powerlines. Federal and state agencies reported 300-plus eagles had been killed by electrocution in Colorado and Wyoming in the winter of 1970-71. But Morley was alarmed. He'd never heard of thirty-three eagles getting zapped in one location. The following morning, he drove his brown Jeep Wagoneer to Twin Falls to inspect the dead birds. Idaho Fish and Game officers were more than glad to get an expert opinion.

As he examined the eagle carcasses, Morley discovered that more of the eagles had been killed by gunfire than by electrication. "Most of them had been shot full of holes," Morley says. "The ranchers were picking them off the power poles."

Still, he was concerned. He'd been aware that some eagles and large hawks got electrocuted on three-strand powerlines, but he did not see it as a significant problem, compared to wanton shooting. By the early 1970s, however, Morley and the National Audubon Society already had done everything they could to change federal laws to protect birds of prey from wanton shooting. Congress had amended the Migratory Bird Treaty Act in 1972 to protect all birds of prey from indiscriminate shooting. Golden eagles had been protected by law in 1962. Enforcement and education were the remaining gaps in remedying the shooting problem, and Morley worked on public education every chance he got.

But the electrocution issue was brand new. It came at a time when populations of bald eagles and peregrine falcons were disappearing faster than anyone thought possible because of the widespread infiltration of organo-chlorine pesticides into the food chain. A new environmental consciousness was afoot in the United States, on the heels of the first Earth Day celebration. Americans were concerned about polluted water, industrial smog and endangered wildlife. A growing distrust of the federal government had emerged over civil rights battles and the Vietnam War. The public wanted to see corrective action taken.

Officials in the electric utility industry were concerned, too. Both the bald eagle and peregrine falcon were listed as endangered species. If any endangered birds were killed by powerlines, companies could be subject to fines and bad press. The Eagle Protection Act made it illegal for anyone to kill a bald eagle or golden eagle, too. No one knew how widespread the problem was, but a few savvy individuals inside Idaho Power Co. thought they should start investigating the issue to see what they could learn.

"We knew we had to get this problem solved before it blew clear out of sight," says Logan Lanham, former vice president of public affairs for Idaho Power Company.

Soon after the incident, Lanham was at a meeting with several legislators in Sun Valley, Idaho. Former state Senator John Peavey, a Democratic sheep rancher from Carey, Idaho, urged Lanham to be proactive about the issue. "You can correct that," Peavey said. "Your company should have a way to do that."

Lanham was open-minded about the issue, and he nodded his head. "We'll see what we can do," he told Peavey.

When he got back to Idaho Power headquarters in Boise, he paid a visit the company's vice president of engineering. Lanham pitched the notion of hiring Morley as a consultant to learn more about how birds of prey get electrocuted by powerlines, and what it would take to fix the problem.

"OK, get Nelson," the engineering chief said.

Morley's reputation was golden. By now, he was considered an international authority on birds of prey. Morley's credibility was rock solid with Idaho Power because of his snow survey work and accurate streamflow forecasts, which the company depended on for the best possible management of its hydropower operations on the Snake River. Morley's golden eagles and large hawks would provide a wide variety of birds to experiment with. Morley also made a habit of filming nearly everything he did related to birds of prey. Those skills would be valuable, too.

Lanham called him in March 1972.

"Morley was honest and knowledgeable. We had confidence he would look the situation over and give us an honest report on what to do about it," Lanham says.

Morley eagerly accepted the offer. He had just retired from his snow survey job in December 1971. He was ready to take on the next challenge. In this case, he'd actually get paid as a professional birds of prey consultant, just as he got paid to work on movies.

Idaho Power opened its labs to Morley for testing the conductivity of hawk and eagle feathers, when wet and dry. The company erected several powerpoles in his backyard for testing different designs with a variety of eagles and hawks.

Morley quickly discovered that single-pole lines equipped with a single-phase line or three-phase lines—both of which carried less than 69,000 volts of current—were particularly deadly to eagles and large hawks. If a bird tried to land on a powerpole, and its wings touched a ground wire and a phase (hot) line at the same time, it could get zapped, particularly if its feathers were wet. The distances between the wires had been configured by engineers many years before to create an efficient way to provide electrical power to customers at the lowest cost possible. Concerns about birds of prey had never been ramped into the equation—until now.

The basic cause of the problem was "too much bird in not enough space," says Alan Ansell, Idaho Power environmental supervisor.

Meanwhile, Bob Turner, western regional representative for the National Audubon Society, based in Boulder, Colo., had written a letter to Pacific Gas & Electric Company, expressing concern about the electrocution problems reported in the media. A copy of Turner's letter was routed to Richard Thorsell, environmental affairs coordinator for the Edison Electric Institute, a national advocacy organization for the nation's investor-owned electric utility companies, located in New York City.

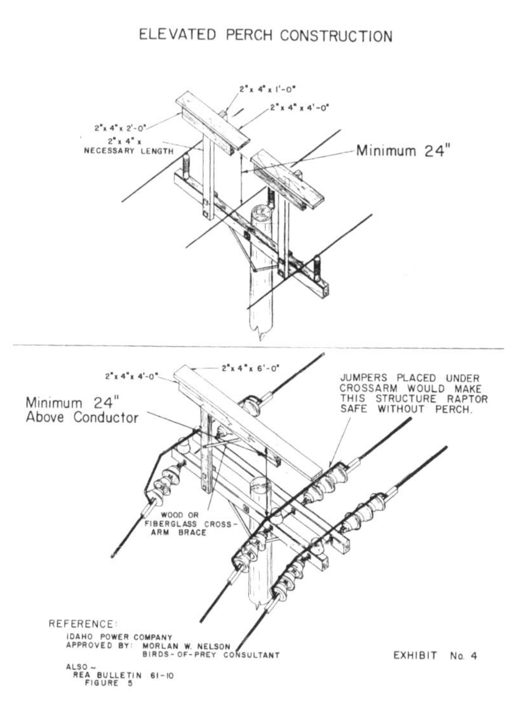

Avian Power Line Interaction Committee (1975, 1996)

These designs and others for protecting birds of prey from powerlines were the fruit of experiments by Morley Nelson and Idaho Power Company engineers. They were published in "Suggested Practices for Raptor Protection on Powerlines."

Thorsell was not an industry insider. A lifelong birder and an Audubon member, Thorsell wanted to do the "right thing" to protect birds of prey, but he knew he had to do so in a way that had full buy-in from the electric utility industry. One of his early mandates was to find "some project where the antagonists (environmentalists), industry and government could work together to solve problems," Thorsell says.

The raptor-powerline issue had potential for a cooperative solution. Thorsell offered an olive branch to the Audubon Society, suggesting that a regional meeting could be organized to discuss the issue. "I have the distinct feeling that (Audubon)

companies wanted to fix the problem for another reason—raptor electrocutions usually caused power outages, and outages cost money, sometimes $20,000 a pop. Utility companies sent their top people to Denver to address the issue.

"Many groups within the industry stepped up to accept the fact that this was a problem, and they wanted to find solutions," says Turner of Audubon. "They were great to work with. They put their best people on it, and put their best minds on it, and said, "Let's figure this out."

Inside the environmental community, Turner says he had to vanquish more militant groups. "There were a few environmental groups saying, we can't work with the industry; let's sue the bastards," he says. 'We had to snuff them out.'"

Near the conclusion of the meeting in Denver, somebody mentioned that Idaho Power Company was looking at raptor-electrocution problems and solutions with a birds of prey expert. Thorsell followed up and called Idaho Power. He confirmed that they were, indeed, working on the problem, and that Morley could be trusted to develop the best solutions for birds of prey.

On April 6, 1972, Thorsell invited the same group of environmentalists, industry and government officials to Denver to strike an agreement on the future course of action. Everyone agreed to work on solutions to the raptor electrocution problem that would produce sound results for birds of prey that were economically feasible. The Fish and Wildlife Service agreed to establish and administer a raptor mortality reporting system. And the REA distributed the first document on potential solutions to raptor-electrocution problems, REA Bulletin 61-10, "Powerline Contacts by Eagles and Other Large Birds."

Solutions embodied in the REA bulletin, Turner says, were really just the beginning. The utility companies agreed to examine problems in their service areas, and develop solutions. But everyone eagerly awaited the results of Morley's research with Idaho Power.

Morley started his research by taking several eagle feathers into the research lab at Idaho Power to check the conductivity of wet and dry feathers. A dry feather could withstand 70,000 volts without arcing. But a wet feather arced at 5,000 volts. Morley shot close-up footage of electricity shooting across the surface of a wet feather. Clearly, the birds would be in most danger during wet weather, either rain or snow.

To determine what kind of modifications would be needed in powerpole and powerline design, Idaho Power's engineers built three different powerpole configurations in Morley's sideyard, next to the Boise foothills. "These mockups are an exact duplicate of the actual powerlines in the field, so they will be a valid test of what happens to the eagle when she comes in and lands," Morley said at the time. "I think once we have flown the eagles over and over to these various configurations, we will have very accurate determination of the problem and how we can solve it."

Morley got permission from the Fish and Wildlife Service to capture a juvenile golden eagle from a powerpole nest in the Snake River Birds of Prey Natural Area. He wanted to use a young golden eagle for the experiments because preliminary data from electrocution surveys had shown that most of the dead eagles were juveniles. Being young and inexperienced, they were sloppy fliers. When they landed on a powerpole, they didn't fold their wings as quickly and expertly as mature birds.

Morley's assistant, David Boehlke of Boise, manned the eagle over a month-long period, and trained the eagle to fly to a T-perch. Then it was ready for the experiments to begin. Boehlke would place a piece of meat on top of one of the powerpoles, walk 100 yards to the other side of the field, and release the bird. Morley set up a 16mm Millikin high-speed camera on a tripod next to the powerpole, and filmed the eagle as it came in to land.

He and Boehlke conducted experiments over and over again, flying the bird from different angles to see if it landed any differently. Then Morley sat down in his film lab, watched the footage in super slow-motion and jotted down ideas on alternative designs.

Morley took his solutions to the powerline engineers at Idaho Power, and between the two points of view, they came up with some new configurations. The solutions included:

- Installing a four-foot wooden perch two feet above the lines to give eagles a safe place to land.
- Raising the center pole of three-phase lines 48 inches to prevent an eagle from being able to touch the center wire with an outside hot wire.
- Placing tubular insulators around wires on powerpoles.
- Putting a two-inch gap in ground wires in strategic locations.
- Installing a triangular barrier between two hot wires on a cross arm of a powerpole to prevent a bird from landing there.
- Using single-pole lines with no crossarms, a preferred method for new lines.
- Burying electrical lines to eliminate the conflict.

At the same time, Morley made notes about what utility companies could do in the field to isolate powerpoles that might cause problems. From experience, he knew that eagles used powerpoles for perching because they were located in the highest, most commanding position over the landscape below. Eagles always sought out the highest cliff or a tall tree near their hunting grounds. If those two features weren't available, they'd select the tallest powerpole that was located cross-wise to prevailing winds. Morley suggested that power companies could hire a wildlife biologist to survey their territory for obvious perching poles, and fix the poles to ensure that they could accommodate eagles and large hawks.

"By examining the poles, it's fairly simple to identify the ones that have been selected by eagles as preferred sites," he says. "Their droppings on the cross arms, the remains of rabbits, fur and feet, their regurgitated castings underneath, all identify these poles."

When Morley walked some of Idaho Power's three-phase lines, he'd find a pole preferred by eagles, and see all kinds of evidence of use. Then he'd walk several miles without seeing any evidence of eagle activity. "Obviously, the birds' selectivity simplifies corrective efforts by drastically reducing the number of poles requiring modification," Morley said. "Of the number of Idaho Power lines studied, it was estimated that ninety-five percent of the electrocution problems could be prevented by correcting two percent of the poles."

Morley's own preferred solution was to place an elevated perch about two feet above the wires. He put the perches on his poles in his yard, and found that the eagles took to them immediately. "The eagles were eager to land on the perches and immediately adopted them as preferred landing sites," he says. "They required no training or encouragement to use them and could land or take off in any direction."

As part of Morley's research, he concocted an idea for creating nesting platforms for eagles and large hawks on large transmission lines exceeding 69,000 volts. These towers were often more than 100 feet tall. They attracted birds of prey because of their height and commanding view of the surrounding landscape. But the towers did not feature any surfaces large enough for birds to fashion a nest. So Morley experimented with different configurations of nesting platforms. He settled on a four-foot-square platform, with two vertical side plates.

"I saw the birds were trying to live on those towers, in places where they didn't have any shade or protection from the wind," he says. "And I thought, well, we ought to create something they can nest on that can provide shade and protection from the wind. We put one up on one of Idaho Power's transmission towers, and a pair of eagles moved into the nest the first year we put it up. The eagles had two young, and they both did great."

When Morley finished his recommendations, Idaho Power Co. invited Thorsell to visit Boise, meet Morley, and go over the material. He was impressed—not only with Morley's recommendations, but also with Idaho Power's engineering drawings that illustrated a full menu of solutions. Thorsell distributed Idaho Power's recommendations to the ad hoc committee of utility companies, environmental groups and government agencies. Dean Miller of Public Service Co., with the assistance of Erwin Boeker of the U.S. Fish and Wildlife Service, Thorsell, and Richard Oldendorff of the BLM, wrote a report that would apprise the electric utility industry of the latest solutions to raptor-powerline concerns.

The report was called "Suggested Practices for Raptor Protection on Powerlines," issued in June 1975. Eighteen people, including Morley, from seven utility companies, government

Idaho Power Company

A golden eagle flexes its wings in a powerline roost designed by Morley. The roost prevents large raptors from touching positive and negative wires at the same time and has saved thousands of birds from electrocution.

agencies and environmental groups contributed to the document.

"Obviously, the work Morley was doing for Idaho Power was very original, and it led to the first guidebook on the prevention of raptor electrocutions. It was applied nationally," Thorsell says. "I mean the minute we had it done, we distributed it throughout the industry."

Still, environmental groups wondered how utility companies were going to implement the new standards, and how quickly they were going to act. Thorsell turned to the Avian Powerline Interaction Committee as a way to show environmentalists and government agencies how the utility companies were going to follow through. Participating utilities would have to hire a biologist to work on the issue, and each one would become a member of the Avian Powerline Committee. The panel would meet periodically each year, and share information on raptor-electrocution problems and solutions.

In 1975, the list of participating utilities included Idaho Power, Utah Power and Light, Public Service Company of

Colorado, Pacific Gas and Electric, Pacific Power and Light, Southern California Edison, Public Service Company of New Mexico, and Sierra Pacific Power. All of the participating utilities were from the western United States because that's where golden eagles and large raptors live. As time went on, the list of participating organizations grew to include utilities from throughout the United States. As osprey and bald eagle populations grew, powerlines in the eastern United States would have to be modified as well.

It was not that difficult to convince utility companies to get involved, Thorsell says. "The threat I used was the prospect of Congress mandating industry-wide standards," he says, chuckling. "I told them if you don't participate, you'll have REA standards inflicted on you. If you don't believe that, consider that environmental groups could sue, using the Eagle Protection Act and the Endangered Species Act, and they could force us to rebuild the whole distribution system in the West. I mean, we could have been forced to rewire the West. We did some back-of-the-envelope calculations, and we came up with cost estimates of $2 billion to $3 billion. If nothing else got their attention, *that* got their attention."

Lanham says it was not a difficult sell inside Idaho Power. "There wasn't any argument," he says. "We knew it would benefit us and it would benefit the eagles. It was the right thing to do."

For EEI and the utility industry to take a national leadership position on the eagle-electrocution issue convinced utility executives about the value of good P.R.—being the guys wearing the white hats. "Thanks to Morley's very original work, we came up with a fix and got the industry out of a sludge pit," Thorsell says. "We came out looking like knights in golden shining armor."

In early October 1975, Morley and Pat flew to Vienna, Austria, to present a paper on ways to reduce or eliminate electrocution risks to birds of prey on powerlines at the International Council for Bird Preservation Conference on Birds of Prey. Morley had a set of engineering drawings that helped explain powerline modifications, and he showed a 10-minute film that he made for Idaho Power called *Powerlines and Birds of Prey*.

The conference presented an ideal setting for Morley to spread the word on raptor-powerline problems and solutions. Birds of prey advocates from throughout the world attended the meeting, and many of them found the information about raptor-powerline issues to be extremely informative and helpful. When they saw Morley's golden eagle trying to land on the powerpole in slow-motion, it became obvious how a large raptor's wingspan could easily touch the center ground wire and outside hot wire on three-phase powerlines. In Africa, large raptors such as Cape vultures, Egyptian vultures, Martial eagles and black eagles encountered similar problems in rural areas. Bonnelli's eagle and the Spanish imperial eagle were subject to similar dangers in Spain. Now officials from those countries could take Morley's information home and try to apply it.

Morley and Pat enjoyed the international birds of prey meeting in Vienna because it featured a great deal of pomp and circumstance regarding the historical reverence with which the people of Austria embraced the ancient art of falconry. Conference participants were invited to observe a falconry exhibition, complete with falconers on horseback decked out in Sixteenth century garb, with falcons on their fists and eagles on T-perches, hunting at an old king's castle and estate. Falconers chased pheasants, European hare, Hungarian partridge and roe deer with peregrines and gyrfalcons, using hunting dogs to flush the prey. After the hunt was over, a special ceremony commenced to honor the quarry that had been taken, which was placed in the center of a grassy square. Pine boughs were laid out around the edges of the square.

Men rode up on horseback and played unique notes on a hunting horn to honor a particular prey species. Each type of prey received a special tune.

"That was something else to see those Austrian falconers work with the horses and the dogs just like their ancestors had done for thousands of years," Morley says. "Their peregrines and gyrfalcons were well-trained, and they could take anything in the air."

After Morley returned to Boise, he assisted Idaho Power and Utah Power and Light officials with erecting forty-two nest boxes on tall, high-voltage lines in parts of southern Idaho, northern Utah and western Wyoming. He and assistant Pat Benson helped utilities in the Northwest survey three-phase lines and identify problematic powerpoles.

At the urging of Thorsell and Nelson, utility companies invited the media to cover the story about raptors and powerlines. They took reporters on field trips to areas that needed attention, and provided photo opportunities of powerline engineers installing new equipment on powerpoles, and Nelson installing nest boxes on tall transmission lines. The strategy worked, resulting in dozens of newspaper and magazine stories and TV news reports throughout the region.

"Utilities learned that this is good P.R.," Thorsell says. "They quickly realized that fixing powerlines put them on the right side of the issue where they were the ones wearing the white hats."

Idaho Power ran print advertisements educating the public about the issue as well. One of them said, "Because eagles don't know about electricity, we got help from Morlan Nelson, who knows about eagles."

"Idaho Power is making remarkable progress in its eagle protection plan," Morley says in the ad. "A far greater danger to the birds than electric lines are irresponsible shooters. I wish we could correct them as easily as we did the power poles."

The ad concludes with the message: "Idaho Power Company. A Citizen Wherever It Serves."

Morley's involvement in the raptor-powerline issue, on top of his recent achievements in protecting the Snake River Birds of Prey Natural Area, created more interest by prominent national journalists. Don Moser, a former assistant managing editor of *Life*, interviewed Morley for a lengthy feature in the November 1974 issue of *True*, a national magazine geared toward the American male audience.

A first-class writer, Moser probed many facets of Morley's life at the time, including his crusade on fixing powerlines, falconry, film projects, the Snake River Canyon, raptor-rehabilitation and more. Perhaps more than anything else, Moser captured the frenetic pace with which Morley faces every day, even though he was technically "retired."

"An hour or so in the Nelson home illustrates the point that if you are good enough at something, no matter how specialized or obscure, the world will jam your telephone line," Moser writes. "The phone rings: a film producer is calling to request his services. (Nelson and his sons have a small film company specializing in footage of birds of prey). The phone rings: a power company executive solicits his advice. (Nelson has developed a procedure for modifying high-tension poles, which are attractive but dangerous perches, and mete out capital punishment to eagles at an apocalyptic rate.) The phone rings: a man whose name *is* a household word is calling to see if Nelson might possibly take on a problem child as a kind of falconer's apprentice for the summer, and maybe straighten the kid out. The phone rings: a distraught woman, a member of a local bird club, tells Nelson that a lovely pair of sparrow hawks are nesting in a tree on a residential street, and some tree surgeons are about to fell the tree. Can he help? she asks. Does a cat have whiskers? Nelson heads for his Jeep and speeds to the rescue."

Moser also is entranced by Morley's mews. "Indeed, his household resembles a kind of raptorial M*A*S*H, where at a given moment one may find a waterlogged eagle drying out in the oven or a gunshot falcon recovering in the laundry room. Most of the birds in Nelson's hawk house are recuperating from something or other—mostly acts of human violence. Every year, from all over the country, 40 or 50 birds are sent to him for repair."

Good press continued to pour from national and regional media while Morley continued to work on the raptor-powerline issue in the field, and he gave more talks to service clubs and community organizations. The Smithsonian Institution invited Morley to Washington D.C. to give a talk about the raptor-powerline issue, and he brought a golden eagle with him for the talk.

Former Senator James McClure, R-Idaho, recalls accompanying Morley as he walked down the street in Washington D.C. to the Smithsonian, with a golden eagle on his fist. "Talk about a crowd-stopper," McClure says. "Everyone slowed to a stop to stare as we walked down the street. Eventually, Morley gave a talk about birds of prey, and showed a movie. It was a very successful speech."

In the fall of 1977, Morley and Pat were invited to give their paper on raptors and powerlines to a worldwide conference on "Bird of Prey Management Techniques" in Oxford, England. All of their expenses were covered. Several years after the initial guidebook on raptor-powerline issues had been released, Morley had been impressed by the response by utility companies to embrace solutions and install nest boxes. In media interviews, he seized on opportunities to congratulate the utility industry for being so proactive.

In a United Press International article on the subject that ran nationwide on September 28, 1977, writer Robert Van Buskirk reported that the Bonneville Power Administration, a federal power broker in the Pacific Northwest that oversaw thousands of miles of high-voltage transmission lines in the region, had joined the group of cooperating utilities on the raptor-powerline issue. Morley said the positive response by utility companies marked a new era of cooperation between industry and the environment.

"It has set a precedent of cooperation between conservation and power interests to inform the public at large of the potential for increasing the habitat for birds of prey in general," Morley said. "And it has won international attention."

Beyond all of the hype, utility companies were making strides in installing perches on powerlines and changing pole configurations in the field during the second half of the 1970s. The Avian Power Powerline Interaction Committee did not keep track of exactly how many poles were fixed or how many nest boxes were installed, but committee members say the number was high.

"At Idaho Power, we've modified hundreds of poles, and built a lot of new lines that are built to raptor-safe standards," says Ansell. "We've invested hundreds of thousands of dollars in it."

Industry wide, no statistics have been compiled to understand how many miles of powerline have been addressed to make them raptor-safe, or how much money has been invested. EEI officials say the investment, industry-wide, is in the magnitude of millions of dollars. Most companies install raptor-safe powerlines when they replace old powerlines.

But problems still occur, Ansell says. A powerline between the Thousand Springs area near Twin Falls, Idaho, and Jarbidge, Nevada, killed sixty-nine golden eagles over a twenty-mile area, he said. By gapping ground wires on poles, powerline workers were able to resolve the problem. "The problem was identified and the company responded to it," he says. "That's the way we're dealing with this issue."

In 1981, the Fish and Wildlife Service recorded 175 eagle deaths from electrocution, including 139 in Colorado, Utah, Montana, Kansas and South Dakota. The eagle deaths indicated that the powerline problem continued to pose a challenge to the industry, but eagle advocates were satisfied that utility companies were responding to the problem areas and fixing them, according to an August 8, 1982, article in *Empire* magazine.

"Up to ninety-five percent of the eagle electrocution problem is solved," said Bob Turner of the National Audubon Society. "This is a real success story."

"I feel so good and positive about this," Morley said. "I can show you fifty goldens within fifty miles of Boise—and of Denver."

Dean Miller of the Public Service Co. of Denver put it this way. "We haven't killed an eagle in the past five years on the 25,776 miles of the PSC system. Ten years ago, biologists knew nothing about electricity, and electrical engineers knew nothing about eagles."

In ten years, a remarkable change had occurred, not only in the United States, but throughout the world. "We've got a heck of a story going here," Morley says. "This thing we started with Idaho Power has spread clear around the world."

Anticipating the need to spread the word about raptor-powerline conflicts and solutions in his favorite medium—moving pictures—Morley and his boys at Echo Films produced an educational movie for the Edison Electric Institute called *Silver Wires and Golden Wings* in 1981. The thirty-minute color movie contains a number of key slow-motion shots that Morley took of his golden eagle landing on a three-phase line, dramatically illustrating the potential problem by freezing the frame when the eagle's seven-foot wingspan lined up exactly with the center ground wire and an outside hot wire on a powerpole crossarm.

With Morley's extensive movie footage of the whole experiment, start to finish, viewers can see the lab tests of high-voltage electricity getting cranked through dry and wet feathers. Stock footage shows Morley removing a baby eagle from a powerline nest for research purposes. Morley films his assistant David Boehlke manning the eagle, and training it to fly to a baited T-perch. Solution-oriented footage shows how Idaho Power could correct the problem by changing powerpole configurations or with elevated perches. Morley climbs tall high-voltage towers to install his custom nest boxes.

Steve Eaton, a professional musician from Salt Lake City, wrote an original score for the movie. It was produced for EEI, which in turn, distributed the movie as far and wide as possible—not only to utility companies around the world, but also to public groups that were interested in the issue.

In the film's conclusion, a golden eagle flies into a setting sun, a big golden bulb in the center of the frame, outlining a silhouette of an eagle. "We must have the silver wires to bring us light, warmth and power, but we also need the golden wings soaring across the sun, bringing us the beauty of poetry in flight, and the symbol of freedom in a world that needs reminding of freedom's blessing."

"*Silver Wires and Golden Wings* was a way to show everyone else what we already knew," Boehlke says. "Morley and I would smile at each other. We knew we had to do the film to show everyone what's necessary and make them understand, but we could have made the recommendations just about as easily without the film part. The film demonstrated it. Morley's film-making ability helped him present things real clearly to make a point. It's harder to skew things when you've got it on film."

Morley and his boys received three awards for the film, including the Gold Award from the Houston International Film Festival, an award from the Public Relations Society of America, and a "More than a Million" Film Achievement Award from the Modern Talking Picture Service organization in August 1985.

"The More than a Million award meant that *Silver Wires and Golden Wings* was seen by more than a million people," says Tyler Nelson, producer of the picture. "It was done in 16mm format, so that's a pretty major deal for a film to receive that many private showings at service clubs, utility companies and community groups. I'm telling you, that film was seen all over the world."

When *Silver Wires and Golden Wings* was first released, Thorsell threw a big party in Washington D.C. for Morley and Pat. He invited more than 100 people to his house for a reception and a special viewing of the film. Morley had a golden eagle on hand for the festivities, of course.

Thorsell went deep below the surface when he introduced Morley, who was now not only a business associate, but a good friend. "Nelson may have been the reincarnation of a golden eagle," Thorsell said. "He saw the King Eagle in his kingdom, and the Eagle God said it was OK for him to come back to earth as a human, as long as he had the heart and soul of an eagle. Well, there isn't a person alive in the world who doubts that Morley's got the heart and soul of an eagle."

Chapter Fourteen

Saving the peregrine
1965-1999

On a chilly, breezy spring day in April 1965, Morley drove his red International Scout to the Black's Creek area, about ten miles east of Boise, to fly his peregrine Blacky after work. A gusting wind bowed tall strands of bluebunch wheatgrass against the white, chalk-like soil on the ground, and elicited high-pitched mournful sounds as it whipped through five-foot-high bitterbrush. Up in the sky, angular banks of salt-and-pepper clouds swept across the Snake River Plain, casting shadows on the landscape below.

"Jump up big hawk," Morley says, taking Blacky from her perch on the top of the back seat.

Morley brought Blacky out into an open area and faced her into the wind. Gawd, you're a beautiful bird, he thought to himself, looking into her yellow-ringed jet-black eyes. Blacky stood completely erect, her white salmon blue-barred breast reflecting a bronze glow of western, evening light. Her eyes darted as she surveyed the landscape. She was a big "rock" peregrine that Morley received as a gift from Jim Fox, an East Coast falconer, in 1949. She'd been a trusted companion for more than 15 years.

"You go anytime you're ready, old girl," Morley says softly.

Blacky leaped off of Morley's fist. Her wings beat rapidly as she flew into the wind, darting one way and then another, twirling in a tight circle, gaining altitude rapidly. Like all mature peregrines, Blacky cut such a perfectly sleek form. Her angular slate-blue wings were immaculately trimmed for speed, as was her bullet-like body. Today, Morley was going to let her fly for a while and enjoy the aerial show before he swung the lure to bring her back. She gained more height, and pretty soon, Morley couldn't see more than a tiny silhouette in the western sun.

Then he heard the unmistakable courting call of a male peregrine. I'll be damned, he thought, another peregrine is passing through on its migration back to the north. A tiercel. He listened to them chatter. "I understood everything they were talking about," Morley says. "When he started calling to her in a really sexy way, she said, well, all right, I'll go with you."

The newly bonded peregrine pair flew together overhead, banked into the cool north wind, and headed for the tree-cloaked mountains above the Boise River. Morley didn't even bother to swing the lure. After all these years, she had found a mate. It was time for her to go. Watching them fly to the north, Morley's eyes watered as he felt an emotional tug in his heart, knowing one of his favorite hawks was returning to the wild, where she truly belonged. It felt similar to the feeling of watching his son Norm leave for college the previous fall. The difference was, he'd never see Blacky again.

"I felt proud of her, and glad for her that she was finally going to have babies of her own," Morley says. "But it was a sad moment, too, because she was such a spectacular hunter. I knew that I'd miss her."

Morley wondered where the pair would go to establish a nest. He had not seen any peregrines nesting anywhere in Idaho for several years. Most of the peregrine aeries he had visited over the last five years in the Northwest were vacant, except for an aerie that Morley always checked each summer on the Oregon Coast. There was so much food available there, in terms of shore birds, pigeons and songbirds, that Morley couldn't ever imagine that peregrines would abandon that territory.

However, he had observed prairie falcons taking over a number of peregrine falcon aeries in Utah, and in eastern Idaho, and that made him curious. Could it be that a prolonged dry cycle was reducing the prey base for peregrines, and forcing them to live elsewhere?

A couple days after Blacky took off, Morley received an unexpected call from Professor Joseph J. Hickey at the University of Wisconsin-Madison. Hickey was planning a global conference on peregrine falcons, and he wanted Morley to speak about the status of peregrines in the Pacific Northwest.

Hickey had chilling news about the plight of peregrines in the eastern United States. Recent examinations of 236 historical peregrine aeries east of the Mississippi River did not turn up a single nesting pair.

"Something terrible is going on that's wiping out the peregrine," Hickey says. "We've got to find out what it is, and how we can bring them back."

"I've got a few ideas on that," Morley says.

So did Hickey. Rachel Carson, author of the environmental ground-breaking book *Silent Spring*, warned Americans about the unintended consequences of the widespread use of DDT and other insecticides and pesticides in 1962. Most of the conference attendees, including Hickey, Morley and Cade, had read the book. A former marine biologist for the U.S. Fish and Wildlife Service, Carson sounded the alarm with incisive prose.

"Since the mid-1940s, over 200 basic chemicals have been created for use in killing insects, weeds, rodents, and other organisms described in the modern vernacular as "pests," Carson wrote. "These sprays, dusts, and aerosols are now applied almost universally to farms, gardens, forests, and homes—nonselective chemicals that have the power to kill every insect, the "good" and the "bad," to still the song of birds and the leaping of fish in the streams, to coat the leaves with a deadly film, and to linger on in soil—all this though the intended target may be only a few weeds or insects. Can anyone believe it is possible to lay down a barrage of poisons on the surface of the earth without making it unfit for all life? They should not be called "insecticides," but "biocides."

"The whole process of spraying seems caught up in an

endless spiral," Carson continued. "Since DDT was released for civilian use, a process of escalation has been going on in which ever more toxic materials must be found. This has happened because insects, in a triumphant vindication of Darwin's principle of the survival of the fittest, have evolved super races immune to the particular insecticide used, hence a deadlier one has always to be developed—and then a deadlier one than that. It has happened also because, for reasons to be described later, destructive insects often undergo "flareback," or resurgence, after spraying, in numbers greater than before. Thus the chemical war is never won, and all life is caught in its violent crossfire."

DDT, an acronym for dichloro-diphenyl-trichloro-ethane, was the first chlorinated organic compound invented to control insects. Doctor Paul Muller of Geigy Pharmaceutical in Switzerland received the Nobel Prize in medicine and physiology in 1948 for discovering the ability of DDT to control malaria. The use of DDT exploded worldwide after World War II. In the tropics, it was widely used to spray mosquitoes to prevent the spread of malaria and to kill lice that carried typhus. According to the World Health Organization, an estimated 25 million lives were saved over a fifteen-year period. But as Carson noted, mosquitoes became resistant to DDT over six to seven years, and new chemical agents had to be devised. Eventually, antimalaria drugs were used more predominantly than insecticides to control malaria in the tropics.

In the United States, DDT was used primarily to control insects on farmland and forests. "From small beginnings over farmlands and forests the scope of aerial spraying has widened and its volume has increased so that it has become what a British ecologist recently called "an amazing rain of death" upon the surface of the earth," Carson wrote.

Millions of acres of land were sprayed nationwide to control farm pests, as well as insects that preyed on trees, such as the gypsy moth and tussock moth, and many others. The production of synthetic pesticides escalated from 124,259,000 pounds in 1947 to 637,666,000 in 1960—a five-fold increase.

In Carson's chapter, "And No Birds Sing," she recounts dozens of anecdotal accounts of songbirds and raptors disappearing from communities and well-known territories, the unintended consequence of relentless spray campaigns to control forest and farm pests. Everything from robins to bald eagles were affected.

Charles Broley's outstanding volunteer work banding bald eagles over a twenty-year period showed a chilling drop in eagle reproduction from 1939 until his death in 1958. When he started banding birds, Broley worked a series of 125 bald eagle nests on the west coast of Florida, from Tampa to Fort Myers. In 1947, he started to notice a decline in eagle production. Between 1952-57, 80 percent of the bald eagle pairs failed to produce young. In the last year of his banding work, in 1958, "Mr. Broley ranged over 100 miles of coast before finding and banding one eaglet. Adult eagles, which had been seen at 43 nests in 1957, were so scarce that he observed them at only 10 nests," she wrote.

In *Silent Spring,* Carson raised the specter that the U.S. Department of Agriculture and American chemical companies had unleashed a wrong-headed insecticide campaign that must be stopped.

"As crude a weapon as the cave man's club, the chemical barrage has been hurled against the fabric of life—a fabric on the one hand delicate and destructible, on the other miraculously tough and resilient, and capable of striking back in unexpected ways," she wrote in the concluding chapter, "The Other Road."

"The "control of nature" is a phrase conceived in arrogance, born of the Neanderthal age of biology and philosophy, when it was supposed that nature exists for the convenience of man. The concepts and practices of applied entomology for the most part date from that Stone Age of science. It is our alarming misfortune that so primitive a science has armed itself with the most modern and terrible weapons, and that in turning them against the insects, it has also turned them against the earth."

Prior to the beginning of his conference, Hickey knew about the deleterious impacts of DDT on songbirds and birds of prey. His students at the University of Wisconsin tracked the impact of DDT on robins in the Midwest during a spray campaign intended to combat Dutch elm disease. Hickey also had watched peregrine populations plunge in the eastern United States. In 1942, he had conducted the first survey of peregrine falcon aeries in the eastern United States, when peregrine populations were stable, and few environmental threats existed, aside from careless shooting by slob hunters.

When no peregrines could be found nesting in the wild in the eastern United States in the early 1960s, no one was more concerned than Hickey. "During the years 1950 to 1965, a population crash of nesting peregrine falcons occurred in parts of Europe and North America on a scale that made it one of the most remarkable recent events in environmental biology," he said in his opening remarks.

By bringing together the world's best experts, he hoped to compare data on declining peregrine populations, discuss the reasons for the decline, search for solutions and develop ideas for future scientific rcsearch.

Who's who of the birding world was there, including Audubon bird expert Roger Tory Peterson, Frances Hamerstrom, the grand dame of falconry from Wisconsin, Tom Cade, then a professor at Syracuse University, Walter Spofford of the University of New York Medical Center, Frank Beebe, co-author of *North American Falconry and Hunting Hawks,* of Victoria, B.C., and Morley. Raptor authorities from western Europe also attended.

As discussions began on the status of peregrines in North America, bird of prey authorities from Alaska and British Columbia starting things off on a positive note. One researcher revisited the middle section of the Colville River, which Morley had floated with Cade in 1959, and he found two more pairs (13) than Cade had observed. Other areas of Alaska featured healthy peregrine populations.

In British Columbia, peregrines were thick as fleas. In the Queen Charlotte Islands, west of Prince George, Beebe estimated there were at least 100 pairs of peregrines.

Now it was Morley's turn to give his perspective on the status of peregrines in the Northwest. "The peregrine has always been a rare bird in the Intermountain West," he said. "At the same time, to those who knew the birds intimately, several

Nelson family collection

Morley with some of the bird handlers he trained over the years.

nesting sites could be found within a 100-mile radius of any of the major cities. In 1938, a trained student of falcons could locate the peregrine falcon's aerie without too much trouble whether he was in New Mexico, Utah, Idaho, Oregon or almost any other western state . . ."

In 1938, Morley's first year in his Soil Conservation Service job in Utah, he found three peregrine aeries. And he watched prairie falcons attempt to take over the nest sites. "The aerial battles were spectacular with a fierce and awesome beauty. The prairie falcons seemed to win these battles, much to my surprise," he said.

During the period 1939-42, he observed prairie falcons take over 50 percent of the peregrine aeries in Utah. "The prairie falcons were steadily taking over the nesting sites of these birds, often without a fight of any sort," he said.

Morley wasn't the only person to observe this trend. Allan Brooks of British Columbia watched ten peregrine aeries get overtaken by prairie falcons near Okanagan Lake, B.C., between the early 1900s and 1929.

In 1948, the year Morley moved to Boise for his snow survey

job, he received a list of forty known peregrine aeries from Richard M. Bond of the SCS in Portland. Bond was a falconer, too. Morley always checked the aeries when he got a chance to keep an updated tally on occupied peregrine nesting territories. He noted that three aeries in the Malheur National Wildlife Refuge were gone by 1952, perhaps coinciding with the reduction of the lake's surface area.

As part of his work for the SCS, Morley had been conducting research on the water levels of twelve large, landlocked lakes in Oregon, Utah and Nevada to track long-term precipitation trends. The results were stunning. From 1896 to 1963, all of the lakes had shrunk considerably, and several of them were almost dry. The Great Basin was in the midst of a drought.

The Great Salt Lake, for example, dropped in surface area from 2,400 square miles to 950 square miles, a reduction of 60 percent. Malheur and Harney lakes, the backbone of the Malheur National Wildlife Refuge in southeast Oregon, had shrunk from 125 square miles in the late 1800s to one square mile in 1961, a ninety-nine percent change.

He was certain that the shrinkage of the lakes had dramatically reduced the presence of waterfowl such as ducks, geese and swans, not to mention shore birds peculiar to wetland environments. Any peregrines that depended on those habitats for survival had to go elsewhere to make a living.

Morley confirmed three active aeries on the Oregon Coast in 1958. By 1964, two pairs had disappeared. This was curious because it was not a competition issue with prairie falcons. In the Columbia River gorge, only one or two pairs existed out of the thirteen pairs once observed. About eighty to ninety percent of the nesting birds in Utah, Idaho, Oregon, Washington and western Wyoming had shifted to "some new location" from their former nesting sites, Morley said.

"There is considerable evidence that a drop in nesting populations of shorebirds in the Intermountain West has been similar to that of the peregrine when considered from 1930 or earlier, to the present," he said.

A long-term general increase in temperature (1.6 to 2.4 degrees Fahrenheit) could have increased the mortality of peregrine chicks in the nest due to a reduction in cool temperatures, he postulated. "The direct rays of the sun on birds of prey, just out of the down and into feathers, can kill them in less than one-half hour of exposure between 10 a.m. and 5 p.m. The temperature must be 90 degrees or warmer for this to occur, and in May and June, these temperatures can be common below 6,000 feet," Morley said.

"It was my grim experience in June 1962 to watch three golden eagle aeries lose all of their birds in this way. At the Castle Butte site, my son Norman and I watched the adult eagle stand over the young with her wings spread and her beak open until she could stand it no longer . . . The two young died before we could go down to the aerie. The temperature was between 98 and 102 degrees on that day. We estimated that 30 percent of the young eagles at twenty-eight aeries we observed were killed by the sun during the hot period.

"Below the 6,000-foot elevation in the mountain states, peregrines are doomed to nesting failure in May, June or July if exposed to the direct rays of the sun . . . . We should consider the possibility that the peregrines moved north or up in elevation … at a time when precipitation started to drop and average temperatures climbed steadily, even up into the Arctic areas."

Morley noted that there was no pesticide problem when the decline or shift started to occur. Hickey found Morley's thoughts illuminating

"It seems to me that Nelson has put forth evidence for a population change which we must agree has taken place in some fashion," Hickey said. "I believe Nelson should be commended for giving us some ecological evidence to explain what until now has been an extremely puzzling phenomenon in peregrine population biology."

As the meeting continued, it became clear that what many raptor experts had seen as a local population decline was much bigger than that. "The problem was, everyone was seeing this phenomenon take place locally, but they all thought it was just a local deal," says Cade. "At the same time, there was a lot of correlative evidence put forth implicating DDT as the problem. The proof didn't come until about five years later."

Derek A. Ratcliffe, a scientist with The Nature Conservancy in Huntingdon, England, was credited with being the first person to identify scientific correlations between a decline in peregrine falcons in Great Britain, and the presence of DDE, a metabolite of DDT, and other organochlorine pesticides. Ratcliffe and others had seen a stable population of peregrines in Great Britain drop precipitously. Researchers had found many nests unoccupied with broken egg shells present.

"The pattern of peregrine decline not only matches the intensity of pesticide use geographically, but also in timing, showing close agreement with the development of organochlorine compounds as the principal insecticides of British agriculture," Ratcliffe said in his presentation. "Unhealthy symptoms, in the form of frequent egg-breakage and reduced nesting success, had been widespread south of the Highlands since 1950, but actual population decline seems to date from about 1955, and had become severe by 1961. The first symptoms correlate with the general introduction and widespread use of DDT and BHC; but the population "crash" corresponds with the advent and rise to prominence of the most toxic chemicals, aldrin, dieldrin and heptachlor, and also follows the time pattern of the widespread and spectacular kills of wild birds (many of them prey species of the peregrine) which were caused by the use of these substances in particular."

In Great Britain, where the peregrine decline was well-known earlier than it was in the United States, researchers had been checking peregrine eggs for evidence of pesticides. In one study, eighteen peregrine eggs obtained from thirteen peregrine aeries all contained residues of four major pesticides, DDE, dieldren, heptachlor and BHC. The carcasses of four peregrines found dead in the wild contained the same residues, Ratcliffe said, and fifty prey species suffered from the same contamination.

"This being so, virtually all British peregrines are "at risk" to contamination, and the actual degree of contamination for any individual will be largely a reflection of that for the total prey population with which it has contact," Ratcliffe said.

Despite the research findings thus far in Great Britain, Ratcliffe recommended that more study was needed to determine what levels of pesticide contamination could be lethal or harmful to reproduction.

Summing up the findings of peregrine experts worldwide, Hickey said, "The persistence of DDE in the world's environments and its concentration at the tops of certain ecosystems have led to extracontinental phenomena that bind peregrine falcons, Scottish golden eagles, sparrow-hawks, American ospreys and bald eagles together in a new process of physiological deterioration. Many other species are almost certainly involved—and in regions far removed from the original points of environmental contamination."

Still, much more research was needed to make an iron-clad scientific case about the causes and effects of pesticide contamination in peregrines, bald eagles and other species that ate insect-eating birds and fish, key carriers of pesticide contamination.

In the meantime, Morley shared his ideas with the conference attendees on the need to explore further the possibilities of captive-breeding of peregrines to rebuild populations throughout the world. Naturally, Morley thought back to the time when Blacky laid infertile eggs. Blacky "became so much a part of my family that she began to lay eggs, and eventually it became obvious that the peregrine could be raised in captivity," Morley said. "I knew that it had been done in other countries on rare occasions. This old peregrine laid many eggs and raised two young prairie falcons when her own eggs didn't hatch. However, at the time, it seemed to me such work was 100 years ahead of our needs in falconry. Now, as this conference implies, we may be 30 years behind time for the public at large and falconers."

Morley suggested that more research should be conducted on finding ways for peregrines to be bred in captivity, and in the meantime, healthy populations of peregrines could be used to restock areas where the birds had disappeared. Morley laid out a 10-step process for reintroduction efforts.

Before the birds are reintroduced to old territories, Morley said it was critical that all people who support birds of prey work for improved public acceptance of raptors.

"Laws alone will not protect these birds, as we already know," he says. "We must have the active participation of educators, conservation clubs, sportsmen's clubs, state and federal organizations, falconers, and finally the majority of the public at large. . . . If the public is not convinced of the value of this work, it would be impossible for us to accomplish it.

"The hawks, eagles and falcons have been an inspiration to people of all races and creeds since the dawn of civilization. We cannot afford to lose any species of the birds of prey without an effort commensurate with the inspiration of courage, integrity and nobility that they have given humanity with very little but persecution in return. If we fail on this point, we fail in the basic philosophy of feeling a part of our universe and all that goes with it. Every person needs this now, more than ever in history."

Morley's speech drew a standing ovation and thunderous applause. Every participant in the meeting felt exactly the same way. Now it was time for action. Following Hickey's ground-breaking conference, all of the participants hit the ground running in the following months to act on recommendations for more peregrine surveys, DDT research and more.

"It revved up everybody to start doing things," Cade says. "It got everybody out in the field to at least survey all of the nesting territories they knew of. I went up to Alaska, to the Colville River, and we looked everywhere else in North America. Very quickly, we discovered that there were practically no peregrines left anywhere, particularly in the western United States and southern Canada. The population was still looking good in the Arctic, and in the coniferous forests in Canada and Alaska, but everywhere else, things looked pretty grim."

Followup research on the effects of DDT on peregrine eggs showed that every egg that had been laid after 1947 in North America was contaminated with varying levels of DDE. As a result, many of the eggs were close to twenty percent thinner than normal, causing the eggs to break prematurely and the embryos to die.

Research findings, combined with public outrage incited by *Silent Spring*, led the National Audubon Society and National Wildlife Federation to lobby Congress and the Nixon administration in hopes of banning DDT. Congress did not act, but

William Ruckelshaus, the former director of the Environmental Protection Agency, outlawed the domestic use of DDT in 1972.

Morley, meanwhile, focused on drumming up publicity about the plight of peregrine falcons in hopes of getting the American public behind the cause of banning DDT and restoring raptor populations. Morley worked his contacts in the movie business to find a producer to do a film on endangered peregrines. Repeated inquiries with the Disney Company came through, and Roy Disney decided to pursue the project personally as executive producer.

The end product was *Varda the Peregrine Falcon,* an entertaining story that traces the life of a peregrine named Varda. She's hatched on a cliff in Alaska, and has numerous close calls with predators. She gets trapped by a falconer in Florida, but being a wild bird, she is determined to fly free, and returns to the wild. The story line also touches on the threat posed by DDT on peregrine falcons while Varda migrates south from Alaska to Florida for the winter. "We weren't just into entertainment, we were trying to show the truth," says Roy Disney. "At the time we did the movie, there was a lot of evidence of crushed eggs in the nests. The point was that DDT was having a deleterious effect on the birds."

Morley and Tom Cade had input on the development of the script, and Disney plucked other ideas for the story line from the book, *The Peregrine Falcon*, by Robert Murphy. A different falconer, Dennis Grisco, handled the bird behind the scenes, and a Seminole Indian, Peter De Manio, played the role of the falconer in the film. Morley's son, Norm, worked as a photographer's assistant for Frank Zuniga, director of photography, a key step in launching Norm's film-production career.

Primarily, *Varda* acquainted Americans with an up-close-and-personal portrayal of peregrine falcons, including fine footage of peregrines flying and hunting for ducks. Most people had never seen a peregrine up close before, and the movie helped educate Americans about a bird they would hear much about in the years ahead. TV ratings showed that *Varda* received a forty-one percent share of the American households tuned in on the evening of the broadcast premier. Disney was pleased.

Captive-breeding experiments with peregrine falcons presented unique challenges. By the late 1960s, twenty-three species of raptors had been bred in captivity, including hundreds of American kestrels. Peregrine breeders wouldn't have to reinvent the wheel completely, but no one had ever bred large numbers of peregrines in captivity.

There were two major motives that pushed the experiments, Cade says.

"One was the desire of falconers to ensure a continuing source of peregrines available for training and hunting; the other involved the conservationists' goal to preserve the peregrine as a wild species and if possible to repopulate vacant range by releasing captive-produced falcons. Regardless of how one evaluates the first motive, it is highly unlikely that the conservationists' objective could have been achieved without the intensely personal and essentially selfish desire of falconers to be able to continue possessing and using peregrines in their sport, because this drive, more than anything else, was responsible for the successful breeding of peregrines in captivity."

It should be pointed out that the two influences were not mutually exclusive. Many of the falconers who worked on the front lines of the captive-breeding effort were just as concerned about restoring peregrines to the wild as they were about creating a surplus of birds for falconry. In the background of the whole peregrine recovery movement, certain National Audubon Society officials cast a dim view on falconry because of the historic practice of removing birds from the nest. But Audubon purists didn't realize that it was the falconers' knowledge and skill that would lead to a miraculous recovery for peregrines.

Larry Schramm, a falconer from Portland, Oregon, was the first American to raise a young peregrine to the point where the bird fledged from an artificial nest and flew on its own in 1968, Cade wrote in the book *Peregrine Falcon Populations: Their Management and Recovery,* co-edited by Cade, James Enderson, Carl Thelander and Clayton White (The Peregrine Fund, 1988).

Several years later, Doctor Heinz Meng in New York bred a pair of Peale's peregrines and hatched a single bird using artificial incubation. The bird became famous as "Prince Philip," Cade says. The following year, Meng hatched and raised seven young peregrines from the same pair of adults. He increased production by taking the first clutch of four eggs away from the birds, so they would lay a second clutch. He also used artificial incubation to ensure the eggs did not break. He ended up with seven peregrine chicks.

Meng's success "was a great boost to breeders, because it confirmed the notion that large-scale production of peregrines in captivity could be achieved with a relatively small number of breeding pairs," Cade says.

Meng donated Prince Philip to Cade and three partners who created a non-profit corporation called The Peregrine Fund at Cornell University in 1972. Through the process of artificial insemination, Prince Philip and his mate produced twenty-three young in the next five years, laying the foundation for what would become a large-scale captive-breeding program by The Peregrine Fund.

Cade created The Peregrine Fund as a way to raise tax-deductible funds for the peregrine-recovery effort because traditional university sources of money were far too limited for the task at hand. "I had to start the Fund because Cornell's development office had its own priorities," he says. "If we found a large donor, and I had an inside track with the money, I didn't necessarily get the money. So to make a long story short, I got fed up with that. I just decided, well, I'll incorporate.

"So we just did it. I didn't ask permission from the university. Jim Weaver and I, and Bob Berry and Frank Bond, all falconers, all guys trying to breed birds in captivity, formed the corporation in 1972. And it just grew from there. We were the only directors for several years. We brought in Morley and Kent Carnie pretty early on. Once we got a little publicity, people wanted to give us money. That was one reason I thought we should incorporate, the tax-deductible aspect of giving.

"It was funny how things evolved in the fund-raising arena pretty quickly. Cordelia Scaife May and a Mrs. Smith, who had given us $1,000, stopped by Cornell. We didn't know much about Mrs. Smith, except that she was the wife of the U.S. ambassador to Canada and an avid birder, and we knew even less about Mrs. May.

"They came into the hawk barn, and Jim Grier had an imprinted golden eagle that had been artificially inseminated, and the deal with this eagle was, she wouldn't get off the eggs to eat unless Jim went in there and "sat" on the eggs. (She was an imprinted bird; she thought Jim was her mate). The ladies got to see this whole operation, with Jim and the female eagle, and they were just totally blown away by it. So they went on their way, and it turned out Mrs. May was the head of a family foundation, and she gave us several nice contributions. That's how we kept going."

Cade had a mixture of non-academic falconers and graduate students, nearly all of whom were falconers, working on captive-breeding experiments. Guys like Weaver were handy because he was a falconer—he knew how to handle birds, and he knew how to build things and fix things. "I tried to run it with graduate students, but that didn't work, I needed people who were totally dedicated to the birds," Cade says.

Weaver convinced a lot of falconers to donate birds to The Peregrine Fund for experimentation with breeding. About half of their birds came from the falconry community, and the others came from Alaska, Canada, northern Mexico, and Spain.

Experiments at The Peregrine Fund, combined with the knowledge of other captive-breeding research projects elsewhere in the United States and abroad, helped narrow the techniques. "There's more than one way to skin a cat," Cade says. "You can go with imprinted birds (falcons that think humans are their mate), that's one way to go, but then you're stuck with artificial insemination. It works, but it's very labor intensive. And then we had captive birds from falconers that mated in captivity."

For artificial insemination, the semen from male birds could be retrieved by using massage techniques. But it also could be extracted using an ingenious method invented by falconer Les Boyd, a teacher at Washington State University. Working with an imprinted male peregrine, Boyd donned a hat with a rubberized ring on top, and he played the role of a female peregrine. He courted the tiercel to mate.

In an April 1990 article in *Smithsonian* magazine, author Don Moser describes the technique that Boyd invented as performed by The Peregrine Fund's Cal Sandfort. "To be prepared as a stud—called a hat bird for reasons that will become clear—a young male is reared and hand fed for between thirty and forty days, by which time the bird regards its handler as another falcon or itself as a human, and would regard another peregrine as an alien being," Moser wrote. "When the bird reaches sexual maturity at two to three years, courtship begins. In the hawk barn, Cal enters a chamber where an imprinted male is perched on a feeding ledge. Cal kneels to be on the same level as the falcon, and proffers a piece of quail breast. The falcon accepts it, plucks a few feathers from it and offers it back. Cal holds the meat for a moment, and then returns it once more, reenacting the food exchange that is part of peregrine courtship ritual.

"The falcon then bows its head and utters a series of piercing cheeps. Cal bows and cheeps in return, just as a smitten female would, and for the next few minutes, the two of them provide an amazing spectacle, man and bird bowing and cheeping, affectionate lovers arousing each other. Finally, the hat—a nondescript fedora with a rubber dam around the crown to catch the semen. With the male fully aroused, Cal will don the hat and turn his back (the male peregrine mounts the female from the rear), the male then flies to the hat, and with much cheeping and fluttering of wings, copulates with it."

Peregrine Fund bird-handlers would take the semen and insert it into a female peregrine at the appropriate time. Eventually, about fifty percent of the birds released by The Peregrine Fund were produced via artificial insemination. Continued experimentation showed that some adult peregrines would mate in captivity. Often times, the bird-handlers had to move adult birds around in the Cornell hawk barn until a pair bred successfully.

"We had the best luck with birds that had had some handling, but were not imprinted on humans," Cade says.

Repeated experiments showed that wild-trapped peregrines

would not mate in captivity—even if they were trapped before they mated in the wild. "No one knew it was a mistake at the time, but you could keep wild birds for ten years, and they'd never breed," he says.

The U.S. Fish and Wildlife Service tried to develop its own captive-breeding program for peregrine falcons at the Patuxent Wildlife Research Center in Maryland, at the same time The Peregrine Fund was cranking up. Research specialists at the Patuxent facility trapped their breeding stock on the East Coast as wild peregrines migrated through the area from the Arctic. However, the FWS ran into the same problem The Peregrine Fund did—wild-trapped birds would not mate in captivity. The Patuxent Center never produced a single peregrine in captivity.

"The feds didn't get it done because it's duck hawks this week, and next week, it's whooping cranes," says Kent Carnie, an early board member of The Peregrine Fund and a longtime officer in the North American Falconers' Association. "There wasn't the passion, and the fire for the species like there was among the falconers for the peregrines. That's what kept it from getting stale. The P Fund kept working at it and working at it until they got it done. They did it because they were falconers. One, they had an insight into the bird from their association with the bird in the sport, and two, they had the passion, because the bird really meant something to them. That combination is what made the P Fund succeed."

The Peregrine Fund, aided by more experience and intimate knowledge of birds of prey through falconry, focused on the methods that worked. By 1973, twenty young were produced from three mating pairs. Cade began to devise plans for a coordinated release program in the eastern United States.

In the meantime, James Enderson, a biology professor at Colorado College, had been working on raising peregrines in captivity at his home in Colorado Springs. A falconer, Enderson had attended the Madison meeting and was enthusiastic about assisting the peregrine recovery effort. In 1973, he produced three young Rocky Mountain *anatum* peregrines, the first of that subspecies to be produced in captivity.

In July 1974, Bill Burnham, a Colorado native, falconer and a graduate student at Brigham Young University, met Jim

Weaver while working on a peregrine banding and population survey project in Greenland, along with project leader Bill Mattox and Scott Ward. Burnham had expressed interest in The Peregrine Fund's captive-breeding program, and the Fund was looking for someone to establish a breeding center in the West. Before they left Greenland, Weaver asked Burnham, "Oh, by the way, we're looking for someone to run the project. Are you interested?"

Burnham accepted, and in December 1974, he moved into a facility at the Colorado Division of Wildlife field station in Fort Collins, Colo., on Christmas Eve, with his wife, Pat. Enderson offered the use of his captive pairs of peregrines to begin the production effort in Fort Collins.

Mattox, deputy director of the Ohio Department of Natural Resources who had studied gyrfalcons and peregrines in Greenland for more than twenty years, says Burnham was well-suited for the job. "Bill was a very serious, devoted, hard-working guy. He was a very focused individual, and he was very enthusiastic about birds of prey."

In the spring of 1975, Cade, Weaver and the gang at Cornell had raised enough peregrines in captivity that they were ready to begin releasing birds in the wild. They used a method called "hacking," an age-old falconry practice of using a man-made nest structure of some kind to provide a safe haven for young peregrines before they're ready to fledge the nest. The Peregrine Fund hired hack site assistants to keep watch over the birds and feed them in a way that they did not associate food with humans. Vertical bars on the front of the hack boxes prevented predators from raiding the nest and killing the valuable peregrines. Great-horned owls were the biggest concern.

That year, The Peregrine Fund released 16 young peregrines at four hack sites, including one located on property owned by the Massachusetts Audubon Society. Twelve of the birds returned to the same locations the following year. "That was really encouraging—I mean who's going to stop after that?" Cade says.

The Peregrine Fund used skyscrapers in a number of eastern cities for release sites as well. The buildings provided shelter for the birds, and there was an ample supply of food in the form of pigeons and other small birds. A key benefit was that great-horned owls didn't live in cities and wouldn't pose a threat to the newly released birds. Cities present their own set of hazards, however, such as heating vents, electrical wires and mirror-glass. But over time, urban areas proved to be a valuable place for hacking peregrines. And some birds that were released in wild settings ended up taking up residence in cities.

By 1989, for example, ten pairs of peregrines had established residence on skyscrapers or bridges in New York City alone. Overall, more than 150 pairs were established in the eastern United States after ten years of captive-breeding work by The Peregrine Fund at Cornell.

In the West, Bill and Pat Burnham cranked up production in Fort Collins by following the same techniques developed through trial and error at Cornell. They used Enderson's breeding pairs, a pair of birds that Burnham had been working with, and three pairs of birds sent from Cornell. Once again, freshly laid eggs were removed to cause the adults to produce a second or third clutch. Eggs were meticulously monitored under incubators. Temperature and humidity were tightly controlled. The eggs were weighed every third day to ensure that they slowly lost weight as the chick grew inside—five grams over thirty-one days. Eggs could be moved to more humid conditions or rubbed with crayons to reduce weight loss, or they could be rubbed with sandpaper to increase it. All of the monitoring required a great deal of attention and time.

In the first couple years of production at the Fort Collins facility, five peregrines were released in 1975 and 1976. In 1977, higher production began to occur, allowing the release of twenty-three birds that year, and forty the following year.

Bill Heinrich, a falconer and a Colorado native, was the second person in Fort Collins who was hired full-time to work for The Peregrine Fund in 1976. Heinrich remembers that Bill Burnham was extremely busy.

"Bill was doing everything single-handedly," Heinrich says. "He didn't have a lot of time to learn. They just gave him the incubators, and he got the whole operation off the ground. He did an excellent job."

Heinrich helped build the hawk barns, while he raised quail

as a meat source for peregrines. Then he was placed in charge of the release program in the West. By the late 1970s, the Colorado facility released about fifty birds a year, primarily in Colorado. In the early 1980s, production doubled, and they released more than 100 birds a year. Because of hazards in the wild environment, only thirty-five percent of the birds survived to breeding age, about three or four years old. It would take a long-term campaign to restore peregrine populations fully.

In the late 1970s, the Santa Cruz Predatory Bird Research Group joined the peregrine recovery effort as a partner with The Peregrine Fund. Brian Walton, coordinator of the group, was a recent college graduate at the time. He and a faculty member at the University of California-Santa Cruz offered to begin a peregrine recovery program in California.

"We had a wild idea that maybe we could start a program on the West Coast," says Walton, a falconer. "It was clear that Tom Cade couldn't cover the whole continent, and we had a lot of support from the university and a lot of interest in California. So we decided, yeah, let's try it, and it took off from there."

The Santa Cruz Group obtained adult peregrines from falconers in California and The Peregrine Fund and began a breeding program. The first releases occurred in 1977.

Ten years into the peregrine-recovery effort, The Peregrine Fund was producing outstanding results—surpassing most people's highest expectations. Cade and founding board member Robert Berry, an insurance executive, were steadily making gains in fund-raising. As time went on, Cade wanted to add new board members to The Peregrine Fund who had wealth and influence. It was vital to the organization's future. Morley had been in continual contact with Cade and others through the 1970s, and he was officially added to the board of directors in 1981 due to his stature as a falconry legend and his influence in the film business. Morley didn't have any money to add to the cause, but he would influence others to contribute.

To assist in fund-raising and public education, The Peregrine Fund needed a promotional film about the peregrine falcon captive-breeding and reintroduction effort. Naturally, the board turned to Morley and his boys at Echo Films to produce a thirty-minute film called *Peregrine*.

Against the backdrop of the Rocky Mountains in central Colorado, Tyler Nelson shoots adult peregrines flying and soaring in classic acrobatic maneuvers against a clear blue sky. "In the judgment of many naturalists, the peregrine is the most accomplished and superbly developed flier on earth, combining to a marvelous degree the highest powers of speed and aerial agility. Stooping at speeds of over 200 mph, she is the winged embodiment of Diana, the goddess of the hunt, the hallmark of nature's perfection," the narrator says.

Tyler: *We had to pack all of our film gear, our camping gear, tent, sleeping bags, a box of quail and enough food for three days —3,600 vertical feet to the top of that cliff to get that shot. Just our camera gear alone was more stuff than anyone would want to carry, and then we had to carry our camping stuff on top of that, and the box of quail to attract the birds for close-up action. And we did it all in one load. We damn near starved to death eating that freeze-dried camping food for three days, and when Bill Burnham and Morley came up to get us, we said, we're so hungry we're going to eat you. But no, it was a really cool spot and I was really happy with the footage we got.*

"And then came another bird," the narrator says, "a man-made bird spreading poison on the land."

Morley shoots a helicopter flying down the throat of a forest canyon, with a 100-foot spray bar attached, showering the trees with DDT.

"The reason DDT is so deadly is because many of the small birds that peregrines eat have eaten contaminated insects," says Jim Enderson, interviewed in a lab at Colorado College. "The cumulative effect of this is that peregrines lay thin-shelled eggs, which don't hold up well on rock ledges under an incubating mother falcon."

The Nelson brothers shoot a close-up shot of Enderson squeezing the edges of a half-broken peregrine egg shell, showing its fragility.

Next, the film segues to a scene at Cornell University's Laboratory of Ornithology, where Tom Cade gives a tour of a hawk barn. He peers into a room full of peregrines used in the captive-breeding effort through one-way glass. He lifts the cover of an egg incubator, revealing neat rows of brown peregrine eggs lined up on an aluminum rack.

"The first challenge we had to figure out was how to raise young in captivity," Cade says. "And then the second challenge was to find out if it was practical to release captive-raised birds in the wild."

In a smooth transition, the scene dissolves to The Peregrine Fund's Western Project in Fort Collins, showing director Bill Burnham feeding young peregrine chicks on a picnic table outside with a peregrine puppet on his hand. Steve Sherrod, a Peregrine Fund employee in Fort Collins, introduces a male peregrine named B.C., who has provided semen via Les Boyd's hat trick for the artificial insemination of many females. "This bird has fathered more falcons that any other falcon in the world," Sherrod says.

By now, the narrator says, The Peregrine Fund has hatched more than 1,000 peregrines in captivity. Clearly, the Fund's core of expert bird-handlers have the process down to an exact science. Cameras pan to a room inside the Fort Collins hawk barn, where Bill Heinrich monitors a bank of eleven TV screens to watch how adult peregrines are behaving in captive chambers. "We do this to determine which peregrines will mate together and which ones won't," he says.

Next, the film travels to the Foxton peregrine aerie where Tyler shot the opening footage. It's an impressive sheer vertical cliff, with blond rocky spires and columns perched above a dense forest. This was one of only six active peregrine nesting territories in Colorado. To be sure that the young were cared for and hatched from the thin-shelled eggs, without the threat of breaking, Burnham and Heinrich had taken the eggs from the wild adults, and substituted plastic eggs in their place. Now they were returning the young birds to their parents. Before the birds were placed in the nest, Heinrich and Burnham placed bands on the birds' legs, while the adult peregrines dive-bomb them, protecting their nest and territory.

Showing the diversity of The Peregrine Fund's hacking locations, the film moves next to Baltimore, where Scarlett, a female peregrine, lived on the thirty-third floor of the U.S.

Fidelity and Guaranty Insurance Company building. She needed a mate, so Cade and his crew at Cornell provided a male peregrine named Rhett. They mated successfully and raised three young. The narrator points out that skyscrapers and bridges have become a favorite nesting location for peregrines in Washington D.C., Philadelphia and New York City.

In upstate New York, Cade removes a tiny radio transmitter from the leg of a peregrine and replaces it with an aluminum band. The Peregrine Fund placed radio transmitters on some of the released birds to track their movements and, at times, to help hack site attendants protect the birds from predators.

Transition to the Snake River Canyon in Idaho, where a "cross-fostering" experiment was tried. Three peregrine chicks were placed under adult prairie falcons to see if the prairies would accept the young birds (they did), and to see if peregrines could reestablish nesting territories in the Snake River Birds of Prey Natural Area, which remains to be seen.

Tyler and Norman Nelson shoot footage of young peregrines performing aerial acrobatics above the Snake River, and the music builds to a crescendo for the final word. It's Morley, dressed in a red flannel shirt and jeans, standing on the edge of the canyon, with a peregrine on his fist.

"The peregrine falcon and all of their races on every continent represent humanity's highest ideals, the protection of their young, courage and nobility in life," he says. "They've inspired every cultural society in the world with their awesome ability in the air. The future holds problems, but concern in almost every nation shines brightly for restoring them to their historic place in our environment. The great cliffs they have occupied for centuries may again be graced by their bold beauty."

Peregrine was shown hundreds of times to enhance public education about the peregrine recovery effort, and it was used frequently by Cade and Burnham to help promote The Peregrine Fund's activities and raise money. The film won a 1983 Blue Ribbon Award at the American Film Festival.

After she starred in *The Eagle and the Hawk,* Nell Newman came back to Idaho numerous times in the 1970s to visit Morley in the summer and to fly birds. "I used to come out for a couple weeks every summer, and I brought one of the birds that I was flying, or I worked with one of Morley's birds," Newman says. "I had a Swainson's hawk, and a Harris's hawk at the time. It was a great time to come out here and just see the birds. It was always better than seeing any nature movie just to watch Morley fly his birds."

"I took her out to see the wild aeries, all over here, and she saw all of the red-tails and golden eagles and prairie falcons when she went down to the Snake River Canyon with me," Morley says.

During the summer of 1976, when Nell was eighteen, she learned about an opportunity to monitor the cross-fostering experiment in the Snake River Canyon. She spent eight weeks in the canyon between April and June, watching adult prairie falcons care for three young peregrines and monitored the young birds after they fledged the nest.

"It was really impressive to watch their flying abilities," Newman says. "You see how they learn to fly, and their ability improves every day. I remember the peregrines were trying to catch cliff swallows, and just to watch them was really amazing. But that was a tough site in the Snake River Canyon. It was in the spring, but it was hot."

During that summer, Paul Newman came to Boise to visit Nell. While he was in town, Morley asked Newman if he would narrate a public service announcement to deliver a conservation message about birds of prey. Newman agreed. The U.S. Fish and Wildlife Service financed the production of the piece, and oversaw nationwide distribution.

Dressed in a powder-blue denim shirt, Newman narrated the 60-second PSA in Morley's backyard. The piece opens with striking footage of a golden eagle soaring in the Snake River Canyon. "These are the sky hunters, top of the line in the bird world—the eagle, the hawk and the owl. Not too many years ago, many people thought these birds were the enemy, killers of livestock and game. Now we know their hunting benefits nature and man, by helping to control destructive increases in populations of rodents and other small animals."

Light symphonic music makes a bold transition to the

musical score of *Born Free*. A prairie falcon stoops in the wind. A great-horned owl peers directly into the camera with its blinking bright yellow and black eyes.

"Despite persecution, pollution and loss of habitat, America's birds of prey still cling fiercely to a way of life that symbolizes wild freedom. Our skies would seem empty without them. All the more reason to remember that all eagles, hawks and owls in this country are protected by law."

Newman looks stern and sincere for the final line. "I think you can see why."

Norm Nelson: *Dad was really nervous before we shot that scene with Paul Newman. He made sure we had an extra camera, extra tape recorder, the whole nine yards. It's like you can't screw this up. You don't get a chance to do something like this with Paul Newman every day. And we're like, yeah, yeah, OK, Dad. We'll get the shot, don't worry. But the best part about it was Newman and Morley had a great time together. They were like laughing and joking the whole time, and so Newman was really relaxed for the narration scene.*

Several years later, Nell Newman served as a hack site attendant in central Massachusetts, while taking a break from college in Bar Harbor, Maine. "It was called White Rock. It was in an old-growth forest, and it was really beautiful," she says. "It was a three-mile hike up there, and from the top of the cliffs, you could see across a huge valley."

Like all hack site attendants, Newman and her partner had to watch the birds continually with binoculars, feed them every day with fresh meat, and watch for predators. Hack boxes were rectangular, made out of plywood. When the young birds were ready to fledge the "nest," Newman removed the protective screen in front of the box and let the birds paddle their wings in the wind. Eventually, they'd take off and everyone hoped for the best.

Newman was one of more than 1,000 people hired nationwide by The Peregrine Fund to serve as hack site attendants. Early on, Heinrich and his friend, author and falconer Dan O'Brien released the birds personally in a seven-state region in the West. As every hack site attendant knows, the first day or so after the protective cover of a hack box is removed, it can be awfully stressful to watch the young birds try to cope for the first time in the wild, knowing a predator could be nearby.

In his book *The Rites of Autumn: A Falconer's Journey Across the American West*, O'Brien writes, "Birds of prey are fragile. They have evolved so that only the very best survive. (Perhaps as few as 10 percent ever make it to breeding age.) The very best are not orphaned; they do not break wings. The very best are perfect and don't need anything that a human can give them except a decent environment. The system that has evolved in nature for the raising and selection of birds of prey is complicated and, in human terms, cruel and severe . . . .

"As humans begin to do certain things, like walk and talk, at certain ages, so do falcons. They begin to eat on their own at about thirty days. They fly at about forty. The peregrines in the hack box beside me were forty-three days old. Eight days before, these falcons had been flown by private airplane to an airstrip close to the hack site. I had picked them up and carried them in a wicker backpack two miles to the top of the cliff. Then I climbed down a rope to place them in the hack box. For eight days, these falcons would live in the hack box, acclimating and readying themselves for freedom. These birds had been raised in a lab. This was their first experience of life in the wild."

O'Brien writes about working a hack site in the Northern Rockies in Montana. The primary concern in that area, in terms of predators, was not great-horned owls, but golden eagles. If a golden eagle spies a freshly released falcon on top of a cliff, the young bird won't have a chance. O'Brien carried a shotgun with firecracker-like exploding shells as a way to spook eagles away from the site, if needed.

"Of course I knew there was very little that I could do to protect the young falcons," O'Brien wrote. "Many die and that is unavoidable. But you never get used to it. The term "survival of the fittest" makes perfect sense when read from a college textbook or discussed over coffee, but the reality of it can be demoralizing. Nothing makes you feel more insignificant than to witness a golden eagle swoop down to pluck a clumsy young peregrine falcon from the air. It is a grisly thing to see and only underscores the fact that nature is extraordinarily unforgiving."

On the first night after removing the protective screen from the hack box, O'Brien heard the peregrines issue a distress call, *kak kak kak*. He saw a dark object streak across the sky. "I knew it was a golden eagle, stooping at perhaps 100 mph toward the young falcons," he wrote. "Shouting, I pointed the shotgun at the eagle and pulled the trigger. It would be wonderful to say that the eagle heard me, that the projectile exploded in his path and diverted it from its prey. But in reality, my voice was lost in the mountain breeze and the projectile exploded 200 feet behind the eagle. I heard the kakking become frantic and then fade to nothing."

O'Brien walked back to camp, where his hack site helpers had watched through binoculars as the eagle snatched a young peregrine named Yellow, as it prepared to make its first flight from the top of the cliff. The eagle grabbed the bird with its talons in one clean sweep and flew down canyon. Now three other young peregrines were at risk in the nest, because the eagle would come back to look for another easy meal.

The wild environment could be a cruel world, as O'Brien eloquently describes. That's why The Peregrine Fund produced hundreds of peregrines each year for release, and worked all the angles to reduce predation at release sites. But there was only so much they could do to reduce the risk. Eventually, the birds had to learn to survive on their own.

In the early 1980s, The Peregrine Fund was forced to move its Rocky Mountain captive-breeding center in Fort Collins to a new location. Anheuser-Busch Co. wanted to build a new brewery in close proximity to the breeding center, and the new plant required a new access ramp on Interstate 25. The ramp "will intrude so close to our location on Frontage Road that it will result in unacceptable levels of traffic and produce other disturbances to our breeding falcons as well as encourage secondary construction of businesses and houses," Burnham wrote in the Fall 1983 Peregrine Fund newsletter. "As soon as Anheuser-Busch learned about our predicament, representatives of the company showed an immediate concern for our program."

In fact, Anheuser-Busch donated $450,000 to the Fund to cover the costs of dismantling the hawk barns and building a new headquarters in a different location. While Colorado political officials including Governor Richard Lamm and Senator Gary Hart urged The Peregrine Fund to keep its operations in the Rocky Mountain state, Burnham and Cade explored alternative locations in Georgia, Texas, California, Oregon, Wyoming, Montana and Idaho.

When Morley learned that Cade and Burnham were shopping for new locations, he started to lay the groundwork for bringing the Fund's operations to Boise. He visited with Democratic Governor John Evans, Mayor Dirk Kempthorne, and Idaho's congressional delegation. He spoke to the state director of the Bureau of Land Management, Clair Whitlock, and local BLM officials in the Boise District.

"I wanted them to come to Boise because the Snake River Birds of Prey Area was already here. It seemed like a natural thing to do," Morley says. "I had known Tom Cade from the time when he was going to college, and I knew he was one helluva good man. As a board member of The Peregrine Fund, I knew Bill Burnham, too, and he was a real go-getter. I knew that if The Peregrine Fund moved here, we could get more peregrines reintroduced to the wild."

Morley was well connected in the Boise business community, having been a longtime member of the Boise Downtown Rotary Club, a group that included many well-to-do business leaders in the community. Morley had solicited contributions to The Peregrine Fund from Boise Cascade Corporation and the J.R. Simplot Company, two major business interests in the city of Boise.

When Burnham and Cade traveled to Boise, they were impressed.

"Morley came through for us in a big way," Cade says. "He was directly responsible for us moving here. He had the contacts with the BLM to get the land that we eventually got from the city. He could walk right into the governor's office and see the governor without notice. He arranged for us to see anyone that we needed to see, and it was quite easy. Everyone was on board with birds of prey, long before we got there, so it was all kind of made to order that way."

Whitlock, who was a real supporter of birds of prey, present-

Peregrine Fund
Bill Burnham accepts a check from Mike Barker and Janet Cool of the Exxon Company in support of peregrine falcon restoration. Morley was there as a board member of The Peregrine Fund."

ed Cade and Burnham with a number of options for a new location, including the Boise Foothills, a site near the Snake River Birds of Prey Area, and a potential spot in Flying Hawk Reserve. The BLM had deeded 500 acres of the reserve to the city as a wildland park in south Boise. But the city was willing to trade the property back to the BLM for land elsewhere, if The Peregrine Fund wanted it. Cade and Burnham liked Flying Hawk Reserve the best.

"Clair Whitlock was really helpful," Cade says. "We got the land for next to nothing—it was like ten percent of appraised value. And we had the money from Anheuser-Busch for the initial construction of the facility."

By the fall of 1983, The Peregrine Fund made an official decision to move to Boise and build a new headquarters on 280 acres in Flying Hawk Reserve. The new facility would be called the World Center for Birds of Prey. Recently, the Fund's board of directors had made a decision to expand its captive-breeding operations to work on the conservation of birds of prey on a global scale, beginning with orange-breasted falcons in Guatemala, aplomado falcons in Mexico and the southwestern United States, Philippine eagles in the Philippines, and Mauritius kestrels on the island of Mauritius.

"It just seemed like a natural progression for us," Cade says. "We had the captive-breeding of peregrines down to a science, and we knew that the same techniques could be applied to other birds of prey. It was time to make a move in that direction."

Soon after The Peregrine Fund established its world headquarters in Boise, a number of beneficial events occurred. Burnham, now the Fund's president, met a number of prominent members of the Boise business community who were interested in serving on the board of directors. Although Boise was a city of about 120,000 people at the time, it had an unusually high concentration of wealthy individuals and corporate headquarters for national and global firms, including Albertson's grocery stores, Boise Cascade, Simplot, Ore-Ida Foods, Morrison Knudsen, and Micron Technology.

Before long, Idaho rancher Tom Nicholson, businessman Ron Yanke, businessman Joe Terteling, Ore-Ida Chairman Gerald Herrick, Velma Morrison and Nell Newman were added to the Peregrine Fund's board of directors. Cade says the business leaders helped raise more funds for the private non-profit corporation.

"Once you get someone like Dan Brimm and Joe Terteling, they know other people with wealth," Cade says. "We've been very fortunate, and we've been very lucky—our board always had a great mix of people on it. As time has evolved, we've had one of the strongest boards in the conservation community, maybe the strongest."

Later, people like Roy Disney, Julie Wrigley, Robert Comstock, Yvon Chouinard, Lee Bass, Hank Paulson and other high-profile people with influence and wealth served on the board. Morley continues to serve on the board, playing an effective role as a media-savvy conservation champion.

Over the years, fund-raising efforts grew substantially to more than $5 million in annual income. Burnham kept administrative costs to a minimum (four percent), allowing most of the funds to be spent on species restoration and conservation programs.

In a feature article about The Peregrine Fund's success in the conservation and fund-raising arena in *The Wall Street Journal,* Morley was the central character profiled. "Mr. Nelson is seeking help to rescue the peregrine from extinction, and the falcon on his fist makes a striking sales prop," writes staff reporter Ken Wells in the October 12, 1983, front-page article. "Even more persuasive, though, is Mr. Nelson's unusually cooperative approach to industry. Unlike some conservationists, he generally avoids criticizing industry for gobbling up wildlife habitat or polluting the environment.

"My point is that if we use a little intelligence and consideration, we can always have these inspirational forms of life like falcons and eagles around, and we can still have industry providing human beings with energy and food and fiber," Morley says. "It doesn't have to be one way or the other."

"That approach has paid off," Wells continued. "With it, Mr. Nelson has lured thousands in cash and other assistance from Idaho timber and mining concerns for his favorite charity, the Peregrine Fund."

A third benefit of The Peregrine Fund's move to Boise was accomplished several years later, when John Kaiser, president of Boise State University, created a graduate program in raptor biology. A number of raptor advocates and scientists urged Tom Cade to move to Boise and serve as the chairman of the new program. The offer came at a perfect time, Cade says, since Cornell's involvement in The Peregrine Fund's operations ended after all of the captive-breeding operations were combined to one location at the World Center for Birds of Prey in Boise.

"All of the birds were cleared out of here by 1985, and I figured I was going to leave, too," Cade says. "There wasn't all that much left there for me to do that was interesting anyway."

A number of Ph.D. level biologists in the Boise area urged BSU to make the move, knowing a natural synergy could exist between The Peregrine Fund's ever-expanding need for graduate students to conduct research around the globe, ongoing research needs at the Snake River Birds of Prey National

Conservation Area and at Boise State. Cade took early retirement and moved to Boise, with his hawks and his wife, Renetta.

On June 29, 1984, the U.S. Fish and Wildlife Service arrested thirty-two falconers in fourteen states in a major crackdown against people allegedly involved in the illegal commercial trade of birds of prey, including endangered peregrine falcons and gyrfalcons. The arrests followed a three-year undercover "sting" investigation, known as "Operation Falcon." Beyond the arrests announced on June 29, FWS officials said they would seek more than 100 federal felony indictments against more than 50 falconers, including prominent members of the North American Falconers' Association (NAFA).

As a longtime NAFA member who had permits to possess an unusually diverse selection of birds of prey for purposes of rehabilitation, education and conservation, Morley was on the FWS's hit list. So was Tom Cade and The Peregrine Fund, and many others. Neither Morley nor Cade were ever arrested or indicted, but the whole experience of being investigated made both of them anxious and upset.

"Now that's a big story," Morley says. "They made a statement to me, boy, I really got mad about that. They said that Tom Cade and I were guilty of taking peregrines from the wild. They had bad data. They didn't know what they were talking about."

A FWS press release elaborated on the allegations. FWS agents contended that young peregrines, gyrfalcons and raptor eggs were being captured in the nest and smuggled across U.S. borders to sell to falconers in Europe and the Middle East. Between 1981 and 1984, up to 400 birds had been allegedly taken from the wild to be peddled in the illegal wildlife trade. Highly prized birds, such as a white female gyrfalcon, were sold for $10,000 to falconers in the United States and for up to $50,000 in Europe and the Middle East, the FWS said. Endangered peregrines were sold for up to $2,000, goshawks $1,500, prairie falcons $800 and Harris hawks $600.

Under the Migratory Bird Treaty Act, the sale or trade of birds of prey had been illegal since 1972.

In the book *The Pilgrim and The Cowboy* (McGraw-Hill,

1989), author Paul McKay described the crackdown. "Armed with 65 search warrants, and a continent-wide arsenal that included pistols, shotguns and Uzi submarine guns, FBI, Customs, Justice Department marshals and wildlife agents burst into hotel rooms, private homes, and 17 falcon breeding facilities in Arizona, California, Colorado, Idaho, Illinois, Minnesota, Montana, Nevada, New Mexico, Texas and Utah. In Canada, raids and arrests were carried out in Ontario, British Columbia and the Yukon.

"When the smoke cleared, the agents had seized mountains of documents, pictures and financial records, as well as 106 birds of prey, an airplane, several automobiles and some counterfeit money. Arrest warrants were issued for people as far away as West Germany, Finland and Great Britain."

Morley began practicing falconry in the 1930s, long before any state or federal permits were required to remove young birds from a nest. He took prairie falcons, a gyrfalcon and golden eagles from nests in western North America for the purpose of training the birds to produce films and to use in the sport of falconry. Morley's personal conservation ethic ensured that he never abused the privilege. He never took more than one bird from a nest, and he never sold a bird in the black market. When he finished making films for which he had trapped and trained a dozen birds, such as *Ida the Offbeat Eagle* or *Rusty and the Falcon,* he released the surplus of birds back to the wild or sent them to zoos, wildlife centers or universities for conservation and education purposes.

When the federal government instituted a permit system for possessing captive birds of prey in 1972, Morley applied for and received all of the necessary documents from the Fish and Wildlife Service. He had a golden eagle exhibition permit for using the birds in educational presentations and movies. He had a rehabilitation permit. He had a special purpose permit for trapping golden eagles for powerline research. In 1977, when the FWS began a permit system for falconry, Morley obtained a falconry permit as well.

The Peregrine Fund had all of the appropriate permits required for possessing endangered peregrines at its facilities at Cornell and Fort Collins. The FWS had been contributing financially to the Fund's peregrine recovery efforts for nearly ten years, but Cade and others were very concerned about getting raided by FWS agents during "Operation Falcon."

"It caused a lot of grief," Cade says. "They'd seize your papers and your permits, they'd seize all of your paperwork and records, and they'd ask you ten to twelve questions. We knew people who had been visited by the undercover agents, and we knew what the questions were. So we took all of our paperwork and stuffed it into somebody's attic. As far as I know, it might still be there.

"We were worried about telephones being tapped and all of that stuff, it was pretty exciting for a while. I think they investigated everybody in the United States who had something to do with birds of prey. Our names were listed at airports with U.S. Customs officials, and they were instructed to "check these guys out." "

Neither The Peregrine Fund nor Morley got raided. Morley did get caught in the cross-fire involving a bird owned by novice falconer Dennis Jeppson who sent an injured female prairie falcon to him for rehabilitation. Later, Morley learned that Jeppson apparently did not have an appropriate permit for possessing the prairie falcon in the first place, and the FWS ordered Morley to hold onto the bird pending a trial. In Morley's correspondence file, he had reams of paperwork about the case, including a subpeona requiring him to testify. But in the whole scheme of things, the alleged illegal taking of a prairie falcon—a common species in the West—was considered minor by FWS authorities compared to the illegal trafficking of gyrfalcons and peregrines.

Ultimately, as McKay points out in *The Pilgrim and The Cowboy,* Operation Falcon turned out to be a giant government boondoggle. When the government's evidence was revealed in the public eye, it showed that not a single endangered peregrine or prized arctic gyrfalcon had been sold to undercover agents by falconers in Canada or the United States. The FWS's own undercover agent, Jeff McPartlin, known as "The Cowboy," had been the *only* supplier of illegal birds by taking them illegally in the wild himself, or by purchasing them from trappers. The giant conspiracy alleged by the government did not exist.

McPartlin did trap a number of unwitting people by offering birds for sale, including several customers in the Middle East. McPartlin's biggest customer, Glen Luckman, known as "The Pilgrim," faced a twelve-count indictment carrying a maximum penalty of 130 years in prison and $220,000 in fines. But federal prosecutors offered him a deal in exchange for his testimony against other defendants. Luckman was a naïve twenty-four-year-old guy who thought he could get rich quick. The plea bargain reduced his sentence to no jail time and a $23,000 fine.

"Incredibly, North America's most notorious falcon smuggler was being fined $23,000, less than McPartlin had paid him for delivering nineteen smuggled falcons and goshawks," McKay wrote. "He had—he later smirked under oath—made a profit off his fines."

All told, the $2 million, three-year undercover investigation resulted in five felony convictions and sixty-one misdemeanor convictions—not even close to the 438 violations alleged on the day of the big bust.

Both Morley and Cade were upset about the whole thing.

"Jesus, they really were out of line," Morley says. "They tried terrible things. I had to threaten them with every god damn thing to get it stopped, including getting a film done on it."

"I don't know what the psychology of it all was, whether they had a lot of money that they didn't know what to do with or whatever, but the whole thing seemed to be designed to kill falconry," Cade says. "But it actually had the reverse effect. The falconers were so united and so outraged by it, they actually ended up in a stronger position socially and politically than they were before the investigation. So they totally failed if that was the government's objective."

On a clear summer day in the southern outskirts of Boise, August 25, 1999, Interior Secretary Bruce Babbitt and Peregrine Fund officers and supporters gathered for the beginning of a weekend-long celebration at the World Center for Birds of Prey in Flying Hawk Reserve. An osprey wind sock soared in the warm breeze next to the Velma Morrison Interpretive Center, a 7,200-square-foot educational facility

Photo courtesy Tom Cade

Tom Cade and friend are interviewed by a NBC-TV crew on top of the Barbizon Plaza in New York City. Cade told viewers high-rise buildings are becoming favorite nesting places for peregrine falcons.

that more than 35,000 people visit each year. Inside the center, people can view a variety of endangered birds of prey from throughout the world, including the Bateleur eagle, California condor and alpomado falcon. A world map shows the impressive reach of The Peregrine Fund's conservation programs across the globe. Special glass-covered displays pay tribute to Morley's and Tom Cade's conservation achievements and their lives.

Babbitt walked up to the microphone: "The peregrine falcon is today officially delisted from the Endangered Species Act," he said. "It's a wonderful story. It's an Idaho story, and it's an American story."

By the summer of 1997, The Peregrine Fund had released more than 4,000 captive-raised peregrines in twenty-eight states. The Santa Cruz Predatory Research Group released 777 peregrines along the West Coast, including 702 in California. Dr. Patrick Redig of The Raptor Center headed up a collaborative program that led to the release of 1,000 birds in the Midwest.

At the time that the American peregrine, *Falco peregrinus anatum,* was removed from the endangered species list, 1,246 animals and plants in the United States were on the list. Only five other species had been restored to the point where they were removed from the list since the act was first passed in 1967.

"This is truly a time for us to celebrate and to reflect on what we have done," said Cade. "It is a human achievement of extraordinary dimension."

"The successful recovery of the Peregrine Falcon is the result of an incredible effort put forth by individuals, biologists, falconers, business people, etc., who have a passion for this species and nature," added Bill Burnham, president of The Peregrine Fund. "They devoted themselves to this effort and it would be short-sighted to credit any particular entity above the individuals who were truly in the trenches making this day a reality. We also must not forget that success would not have been possible without ending the use of DDT, the primary cause of the species' decline."

Local TV stations put the microphone on Morley and asked his opinion.

"I always thought it was possible, but I never thought it would happen so quick," he said.

Bill Heinrich, a supervisor of peregrine release efforts in the West, said, "The most satisfying moment for me was when the birds came back and raised young. And it was like wow, this works! And it's been working ever since."

Cal Sandfort, the wizard behind the scenes in the Fund's captive-breeding chambers, said, "It's a once in a lifetime event. It's basically what I've spent half of my life trying to achieve."

In a celebration at the Morrison Center, Burnham, dressed in a black-tie tuxedo, described what the peregrine recovery means for all people.

"We pass on to our children and generations to come, the wild restless beauty of a peregrine's flight, its defiant cry when defending its young and the embodiment of freedom which it represents. Within the peregrine's dark eyes, we see mankind's reflection. And in this instance, we can feel pride."

Chapter Fifteen

Tribute
2001

Over a span of roughly sixty years, Morley Nelson inspired many people through his participation in movies, conservation activities and his mastery of falconry. Here, a number of friends, associates and falconers share insights and stories about Morley—in their own words—that may provide a better understanding of how he has touched their lives and inspired them to carry on with conservation work of their own.

Roy Disney – Vice Chairman of the Walt Disney Co.

Morley worked on seven films for the Walt Disney Company. He worked with Roy Disney on "Perri" the pine squirrel and later, the two served together on the board of The Peregrine Fund.

I got involved with The Peregrine Fund in the late 1970s and 1980s, and Morley, god bless him, was one of the reasons I got involved. Morley is sort of a force of nature. His interest in birds of prey was really, really important because of his knowledge and his ability as a salesman. He brought a lot of legitimacy to the cause.

When you're around him, you feel like a bit of a disciple. There are a few people in this world who revealed something that was important to know. People like Rachel Carson. When people heard her message, they never forgot it. Jacques Cousteau. He showed you the world beneath the sea and he made you fall in love with it. This is what Morley did with birds of prey.

Morley was so authentic. At the base level, he was a falconer. He knew the birds, he trained them, he knew exactly what kind of moves they were going to make next, and he was always right. One day I took Disney Chairman Michael Eisner up to Morley's house, because I was talking about wildlife conservation and the importance of people paying attention to these things. Eisner brought his 9-year-old boy with him. And we watched Morley fly his white gyrfalcon Thor.

Morley cast him off into the sky, and he started to climb and circle above us. Then Morley let four or five pigeons go. Without a moment's hesitation, Thor saw them and began to climb up above them. The pigeons saw the falcon and they're all like, oh shit, and they dispersed. Thor folded up her wings and came streaking down out of the sky, striking one of the pigeons 20 feet in front of us at eye level. The pigeon exploded. It was an awesome sight. Right away, Michael said, "there aren't any city kids that have ever seen anything like this!" He still talks about it.

I've always felt that Morley could have been a comic book figure. They would have called him something like Morley Man. He's one of kind.

Nell Newman, director of marketing and product development for Newman's Own Organics, Inc.

Nell has stayed in touch with Morley on a consistent basis since they worked together on "The Eagle and The Hawk" in 1971 when she was thirteen.

I really appreciate the education that Morley gave me about birds of prey at a very early age. We were both very perceptive about nature and other animals. Morley also helped give my parents an understanding about, you know, that it was something more than a childish interest. Morley taught them a lot about the history of falconry, which a lot of people don't know. I was talking to a friend recently about flying my bird, and how people used to get their food with falcons before they had guns. She had no idea the history behind falconry.

Falconry gave me respect for the animal world. It helped me understand where the food comes from. Morley and I were always looking for fresh road kill to feed his birds, and I loved it. I'd sit in the passenger seat with the binoculars, really early in the morning, and it was a really good time.

After I finished school at the College of the Atlantic in Bar Harbor, I eventually moved back to California in 1983, and went to go to work for the Predatory Bird Research Group at UC Santa Cruz. I became the executive director of the Ventana Wilderness Sanctuary, which was doing a falcon and eagle restoration program, in Big Sur. Great fun. I learned a lot.

Someone brought in a baby kestrel that was four weeks old, and I adopted it. So I had a kestrel for six years. His name was pipsqueak. His flying weight was 85 grams. He was wonderful.

Later, I did fund-raising for the Predatory Bird Research Group for two years. I was very frustrated with the university system, which made it very difficult to raise money. The foundations were quite happy to fund you when you were putting birds in the wild, but when you asked for money to do monitoring, they weren't interested. I found that really frustrating because peregrines almost went extinct due to the lack of monitoring. It's very crucial.

I decided I wanted to support endangered and threatened species by growing organic agriculture, and giving profits to charity. I've been doing that for nine years. Our business, Newman's Own Organics, has been really booming.

Morley is probably my motivation for doing this. To get the electric utility industry to change the configuration of powerline poles to protect birds of prey, to me, that seemed so remarkable, that you could actually talk the corporations into doing that. And that was the motivation for me to persevere and get into organic agriculture.

As much I can't stand the public eye, I get to be a spokesperson for organic agriculture, and I get a lot of visibility, just like Morley gets, for the same types of reasons. But it serves a purpose – for the betterment of all species.

I find I was really lucky in a way, the things I learned from

Morley, just to persevere, I mean. It's interesting to look back and see the evolution of your life and your career. It's been very important, and that's the direction it sent me, you know, through the bigger picture of what the effects of mankind are on the environment.

Andy Ogden, regional wildlife biologist, Idaho Department of Fish and Game.

Andy learned about training hawks from Morley when he was a teenager in the 1960s. His association with Morley led to a job as a research assistant in the Snake River Canyon, college graduate research work in the canyon and a career as a wildlife biologist.

It was pretty cool doing census work right after high school. I wouldn't have had the opportunity to go to graduate school if it hadn't been for him. Certainly, if Morley hadn't been there, I wouldn't be here working for Fish and Game today, if it hadn't been for that weird turn of events in those early days. When you look back on your life, and think about things that were really important to the way you ended up, and there was no doubt that Mr. Nelson was at the top of my list.

From a pure falconry standpoint, we were reading Frank Beebe's book and the Craigheads' books, and we were trying to figure out how to train hawks, and then to have Morley just pick up a bird, rub his fingers up and down its breast, and talk to it a little bit and be able to tell you what that bird was thinking, we were just in awe.

I had an adult female goshawk that had flown into a picture window, and someone brought it to Morley, and he gave it to me to train for falconry. This was a real prize. I manned her for a few weeks, and we had her coming to me on a creance for twenty or thirty feet, and she was getting close to being ready to fly. Then one day Morley says, let's go chase some pigeons. And I was really nervous about my bird. I wasn't sure if it was ready. We didn't have any radio telemetry equipment in those days, so when you turned your hawk loose you hoped that it was loyal enough to come back and not fly away. I had already lost a bunch of birds.

But Morley didn't lose very many birds, and he knew the time was right. I cast off the goshawk and it flew above the hill about a quarter mile away, and then Morley told me to call it back. He released the pigeons, and when they saw the goshawk, they scattered. And I thought, oh no, my bird's going to fly away, and he said, no, she's going fly after this pigeon over here, and sure enough, she did. Morley knew so much about the birds that he knew what they were going to do before they had even thought of it.

Ted Koch, fish and wildlife biologist for the U.S. Fish and Wildlife Service in Boise.

Ted served as Morley's trainer for a large female golden eagle named Slim from 1990-1994.

In May 1990, I met Morley at a workshop on powerlines and raptors. Morley was the featured speaker. He showed us the powerline stuff, played a short film, and talked about giving papers on it all over the world. Of course, he's very charismatic, and I have a penchant for wanting to be mentored by wise old men. So I made a point of approaching him after the session, and I said I'd like to work with him. He said, well, come on up tomorrow afternoon. We just hit it off.

So the next day he stuck a red-tail on my fist. I truly thought he was kidding. I thought oh my god. After thirty seconds, my arm was killing me, my arm was so tense. The next day, I went up there and he stuck a red-tail on my fist again and he said we got to get you to fly eagles. By September, I got the red-tail to fly on the creance. Morley was really pushing me along, and I was trying to take it easy and pussy-footing around, and then they were flying Pearl (a bald eagle), and Lucy wasn't there yet, and Morley said, you pick up Pearl and fly her today.

Whoa. The first bird that I flew free was a fricking bald eagle! Jiminy Christmas.

I was married, but for four years, every other day, except for one or two, I was at Morley's house. I just loved it and I couldn't get enough.

One day I was up there for one of those nightly sessions, and I asked him, Morley, what made you so sure that I'd be able to handle your birds. You invested so much time and energy in me. How did you make that decision? And he said, well, I'll tell you, when I was in the war, and I was out in the mountains fighting the Germans with bayonets, I had to pick one man to lead ten men around this side of the mountain, and one man to lead 10 men around that side of the mountain, and if I chose the wrong man, I was sending ten people to their death. I never expected that answer. Oh my god. You talk about showmanship and whatever, but there is beef behind that guy. He's got the payola to back it up.

Another thing of note was back in 1994. Morley became aware of the opportunity for corporations to receive wildlife stewardship awards from the Fish and Wildlife Service. So he called me up one day, and he said, we've got to do this for Idaho Power for 20 years of work on powerlines. I wrote a proposal, Morley helped me with the words, and submitted it, and the regional director came out to give an award to Idaho Power. It's an example of how Morley paid them back. He didn't have to do that. He went out of his way to track me down and get that done for Idaho Power.

It is his gregariousness, and his ability to interact with credibility with anyone from a rural rancher to governors and secretaries of Interior, that's exactly the skills I cultivate, and those are the tools that I think you need to be successful. I was oriented that way before I met Morley, but having Morley as a mentor, really helped cement the need to conduct business that way.

Charles Schwartz, a researcher in the Engineering and Manufacturing Technology Department for the J. R. Simplot Company.

Charles sought out Morley in the early 1970s to learn a few things about hooding and manning birds, skills which he applied as a hawk trainer in the Middle East. He later worked with Morley and Chuck Henny on a DDT study in Oregon and Northern Idaho.

When I was going to graduate school at Idaho State University, I trapped a passage prairie falcon, and the bird was wild as anything. I was a typical young falconer—I thought I knew everything but I didn't know anything. I was having a hard time hooding this bird, so I called up Morley to see if he'd help me out. He said sure, of course. Morley treats everybody withan openness that's very uncommon.

I went to Morley's, and in his driveway, in the twilight of an evening in February, he took that hawk—he'd never touched it before—and proceeded to show me how to hood it. He critiqued my technique a little, and told me to practice hooding the bird forty to fifty times that night. And I've never had a hawk that I couldn't hood after that. I could have read a dozen books and not gotten what I learned in ten minutes from Morley.

Later on, I worked in Bahrain in the Middle East for five years. I had to take a very wild captive-bred peregrine to a desert prince in Saudi Arabia. I'd been there for six weeks, and the only Arabic words I knew were "please" and "thank you." I went out to the desert and had to be able to handle that bird like I knew what I was doing. The Arabs don't handle anything but wild birds. They trap their birds as passage hawks before the hunting season and train them. They have full-time falconers who are hired to man the birds, and they put in sixteen hours a day. I had to unhood and hood that nasty falcon in front of twenty Arabs, and I did it perfectly. I silently thanked Morley that day.

When Morley and Chuck Henny needed some assistants to help with a study of the effect of DDT spraying in the national forests on birds of prey, they hired Tom Smith, Craig Campbell and I to help out. Tom Smith had a VW "The Thing" and every time he drove it into Morley's driveway, Morley said it sounded like a German jeep, and it lit him up every time. It takes him right back to the war, and he talks about those god damn Germans.

As part of the study, we found a kestrel nest in a rotten snag near Potlatch, Idaho. We set up on the side of the hill, and I went up the side of the snag with climbing spikes on. We took blood samples of the bird and banded it in the nest. Then all of a sudden, a truck pulls up and this great big monster of a guy with a great big black beard gets out of the truck with a six-gun on his belt.

Morley's eyes got real big and he walked down the hill and started to jabber with this guy, tells him what we were doing, and in less than a minute, he's got the guy eating out of his hand. And Morley says later, you want to get real close to those big guys so you can grab the gun. He's a full major of a man, that's for sure.

In a movie, Morley uses the words beauty, grace and courage to describe falcons. I've often felt that those words describe exactly what Morley is.

Pat Benson, research officer for the Department of Animal, Plant and Environmental Science, University of the Witwatersrand, Johannesburg, South Africa

Pat worked with Morley for more than fifteen years, training birds and working with Morley on film and conservation projects. For the last twenty-one years, he has been studying and conserving cape vultures in Africa.

My first memory of Morley was this crazy guy flying a falcon to a light lure in his back yard. I thought this was amazing. I was somewhere between 4 and 6 years old at the time. He let me fly a kestrel and a red-tailed hawk when I was nine. In the summers, I took care of the birds, cats and dogs when the Nelsons went on summer vacation, usually to the Oregon coast. While they were away I cleaned up the hawk house. The Nelsons were away for eleven days. Morley gave me a dollar a day. I was happy with that. Betty Ann was not. While I was walking up the path on my way home, I remember Betty yelling at Morley, saying the hawk house was cleaner than it had ever been and that he had only given me $11.00. I remember laughing as I walked up the hill.

In 1969 my dad, Morley, Tyler and I all worked on the conservation staff at the National Boy Scout Jamboree in Farragut State Park in Sandpoint, Idaho. Later that summer I went to Alaska with the Foundation for Glacial Research. This was part of a National Exploration Award, which I had received from the Boy Scouts of America and The Explorers Club. Some of the basis of this award was work I had done with Morley. In 1969 I also received the National Youth Conservationist of the Year Award given by the National Wildlife Federation.

In 1970, I worked outside of Cascade, Idaho, with Don Meier

Productions for Mutual of Omaha's Wild Kingdom *as a wild animal wrangler/trainer. This is a job that came through Morley. And 1971 was the year we did* The Eagle and the Hawk *with Nell Newman, Joanne Woodward, et al. Tyler and I flew eagles for that work. It was a great year and the results were great, too. I ended up with a John Denver album cover under my belt. About the same time I took a golden eagle picture for the cover of the first Idaho coffee table book. Morley and I were photographed by George Silk for* Life *magazine.*

I got a lot of flack from guys my own age about flying the birds. It really didn't faze me, though. Years later, one of the guys that had given me a hard time told me that he was very envious of what I had done with birds. I think the whole point of this is that working with raptors is not so much what we do, but more who and what we are. It's a passion. If I weren't a raptor biologist, I would be doing something else with raptors. You don't find many raptor biologists or falconers that haven't been divorced at least once. Something Charles Schwartz once said to me probably epitomizes the whole thing. "You can have one hawk and one wife or you can have two hawks and no wife." We are lucky to have that passion; most people never have that experience. It isn't always easy, particularly when it comes to relationships and family, but it is impossible to change, as it is too much a part of the individual.

Morley has accomplished a tremendous amount. We all have been touched by Morley's passion. I remember at a Raptor Research Foundation meeting in Sacramento, one fellow made some comment about Morley which was not particularly nice. He was immediately surrounded by several guys who would have gladly torn him apart. Morley is loyal, and he instills loyalty.

Morley always had vision and is a good example of the idea of "think globally and act locally." Because of his hydrology/soils background, he understood the basics, what it is all dependent on. He saw the "big picture." So many conservationists become very focused on only their species and don't realize how it fits into the big picture, how it is related to the rest of the world. Morley always had that vision. Though Morley has never been religious, he always was very spiritual and recognized that all things are linked. He has a reverence for all things, not just the "living." I think that is a good lesson, and he was a good teacher. He taught me to believe in certain principles, and the courage to stand up for those things that you believe in. If I didn't stay the course in the conservation field, do what I know needs to be done, I wouldn't be able to look myself in the mirror in the morning. I would be shirking my "duty."

Dave Boehlke, owner of Boehlke Guitar Shop in Boise

Dave trained golden eagles, bald eagles and falcons for Morley for thirteen years. Through Morley's tutelage, he became a master eagle handler and trainer.

I was Morley's first bald eagle trainer. I trained both Pearl I and Pearl II. Berkut was the first of a series of ten-eleven or twelve eagles. It was almost a new golden eagle every year because of the research we were doing. After I worked with the birds for a while, all the myths about the ferocious birds of prey kind of melted away. You see this incredible endearing gentle behavior when they're not threatened. Anybody who sees that will go wow, that totally changes your feelings about predators, because humanity usually sees them defending their young or defending their food or territory or whatever, and you see a pretty vicious being.

Working with the birds every day, you'd see the subtlety of their behavior. You kind of get on the same clock with them. I could go up and see the behavior of the birds, and tell someone, we've got some serious weather coming in, not tomorrow or the next day, but in about three or four days, we've got a big change coming on. They say, what do you mean, well, it's the way the birds are acting. That's what was fascinating to me about, not only manning the birds, but training the birds—to know from one day to the next how much food you've given them and anticipate if a bird was going to be keen enough to do a certain job or not fly away. I mean that's what falconry is all about. A lot of falconers weigh the birds every day. Morley would weigh the birds once in a while, but he already knew that, just from behavior. He didn't have to get down and put them on a scale. He was very intuitive about those kinds of things.

I'd done a lot of different things in terms of learning things.

I've done photography, metal-smithing, electronics in the Air Force, building guitars. Normally, I'd get my hands on every book, every resource I could, to learn those things. But I've never read a book on falconry. It just wasn't necessary. Being there every day, doing it every day, listening to Morley, watching Morley. I was at the top of the chain. That's something I treasure, to have had the opportunity to learn from the grand master.

There's a total honesty about the birds' behavior. With cats, there's the pretense of a cat's affection. A dog defends its territory. But we personify animals, apply human traits to their personalities, the anthropomorphic thing. Morley is animal-pomorphic, where it goes the other way. He's learned from the animals and relates to them better than anyone I've ever seen.

The very best story was with Berkut, the very first year. We were doing a film job, it was something for the BBC, to get a sequence of some kind. Morley said, one of these days, you're going to find yourself in a position where you're not going to have quite enough time to get your T-perch reloaded with a tidbit. You'd try to get several performances of what you're trying to shoot in one afternoon, so you wouldn't give the bird too much to eat. Morley sensed this with Berkut, and he said you're not going to have enough time, you're going to be out in the open, with no where to hide, not a rock or anything, and what you have to do is stand as tall as you can, at the very last second, they'll land at your feet or come at you and try to stab you.

During the fall, when they get really keen, they have a sense of urgency, and they can get aggressive with their feet. This was October. So sure enough, we'd done two flights. She took off the third time, and instead of doing the anticipated thing, she just turned right back around again and came back at me. Like Morley said, I didn't have time to reload the T-perch with meat. There was a rock about this big (size of a large pancake), and the BBC crew claimed I got completely under it. But I had the bird come right at me, and I stood there as long as I could, and then at the last second I dropped, and as I dropped, her talons came down, and I remember I had a brand new shirt on, and it ripped two panels right off my back. For years and years, Morley always said, "I still owe you that shirt," when he said goodbye to me.

Monte Tish, materials and quality assurance technician, Idaho Department of Transportation

Monte has been Slim's personal eagle trainer since 1994. He and his mother, Gen, operate a raptor rehabilitation clinic in Boise.

I remember the phone call distinctly, I was sitting there at home, and Morley was on the phone. He knew I was into rehab, and he asked me if I wanted to learn to fly eagles. That was it, you know, it was like a rocket going off. I was like, of course! Of course!

So I went up there and worked with Ted (Koch) and Lucy and Morley, and I learned so much. I started flying Slim after they made the movie The Vertical Environment. *I've flown Slim for an IMAX picture, the John Denver special, for Disney, and a lot of smaller projects. It's just amazing the things you learn. There is such a great big bond of trust between that bird and me, you know. It took me six months or a year, for her to trust me, how to move, but now it's a great big bond. The first time I was up there flying her, we sent her across the canyon and had her come back. You see an eagle come back at you, and you think about how magnificent and majestic they are, and you see their looking straight at you, they just pierce your soul.*

Morley has taught me a lot. To enjoy what nature has to offer. That's number one. I've hunted deer, elk, pheasants, all that stuff, but being around Slim, and seeing how magnificent and majestic she is, somehow she can touch your soul. You can feel it. This is something I can do to give back to nature. It's a big lesson. You don't need to be always taking away from nature. I appreciate the outdoors a lot more now. I'm trying to preserve nature, that's what Morley and the birds taught me. To enjoy life and relax.

Bill Burnham, president, The Peregrine Fund

Bill made the following statement in a *National Geographic Explorer* segment about Morley.

Morley Nelson is a unique individual to say the least. Trying to describe Morley is kind of like trying to describe the mountains over here that we look at when we got out to fly our birds. You can count and list his achievements, but you can't really list what Morley is or what he isn't, because he's like the mountain.

U.S. Senator Larry Craig, Republican

As part of working with Morley on legislation to permanently protect the Snake River Birds of Prey National Conservation Area, Craig wrote a poem for Morley, and gave it to him as a gift in a matted picture frame.

If you were a gemstone,
I'd polish you
Lift you to the light
And set you high
So all could see you shine
As they walked by.

If you were a brand-new foal,
I'd rub you with clean straw
And brush your coat all bright
And leave you in the sun
So people could delight

If you were a sapling
Green and slender
With new spring leaves
All moist and tender,
I'd shelter you from wind and anger
Until your spring roots found anchor.

If you were an eagle grounded,
Broke wing with flight unsounded
I'd mend you, tend you, help you whole
So nature's heights you could patrol.

My Idaho – I love you, fend you, set you free
From those who bind – including me.
So all can praise you gloriously.

Lucy Nickerson, a Middleton, Idaho, homemaker, falconer and mother of two children.

Lucy has been the sole bird-handler for Pearl, Morley's bald eagle, since 1991. But she had to pay more dues to get the opportunity to fly eagles for Morley than any other falconer. The result has been a positive growth experience for both Lucy and Morley.

I don't remember when I got married but I remember when I met Morley. In 1984, I was a biology major at BSU, and one semester they offered a class on Idaho birds of prey. I had been interested in birds since I was in grade school. I had always thought it would be really cool to have a bird on my fist. I was really timid and shy, but I really wanted to meet Morley after he talked to our class. But there were so many people talking to him after class that I didn't have a chance to say anything. A week later, I felt like this was something I really had to do. I called him that weekend, and it was the same time they were opening up The Peregrine Fund in Boise, and I told him I really wanted to work with him and help him with the birds. He said I don't know, it's really hard, it takes a lot of time, and you've got to pound duck heads and chicken heads and it gets real messy. He said he'd think about it, and told me to call the next weekend. This went on for a couple months. It was really unusual for me to stick with something when people put up barriers. Usually, I'd just give up. But I kept calling. It got to the point where Pat recognized my voice, and she'd say, hi Lucy, I'll get Morley for you.

Finally after a couple months, he told me that a couple people were coming up to see the birds, and if I wanted to, I could come up there and talk to him then. He had quite a few birds at the time. I was just in awe, it was an incredible scene seeing all of those birds that close. I told him I thought it was all really wonderful, and I wanted to help out. He said I could come up and help the other kids who were taking care of the birds, but I had to be there every day. So I did that. I worked there for two years before I had my own bird.

I almost had to beg him to fly a bird. It was a Swainson's hawk that I flew for years. After I trained her, he took her all over the United States to give speeches and stuff. But as far as having an eagle, he said that was a man's job. With Pearl, they didn't want just anyone handling her, they wanted someone who was tall. Morley thought well, you can't do that, you're a woman. I didn't like that, it made me really mad. But he finally gave in. I started out with Slim, a female golden eagle. She tried to grab me all the time. I'd pick her up, and one foot was on my shoulder. She'd grab me in the face. I had blood pouring from my forehead when I was five months pregnant. But after she grabbed me, I wasn't as scared. You build up an idea of how bad it's going to be when you get grabbed, and it isn't as bad as you thought it was going to be.

Even though I knew Pearl, and I had been around her a lot, we had to get used to each other. I was used to dealing with golden eagles. But she didn't try to get me with her feet. She would bite and get my nose and my lips and punch me in the eye. I finally figured out she was taking advantage of me. I started to realize that she was treating me just like another bald eagle. So I started treating her like a bald eagle. Anytime she would get her hackles up, I would do the same thing. Anybody who watched this would have thought that I was crazy. So I started hissing at her and biting at her. And making the same posture, and after a while, she started backing down. Anytime she lowers her head, I lower my head. I learned the same intuition that Morley has with his birds. Finally, I proved myself to Morley.

Morley helped me grow in a lot of ways. When I started out with him, I was extremely shy. I had stage fright. We went back to that same BSU class to give a talk, and the class had close to 100 people. I carried the Swainson's hawk. He didn't know that I had stage fright. I was standing off to the side holding the bird, everybody was concentrating on Morley, and that was fine. Then Morley says, Lucy is going to come up here now and tell you about this bird. And I thought, oh my god! I was too scared to look up at anybody, so I just stood there and talked to the bird. He kept driving me around to these things, and I'd think he's not going to have me talk at this one, and then he'd always ask me to come up and talk about the birds. So I got to the point where I could talk in front of people, and make eye contact with people and do just fine.

Tyler Nelson, vice president, Echo Films
On Morley's wisdom of the wild

We've already talked about the singular, unrelenting passion that Morley has for the birds. The Wisdom of the Wild comes from his overall knowledge of the environment. He's trained in soils, he's trained in vegetation, snow and water. He grew up with horses and bulls and chickens and hogs. And all of that comes together in the wisdom of the wild. He knows more than one element of it, and he knows how all of the elements are associated and work together.

He always says that the universal sign of friendship is a slow movement. That is the wisdom of the wild. He can relate to the bird. He can talk to the bird. That is something that I don't get. He can go over to a house where a Doberman wants to eat me, and it lays down by dad. It's some kind of instinctual thing or a smell or something where they say, this guy is mellow. I'm going to hang out with him. It's the general compassion and understanding he has for the animals. It goes beyond comprehension. It's instinct.

Chapter Sixteen

Epilogue
2001

On a hot, sun-drenched May afternoon, a tour group of 55 people stares at a 400-foot blond and brown cliff face on the south side of the Snake River Birds of Prey National Conservation Area.

"Look for the white blotches on the rocks in the seam between the brown rock and the blond rock, there's three downy white eagle chicks in the aerie," Morley says, speaking into a microphone. "The white spots are from bird droppings. This nest has been occupied off and on for at least twenty years."

"I see it now," says an elderly woman, with a pair of binoculars glued to her face. "Oh my, aren't they just darling?"

Everyone begins to find the eagle aerie, evoking "ooooh's and ah's" from excited tour group guests, all sharing the joy of seeing young golden eagles for the first time. The group is seated in two large blue pontoon boats, tethered together, in the calm waters of Swan Falls Reservoir. Steve Guinn, owner of Birds of Prey Tours, provides boat and bus transportation for day-long tours of the 482,640-acre area during the prime nesting season, April through June. Guests pay $85 apiece to spend the day with Morley, the ultimate tour guide for the area.

The tour starts at 9 a.m. in Kuna, Idaho, about twenty miles west of Boise, and proceeds to the north rim of the canyon, where guests listen to Morley's monologue about the unique soils, vegetation and wildlife habitat in the area, all of which combine to create a perfect home for 600 nesting pairs of birds of prey in the spring. They see Swainson's hawks nesting in trees. They watch short-eared owls standing on wooden perches amid tall grass and sagebrush. They visit red-tailed hawk aeries and golden eagle aeries, and while northern harriers and ferruginous hawks fly overhead.

At lunchtime, the group gathers in a shady park next to Swan Falls Dam to relax. Then they load into the pontoon boats for an afternoon tour of golden eagle and prairie falcon aeries in the steep vertical walls of the Snake River.

"See this cliff over here?" Morley says. "This is where Nell Newman and I rappelled into an eagle aerie for the film *The Eagle and the Hawk*. Over here, see that canyon? That's Sinker Creek, where we filmed *Ida the Offbeat Eagle*. See that little cabin out there on the flat? That was Uncle Billy's cabin in the movie."

The boat drifts downstream, with the motors silent, and Morley continues to chat about the many film projects he worked on in the canyon. "See those black rocks on the top of the cliff over there? That's where I took Jim Fowler and Marlin Perkins over a cliff for a Mutual of Omaha's *Wild Kingdom* program. We tied an eight-pound weight to a golden eagle on top of the cliff, and a national audience could see just as plain as day that the eagle couldn't get off the ground. It was just obvious as hell. The bird flapped its wings like crazy and couldn't go anywhere. That took care of the myth that eagles carried off babies and lambs to the nest."

During the day-long tour, guests not only learn a great deal about why so many birds of prey can make a living in the Snake River Canyon, but also about Morley's many conservation crusades to protect birds of prey. He talks about the days when people shot eagles at will. The group visits a tall high-voltage powerline just below the north rim where a pair of eagles took up residence in a special nesting structure that Morley designed himself. They also hear about Morley's life—how he got interested in falconry as a young boy, and his near-death experiences on the front lines of battle in Italy in World War II.

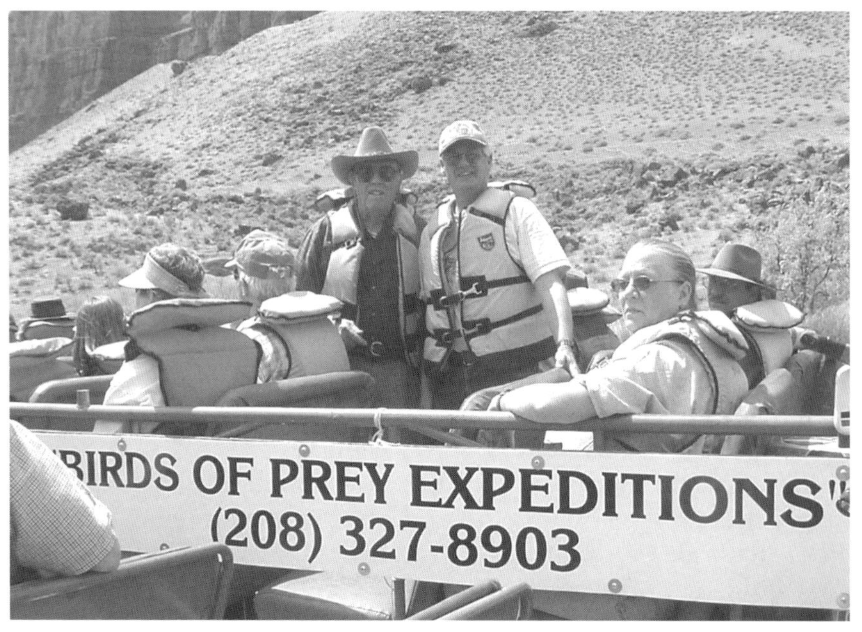

Steve Guinn

Morley, left, and Steve Guinn, teamed up to organize summer float trips through the birds of prey area of the Snake River Canyon.

Most importantly, Morley says, by seeing the birds and the canyon up close, "they can feel a part of the environment."

For Morley to meet Steve Guinn was a real stroke of luck. In the early 1980s, Guinn was a whitewater rafting outfitter on the Payette River, thirty miles north of Boise, and on the Salmon River near Riggins, Idaho. Guinn obtained a permit for a forty-five-mile section of the Snake River that passes through the birds of prey area on a whim, thinking it might expand his business at a time when the other rivers were still snow-bound.

"I'd never heard of Morley's name. I didn't have a clue," Guinn says. "I didn't know anything about what had happened out in the Birds of Prey Area or anything. Hell, I didn't know the difference between a pigeon and a hawk."

But someone from the Republican National Committee called Guinn and asked him if he could take a group of 120 people down the river through the birds of prey area. He agreed to lead the trip. Then Andrus found out about the trip, and asked

Guinn if he could take a group of Democrats down the river, too, on the same trip. He said sure. When the Republicans found out that Andrus was bringing a group of Democrats, they canceled. Andrus assured Guinn that he could come through with 120 guests.

Actually, he showed up with 164 guests for a trip on the Snake River from Grand View to Swan Falls Dam. Guinn had his guides lead the group down the Snake, while he drove a pickup along the dusty roads lining the river, setting up porta-potties.

"We served lunch on Black Butte, and Morley was talking non-stop, and my guide asks me, "Did you hear that old guy talking up there? That was pretty cool."

Guinn didn't think anything more about the birds of prey trip. Later in the summer, he and his guides were packing gear for a week-long Salmon River trip, and Morley called his house. "My guide said it was someone who wanted to talk about birds of prey, and I said, "Tell them to call next year." And my guide says, no, this guy wants to talk to you, he isn't going to go away."

Guinn went to pick up the phone. "It's Morley Nelson here, and I need you to come up and talk to me. We got to talk this thing over and find out what's going on."

"What thing?" Guinn asked. "And he says this birds of prey deal. And he says come on up here, this is really important, and I tell him I'm a little busy right now, I'm on my way to the Salmon River, and he wouldn't take no for an answer, so I told my guides, I don't know what this is about, but I'll be back in a half hour."

He was gone for three hours. Guinn went up to Morley's house, and it took a while to sort things out. "He takes me into his house, and I said, look, I've got this trip leaving for the Salmon River right now and I've got all these people waiting for me. And he says, this is really important, when I set this area up and set it aside, and I'm thinking this was a BLM deal, what's he talking about that he did this? What's the deal with this guy? He's a little bit weird.

"Then I looked over on the coffee table and there was a picture of Gary Cooper holding a golden eagle. I said, excuse me, isn't that Gary Cooper? And he says 'ya ya ya ya, that's Gary Cooper holding one of my eagles.' And I'm thinking, what's he saying about my eagles? I mean I didn't know anyone could have their own eagles. And then I see a picture of Morley and Walt Disney holding an eagle, and I thought, wow, this guy might be the real deal!

"I started to pay a little more attention to what he was saying. He took me downstairs, and he's telling me about trips he's taken to the Middle East and all this stuff, and my mind is swimming by then. I'm not looking at my watch anymore, and he starts talking about these movies and stuff, and how he did these movies for Disney like the *Vanishing Prairie* and *The Living Desert*. Well, when I was a kid growing up, that was the only thing I remembered except for Jack Benny. Then he takes me to the backyard and shows me all the birds, and I'm going holy smokes, this *is* the real deal."

At the conclusion of the meeting, Guinn agreed to work with Morley on future birds of prey trips. "I started to build the trips around Morley's philosophy, and the basic thrust and motivation of his whole life," Guinn says. "He's got that love for the birds, and a really deep passion for the birds. Morley is like an eagle trapped in a human's body."

For the first couple years, Guinn booked a few Saturdays with Morley. As the 1980s progressed, ecotourism—the concept of learning about history, geology, archaeology and the environment as part of recreation outings—gained in popularity. Plus, the World Center for Birds of Prey relocated to Boise. It was a natural for visitors at the World Center to inquire about Birds of Prey Tours on the Snake River. By the late 1980s and through the 1990s, Guinn and Morley were busy every weekend from April through June. They have taken more than 50,000 people on tours of the birds of prey area since the beginning.

Even though Guinn was losing money on the venture for the first five years, he persevered because of Morley's passion. "We looked at it as a way to give back. This was for Morley. He didn't have a desire to own and operate a business to do it. He didn't want to deal with the equipment. We thought, wow, what passion. We'll build the custom boats and make it happen. When I see a man who has so much to say and so much to give, how can I justify not sacrificing myself? What did it matter to me, if I give twenty years to this thing, as long as we can make it fun, and get this story out, it can be nothing but positive. It gets you out in nature, it makes you aware of your surroundings and it shows you the complete system."

Idaho Public Television photo

Actress Lynn Regrave helped promote the national PBS-TV special, *The Vertical Environment*.

Before long, Guinn started to take school groups as well. For $15 per student, he has taken thousands of school kids down the river over the last fifteen years. "This trip might be the only neat outdoor experience they've had their entire life before they leave home, and those kids might turn into conservationists," Guinn says. "That's the reward we get."

"We humans have always looked to the heavens for inspiration. Many have found it embodied in the magnificent birds of prey. Hello, I'm Lynn Redgrave. These raptors are above us only because of what is below us—a complex and fragile ecosystem, a bond between hunter and habitat that just could be the most dramatic environment in the world."

British actress Lynn Redgrave narrated the script for an hour-long feature called *The Vertical Environment*, co-produced by Idaho Public Television and Echo Films. The program was picked up by the Public Broadcasting System and premiered nationally on a Monday evening, September 27, 1993. Norm and Tyler Nelson did all of the photography for the picture, and Bruce Reichert, the award-winning producer of *Outdoor Idaho*, wrote the script. Morley recruited Redgrave for the guest host role. He had met her through her former husband John Clark, who served on the board of The Peregrine Fund.

Morley served as the raptor expert in the film, allowing him to articulate his views about how the vertical environment of the Snake River Canyon created a perfect and unique setting for birds of prey. He started by asking Redgrave to lie down next to the cliff's edge.

"Well, now Lynn, I want to show you how to look at this beautiful cliff. The only way is to lie flat on the rocks here right on the edge."

"Are the high winds a danger?" she asks.

"Oh yeah, it can shove you right over if you're not careful."

Morley lies down next to Redgrave, and they peer into the vertical chasm of the Snake River Canyon. "That's beautiful," Morley says.

"That is gorgeous," she says, giggling.

"Heh, heh, now you can feel like an eagle sitting on the face of a cliff."

"I sure can."

Now Morley describes the dynamics of the canyon.

"The eagles that live in the vertical environment make such a beautiful use of the great walls, the crevasses and ledges," he says in his typical authoritative baritone voice. "The golden eagle is not the most capable in aerial combat, but in its position in the universe, it is the most powerful of all the raptors. The nostrils of an eagle are shaped exactly like the intakes of a jet plane. Regardless of the aerial velocity—over 100 mph—they can still breathe coming straight down for 1,000 or 10,000 feet. They can see a ground squirrel for a mile—way beyond, eight times beyond the capability of humans. The strength and the power of the talons are absolute. They squeeze through the bone and the flesh, and it means life is over."

On the boat tours, Morley explains that prevailing winds blow upriver from the West, creating a perfect updraft for golden eagles, prairie falcons, red-tailed hawks and other birds to ride the wind to the top of the canyon rim, to the "kitchen," on the north side, where they hunt for prey. The birds require little to no energy to climb to the top. After they're finished hunting, they ride the wind back to the nest to feed the young, again requiring little to no energy for the trip.

As the next scene segues to the Oregon Coast, Redgrave

points out that birds of prey thrive in a number of vertical environments in North America. On the coast, peregrine falcons ride the updrafts between hunting for sea birds near the water, to nests high on the cliffs. In cities such as New York, Montreal, Denver and Los Angeles, peregrines nest in skyscrapers, and dive for pigeons for food. The only problem is finding enough wind to return to an artificial nest.

Next, Morley visits an area inside the Snake River Birds of Prey Area that is occupied by the Idaho Army National Guard for tank training. On camera, Morley points out his concerns about tank impacts on the fragile desert soil.

"Morley, we agree that this is a very precious resource," General John Kane says. "That's why we use tracks for our tanks that allow the plants to spring back to life, and we have a program to replant sagebrush. We also have paid for studies out here to see how our activities impact the birds, and what we can do to minimize those impacts."

In the late 1980s, when the National Guard proposed expanding its tank-training activities on the north rim, Morley convinced the Guard to increase the number of proactive measures it performs each year to mitigate its impact on birds of prey, particularly during the spring nesting season. Even so, wildfires caused by lightning, careless campers and from tank ordnance have burned much of the native vegetation on the north plateau, eliminating about fifty percent of the natural habitat. Following the fires, a nasty and prolific annual known as cheatgrass has replaced the shrub-steppe habitat, turning formerly productive wildlife habitat into a thicket of weeds.

The film bounces back and forth from current issues to the natural world, giving viewers a glimpse of the myriad of human and natural forces that persist in the canyon. A set-up scene of Morley's golden eagle, Slim, attacking a rattlesnake made for spectacular slow-motion action. The elusive snake makes multiple strikes at the eagle while Slim dances on the ground with her wings outstretched, waiting for the perfect moment to kill the snake with her talons.

Ted Koch: I was trying to fly Slim to the rattlesnake, and I had the T-perch behind my back in my bare hand. Slim came towards me, and for some reason she was flying right at me and ignoring the snake. Then at the last minute, I threw the T-perch up and she missed it, and she put her halix through the back of my hand and fell down. She wasn't trying to hurt me, she was trying to land. You know when you break a bone or something and you go into mild shock? I went ugh, and I had to go sit down behind the sagebrush and gather myself. Morley in the meantime kept Slim occupied and helped me recover.

Morley is back on the Snake River, leading a large tour group with Birds of Prey Tours. The tour boat is packed full of people, all wearing yellow lifejackets.

"Now there's an eagle aerie forty feet from the top of that big cliff," he says into the microphone. "There she is, mother eagle, she's coming into the aerie. She's going into the wind, slowing down, she's going to go right in and land. She's cutting into the bottom. Beautiful. Just like she's supposed to do. Her mate will be up there guarding the territory while she's feeding the young. There's the male right there."

On the edge of the Snake River Canyon, Redgrave closes the film with a few parting words. "The Vertical Environment is as important as ecosystems of the plain, the desert and the forest. For here, among the desolate cliffs of hardened lava, live the soaring barometers of our planet. Their power, speed and determination place them at the very top of the food chain, among the first to feel man's impact on the earth. We may never truly comprehend the complex environment they live in, but we will always yearn for the freedom they symbolize."

By playing to a nationwide audience on PBS, *The Vertical Environment* sent a heart-warming conservation message to millions of Americans. The film garnered a bundle of awards, including a Golden Eagle award in the environmental category from the Council on International Nontheatrical Events (CINE) in 1993, Best Outdoor Documentary at the National Outdoor-Travel Film Festival in 1994, a Teddy Award (Best Outdoor Documentary) from the Michigan Outdoor Writers Association in 1993, first place in documentaries from the Association for Conservation Information in 1994, and Best of Show at the Huckleberry Video/Film Festival in Sandpoint, Idaho in 1993.

The Vertical Environment was a timely and fitting tribute to the Snake River Birds of Prey National Conservation Area. A month earlier, on August 4, 1993, President Bill Clinton signed Public Law 103-64, which permanently protected the 482,000-acre area as the nation's eighth National Conservation Area.

U.S. Representative Larry LaRocco, D-Idaho, and U.S. Senator Larry Craig, R-Idaho, co-sponsored the bill. Cecil Andrus, who was serving his fourth and final term as governor of Idaho, pushed his friend and party ally LaRocco to permanently protect the Snake River Birds of Prey NCA before his executive order expired in 2000.

"I told LaRocco, hey, we need to get that done, and he was eager to help out," Andrus says.

"This time I think we have a winner," LaRocco said after the bill passed the House on May 11, 1993. "With the work Senator Craig and I have done, we now have a bill that protects this magnificent Idaho resource while allowing for all of the traditional uses such as grazing and recreation."

Craig made sure that livestock grazing and National Guard training activities would continue to be allowed in the birds of prey area. Morley had never had any problem with that, and neither had Andrus.

Plus, a task force of BLM representatives, National Guard officials, environmental interests and recreation groups had been working on the outlines of the final legislation for four years. The continuation of the status quo in terms of public uses allowed in the area was a familiar theme, although continued research studies occurred to monitor the health and population status of raptor populations.

In the mid-1990s, *The Vertical Environment* was the latest film project in which Morley and his boys worked together to send Morley's conservation messages to the nation. Morley didn't know he was doing it at the time, but the fact that he trained his sons to operate a sophisticated Bolex 16mm movie camera when they were in high school, and trained them how to train and fly falcons, hawks and eagles created a fortuitous lifelong partnership that allowed him to reach a wider American and international audience than he could have otherwise.

Before the boys were old enough to help, Morley produced his own educational films such as *Modern Falconry, Nature's Birds of Prey* and *Winter World.* In the 1960s, with Norm's help, he produced *Golden Eagles.* After Norm got his start in the film business by working as a photographer's assistant on *Varda the Peregrine* for the Walt Disney Company, he produced his own film called *Summer's North Face* and launched Echo Films in Boise in 1973. Tyler joined him in 1975, after he graduated from Chico State University.

By then, both of the boys had already worked on the sets of big-time Hollywood film productions as bird handlers in their teen-age years, and later, as photographers. They were ready to blast off on their own. In the beginning, neither one of them wanted to focus on making more pictures about birds of prey.

"I had the illusions of any young film-maker," Norm says. "I was going to make features. And I was going to make dramatic films. But I was going to have a real lean toward skiing, sports, wildlife and westerns."

"Back then we did commercials, grant films, ski films and wildlife films," adds Tyler. "We wanted to be in the film business. And it had nothing to do with birds."

Fairly quickly, however, Echo Films received requests for stock footage of golden eagles, and Norm was able to provide it from his involvement in Morley's educational film *Golden Eagles*, shot in the Snake River Canyon. "That's how we got started with the stock footage people," Norm says. "These people in New York kept calling us up and asking us for our eagle footage, and we ended up making a whole bunch of money selling stock footage. And then they'd call up for some other wildlife thing, and we'd have it, too. That actually helped a lot."

Initially, the Nelson boys shot a number of sports assignments, including World-Cup ski races for ABC-TV. Their experience filming birds of prey—the fastest creatures known to man—transferred well to shooting ski racing. "Here we had been shooting falcons in slow-motion at more than 120 mph, and ABC needed pictures of a full-size man going sixty mph," says Norm, giggling at the thought. "We're like, hell, no problem. We can film the guy's eyeballs coming down the mountain if you want."

In 1978, the Nelson brothers got a fun and challenging assignment to produce a film for Salomon USA called *Winter Wings,* an upbeat twenty-minute picture that blended the sports of powder skiing and surfing with the flying acrobatics of a golden eagle. Morley's good friend Keith Patterson, a former World War II bomber captain, wrote the script. They shot powder skiing footage at Grand Targhee Ski Resort in Wyoming, and surfing on the big island in Hawaii. The film would be used by Salomon's sales representatives in North America and Western Europe to help pump up buyers.

"It was about the control of the turn in all three subject areas," Tyler says. "It's still, to this day, one of the most popular ski films in Europe because it's based around the golden eagle, and the eagle is the national symbol of Austria. It played every night in the apres ski scene in the bars. It's the proudest thing we've done because it was so creative, so different, and so hard to do."

After Echo Films produced the successful documentary *Silver Wires and Golden Wings,* Norm and Tyler received new assignments from the Edison Electric Institute. They produced *Fun Country USA*, a recreation travelogue, and *A Second Chance*, an educational documentary about endangered species, including bald eagles and peregrines. Following the success of *Peregrine*, a fund-raising and educational film for The Peregrine Fund, Echo produced a promotional picture called *The Snake River Birds of Prey Natural Area* for The Nature Conservancy.

Over time, the Nelson brothers worked on a combination of general-interest features and productions that helped spread their father's message. In the 1990s, Echo Films produced an educational film about all birds of prey called *The World of Raptors,* narrated by Joanne Woodward. That film is a modern version of Morley's second film, *Nature's Birds of Prey.* Both of them acquainted the American public with educational information about how to identify all species of raptors, and help people understand the different hunting techniques employed by each one. Plus, Echo produced an Express Mail commercial for the U.S. Postal Service with Morley's bald eagle, Pearl, a rehabilitated bird from The Raptor Center in Minnesota that was trained and flown by David Boehlke.

Nowadays, the phones ring at Echo Films on a daily basis with requests from around the world for stock footage of birds of prey. When film-production companies such as the British Broadcasting Company need particular shots involving birds of prey, they send camera crews to Boise to shoot one of Morley's birds in a wild setting in Idaho.

If needed, Morley always is happy to provide a sound byte for visiting camera crews.

Echo Films has increased the exposure of birds of prey because of the connections they developed in the industry, some with Morley's help, and some on their own. But the boys credit their dad with making it all possible.

"He taught us both how to be cameramen at a young age," Tyler says. "He wasn't dumb. He thought, wait a minute, I can teach these guys how to shoot, then I can go back to being a falconer. Instead of both. He didn't like doing both. He taught me how to shoot, and I shot the football games for Boise High School just for practice."

"Initially, I fought against it because Morley was so authoritative about it," adds Norm. "You know, I was a teen-ager, and no one is very good about listening to their father as a teen-ager. And working that Bolex was so hard for me. I wouldn't have the lens right, or the F-stop right, or the focus was wrong. Sometimes I couldn't get to first base with that camera."

"It's the same camera he taught me on, and you learned so much about cinematography with that one camera in a half hour with him that you felt really low," Tyler says.

The Nelson brothers have many stories that make them laugh and smile with pride from the films they worked on as falconers and photographers. They met dozens of big-name movie stars and producers. They credit their father with making many of those experiences possible.

"Being a falconer on a big movie set really gets you inspired about movie-making," Tyler says. "Look who we started with: Disney! We started with the best."

While the Nelson brothers continue to pitch a variety of new

Steve Guinn

Still going strong

movie ideas to the industry, they still would like to produce films on birds of prey topics, including a documentary on the peregrine falcon recovery effort, American goshawks, the history of the white gyrfalcon, and the definitive story of the bald eagle.

"My favorite is the history of the white gyrfalcon because it's such an exotic bird, and it also has significance between the discovery of North America, the history of Marco Polo and Genghis Kahn," Tyler says.

The main obstacle to selling pictures on the birds of prey today, they say, is that most companies want wildlife pictures on "the big five"—grizzly bears, elephants, sharks, African lions and African leopards. Mega-fauna.

Nevertheless, the boys will keep trying, because when they have produced national pictures on birds of prey, the films have received strong ratings and won awards. "If you do it right, they can be pretty groovy," Tyler says.

In mid-November 1995, Morley and Pat received a visit by her eldest daughter, Shay Hirsch, who lives in New York City. Shay came to Boise to celebrate her mom's sixty-eighth birthday. Pat was not feeling well, however.

"She had this big pain in her chest, and she didn't know what it was," Shay says.

Pat had some tests done to see what the problem was. Morley, Pat, Shay and her brother John all went with her to see the doctor. The news was grim.

"The doctor said she had lung cancer, and it was quite advanced, and there was nothing that she could do. Nothing. There was total silence in the room," Shay says. "And then—and this completely defines my mother—the epitome of her character. She took a big breath and said, 'Well, I'm sixty-eight years old. I've had a wonderful life. I have a great family, and great friends. And I'm lucky.' And that's the way she was to the end."

The doctor said she could live six to eight months, but it would be much shorter than that. Shay helped her mother settle all the details of her personal finances and her will. The Boise Hospice organization was brought to the Nelson's home to make preparations for her death. Shay stayed in Boise. Pat received morphine for her pain.

"Mom never left home again," Shay says.

Pat died three weeks later, on Sunday, December 10, with her family by her side. A memorial service was held. A standing-room-only crowd of about 150 people attended, including many of Pat's friends and Morley's friends. "It was a great celebration of her life. It was more like an Irish wake than a Presbyterian funeral," Shay says.

Shay's husband, George, gave a nice speech, as did Shay's son, Ian, Shirley Ewing and Tyler.

"Pat and I didn't get along so well when she first started living at the house because I was the only kid left at home," Tyler says. "It wasn't pretty. Of course that's pretty typical in family situations following a divorce, but eventually, I came to respect her and she respected me. I learned a lot from her. So I put together a nice speech about how we gained mutual respect for each other."

Following the service, there was a party at Morley's catered by Pat's friends, the Ewings.

In Pat's obituary, her family wrote, "If a family is blessed, there is at least one individual who serves as a magnet, who has a gravitational pull which draws all together in love and laughter, debate and consensus, celebration and consolation, tradition and hope and excitement about the future. She was that kind of person—one who spread her net wide and drew in, not only her husband and children and grandchildren, but her stepchildren and their children, nieces, nephews, friends and neighbors. She will be missed."

Pat had been a lifelong chain-smoker. When she was sixty,

eight years before her death, Morley urged her to quit. She had developed a smoker's hack, a coughing fit in the mornings. "I said it was a terrible thing to be doing, and it will kill you in the end, which of course it did," Morley says.

On the positive side, Morley and Pat had twenty-five years together, and they were always very much in love. "She was a wonderful lady, and a most understanding lady," Morley says. "She was very considerate of my needs and she took care of everything."

Pat was an excellent gardener. She planted beautiful flowers around the Nelson home and down by the swimming pool. Morley called Pat "the commanding officer." She ruled the roost. She helped Morley with many of his presentations, and organized his files. She cooked for him, paid the bills, took care of the taxes. She brought the Yandell family and the Nelson family together for many dinners and holiday celebrations.

For twenty-two years, Morley and Pat took an annual vacation to Mexico. Pat purchased a condominium in Mazatlan. Morley loved going down there to look for exotic birds of prey. Shay and her family often would visit them. So would Morley's friends Forbes and Dee Mack, and Dick and Dot Thorsell. Part of the time, Morley and Pat would go car-camping and look for falcons. Pat kept everything organized, cooked the meals, and brought along a short-wave radio so they could listen to music.

Dee Mack recalls playing a joke on Pat and Morley one time in Mazatlan.

"We were staying at a resort on the beach there, and it was a beautiful night. We had a lot to drink that night. Morley and Pat went down to the ocean to go swimming. They took their clothes off and went skinny dipping. Forbes and I were hiding in the bushes, so we snuck down and took their clothes and hid them. And then Pat came back to get them and she says, our clothes are gone! I know I put them right here! And we're hiding in the bushes and dying laughing, and they finally found them, except they couldn't find Pat's shoes. She had to go out the next morning to find her shoes."

Toward the end of Pat's life, she increasingly grew tired of Morley's frequent visitors in the late afternoon when it was time to fly birds. Many years before, she had built a large addition on the house with her own money to give herself more personal space. But just as Betty Ann had grown tired of the birds and the bird people getting more attention than she did, Pat did, too.

Lucy remembers how Pat's outlook changed over a ten-year period. "In the beginning, Pat liked me because I was the only woman around. I was somebody she could play backgammon with and talk to about Morley and other stuff, and we got along really well," she says. "After a while, we didn't get along very well, and there a couple times that I considered not going back, because I was having such a hard time with it. I think part of the problem was that she was a little jealous of all the time that Morley and I spent together. We'd spend a couple hours flying birds, and then we'd come back to the house and talk about it. We'd watch films or whatever. And she wasn't a part of that."

In the six years that Pat has been gone, Morley has been lonely.

"I think Morley is still grieving," Lucy says. "He's never gotten over it. Morley has never had to take care of himself. Pat took care of him. After Pat died and everybody left, he was on his own. He had to deal with the taxes, the mail, the bills and feeding himself. He didn't like to deal with that stuff, and never wanted to deal with it. He misses her tremendously."

Morley often says, "When I lost Pat, I not only lost my wife, I lost my organization."

On a crisp fall morning in November 2001, Morley chewed on an orange after breakfast in his livingroom. He was watching the national news on TV to check out the latest in the war against terrorism. That week, Idaho Governor Dirk Kempthorne ordered the National Guard to barricade the streets around the state Capitol. Listening to the news, Morley shook his head about the whole thing.

"A lot of people are just evil," he says. "When they get together they can do some really dangerous things that make no sense at all to the average person. I think we can handle this situation if we get organized and find a way to get the people who are responsible for the attacks on our American soil. Boy, it's scary.

When I was fighting in World War II, I could never have imagined something like this ever happening. Never."

Morley's German shepherd, Lady, nuzzled her nose under his arm to get a little attention. He stroked the fur on her back and pondered his achievements and his regrets, looking back over the time in his life since World War II. Morley's living room is smothered with plaques, trophies, sculptures and paintings of birds of prey.

"I think one of the most important things that I ever did was to protect the Snake River Canyon," he says, "but the motion pictures I did for Walt Disney were what made everything else possible. Everybody around the world stopped shooting eagles when they saw *Ida the Offbeat Eagle*. When they saw what an intelligent bird those eagles are, they said I can't shoot those eagles anymore. That was a beautiful thing.

"The snow survey work was for humanity, not for the birds of prey. But measuring the water content of the snow pack in the mountains was one of the most important things that I ever got started in Idaho and the Columbia Basin. That turned out to be a very important and valuable thing. And this year proved that once again after we had one of the worst droughts in the last 100 years. Everyone could see it coming because of the snow survey."

Morley is equally proud of the work he did on reconfiguring powerlines to protect large birds of prey from getting electrocuted. "That work has to be carried on. I gave those talks in Oxford and Vienna and the whole concept spread around the world. I'm very proud of that."

An *Audubon* magazine on Morley's coffee table contained an article particularly critical of several electric utility companies, including the Moon Lake Electric Association, which serves portions of western Colorado and eastern Utah, for failing to install eagle-protection measures on their powerlines. About 170 dead raptors, primarily golden eagles, were found underneath the utility's powerlines. Moon Lake was fined $100,000 after the Justice Department pursued criminal penalties.

"A few companies haven't lived up to the agreements that we put in place twenty years ago," Morley says. "That's always been a problem and still is. But we're working on it. A few companies have yet to develop the proper outlook on this deal. But there's a lot of companies like Idaho Power, Public Service Company and Pacificorp that have been real leaders."

The Fish and Wildlife Service may require a more thorough accounting by utility companies of electrocuted birds along powerlines to ensure better compliance. Up to now, due to the work that Morley did in the late 1970s with the Edison Electric Institute and a number of western utilities, many companies have stayed on top of the problem and worked to reduce eagle electrocutions where possible. The Avian Power Line Interaction Committee in 2001 had nineteen members in North America and Canada, all of whom share information on ways to reduce raptor electrocutions.

"Woof!" Lady sounded the alarm that someone's at the door. It's Monte Tish, one of Morley's eagle trainers who works with Slim, a large golden eagle. Tish is licensed to care for birds of prey that have been wounded and have potential to be nursed back to health. He's a rehabber.

"Hey, Morley."

"Howdy."

"I thought I'd stop by and see you. I got another shot-up golden eagle yesterday. Someone brought it in from the desert. God, we've had a lot of eagles this year, for some strange reason, a bunch of them have been shot."

"Yeah, that's a real problem, isn't it?"

"Yep, I've had three golden eagles come in this year, and two bald eagles whose wings have been blown clear off. You know, I've been hunting all my life. There's no mistaking them for something else. You take them to the vet for an X-ray, and it's not like they're hit with a .22 or a pellet gun. You know it's some guys that are out there hunting and are just shooting at anything."

"In the old days, everybody shot eagles because they carried off babies, and lambs and calves," Morley says. "And the government paid you to shoot eagles. Gawd, that was the craziest thing that ever happened, but I got it all stopped."

"Yeah, but people are still shooting them, and I think they ought to really start sticking these guys to the max," Monte says. "As far as I'm concerned, they shouldn't ever get their hunting license back. It's supposed to be a federal offense. It should be a ten-year jail term and $100,000 fine. Confiscate their truck, their guns, their hunting license and put them in jail."

"Boy oh boy, Monte, you're pretty shook up."

"Oh well, we'll take care of these eagles and see if we can't fix them up."

Morley has been pleased to see so many raptor-rehabilitation clinics emerge in the United States. The Raptor Center at the University of Minnesota has grown into the largest rehab facility in the world, and more than a dozen states have facilities devoted to raptor rehabilitation.

"The problem is, we need more education programs," Morley says. "It takes a big amount of money to do a film that shows it all. My boys need to do a film that will show people all the injured birds that we still have, all around the nation, and the need for more education and enforcement. We've got to get that word out."

"You're right, Morley. I'll see you later."

"Take good care of that eagle now."

Just about every hour or every minute of every day, Morley has a new idea for another film project. His files are thick with reams of paper from film concepts that he's tried to sell over the years. Now his sons peddle the ideas, but Morley wishes that more of his movie ideas could have reached fruition. "I feel confident that we did accomplish a lot, but we need more education, more consideration, more understanding at a greater depth," he says. "The one I'd like to see the most involves the national emblem, our bald eagle. Everyone should know what a wonderful bird that really is. Now that we've got this war going, it might be the perfect time to do it."

As anyone knows who has had a discussion with Morley, the topics can easily spiral from birds of prey to the great issues of the day and circle back to the initial topic at hand. On this particular day, Morley talked about soil, water and overpopulation.

"The biggest problem as far as humanity is concerned is the erosion of soil, and the waste of soil," he says, now speaking from the mindset of a veteran Soil Conservation Service employee. Indeed, about five billion tons of soil erosion occurs

Steve Guinn

Morley was one of 7,200 Americans selected to help carry the Olympic torch to the 2002 Winter Olympics in Salt Lake City.

each year, according to the U.S. Department of Agriculture. About half of that amount ends up in small streams, rivers and reservoirs.

"We're building all of our homes on farmland. We should be building homes in the hills, and leave the level ground for growing crops. That's why I moved up on the hill where I am because I don't want to use up any farmland to build my house, plus it's a heck of a nice place to fly hawks. I saw both things at the same time. But we are losing our soil and hurting our water. All of this affects the carrying capacity of the earth. We must take care of our soil and water. They are vital to our survival."

Naturally, the subject of carrying capacity moves to overpopulation. "That's the biggest problem of all, always has been, always will be," he says. "All we can do is work at it. The systems of birth control we have today are more efficient than we've ever had in history. But the whole world hasn't accepted that. We need more sex education for teenagers. There's no doubt about that. The carrying capacity of the earth is the limiting factor, and we're always pushing the limit. Half the world is starving right now."

As a kid who grew up on a farm, Morley can't believe how dependent all agricultural operations are on petroleum products, mechanized equipment and computers. On his grandfather's farm, everything was accomplished with raw horsepower. "I'll tell you what, boy, if we ran out of oil, we couldn't produce food anymore, because everything is so dependent on gas and oil. We'd all starve to death. But of course, nowadays, for the farmers to stay ahead, they've got to do everything with machines."

"Woof!" Lady signals another visitor. It's Tyler, Morley's son, coming home for lunch. Tyler recently moved into the house to take care of his dad. Since Pat died, Morley has been living by himself. Many of his friends and family members stopped by to visit or take him out to dinner, but he has missed having someone living in the house with him. He's pleased to have Tyler at home.

"It's helped a lot because he cooks for me, and takes care of things," Morley says. "I need to have somebody here."

Although Morley has slowed down of late, he keeps busy. On January 25, 2002, Morley, accompanied by Pearl, his bald eagle, carried the Olympic Torch past the State Capitol in Boise.

Morley was one of 7,200 people nationwide who carried the Olympic Torch on a 13,500-mile, sixty-five-day journey from Atlanta to Salt Lake City. As Morley carried the torch, he thought about his grandson Tim, an elite nordic ski jumper, who has been a member of the U.S. Junior Olympic Ski Team. Tim qualified to be the forerunner on the 120-meter ski jump in the 2002 Olympic Games. He hopes to compete in the 2006 games.

On a bluebird fall day, Tom Cade and Morley stride across a grassy flat in the bottomlands of the Snake River Canyon. Both of them are carrying a falcon, which they will fly for the Nelson brothers' cameras for a scene in *The Vertical Environment.*

"Whoa up!" Morley casts off Zackar into the wind. The tiercel jets off into the sky and forms a series of tight circles as he climbs higher and higher, riding the wind currents, until the radius of his flight grows much wider, waiting on. On this beautiful day, Zackar's flight traces the path of Morley's life in the azure sky. The grand old man of falconry watches below with absolute wonderment. From the time when Morley was a lithe twelve-year-old boy to his twilight years, his passion for birds of prey has taken him on a grand tour that transformed a man, a family, a nation and the world. Clearly, Morley has made a major impact on the welfare of birds of prey, an impact that will resonate for decades to come.

Sitting below an eagle aerie along the Boise River, Morley reflected on his life in this way.

"It is my good fortune to have:

"Felt the gentle strength of the horses eye, their great power in running and the wild expression of the breeding stallion.

"Studied the soil, mountains, water and the atom and am staggered by their symmetry on this planet and in space.

"Been blessed by the noble wisdom in the eyes and action of the falcon, eagle and hawk.

"Been a part of the abject horror of the bullets, bayonets, and blood of war and recovered from their deep sting to hope for a more understanding world.

"Had the honor of working in one of the great missions of humanity, agriculture for food and fiber.

"Cried over the emotions created through the combination of motion pictures, music and life as it really is—fierce, absolute and beautiful.

"Had the greatest opportunity to make lasting friends and hope for my worthiness to them and this nation for such a chance.

"Carried the pride of a wonderful family and its continuance and love of what we have stood for together and for each other.

"Looked for and found goodness, grace and beauty for it is everywhere and infinite.

"Speaking for the raptors and me, thank you for this opportunity."

THE AUTHOR

Mark Lisk

Stephen Stuebner is a Boise, Idaho-based free-lance writer and author. He has written articles for *The New York Times*, *Outside*, *Mature Outlook*, *National Wildlife*, *Horizon Air Magazine* and *High Country News*.

Stephen received his Bachelor of Arts degree in journalism and history from the University of Montana. He was the outdoor/ski writer for the Colorado Springs *Sun* before becoming the environmental writer for the *Idaho Statesman* in Boise, Idaho. He was an adjunct professor at Boise State University and producer for the TV program *Incredible Idaho*.

He has received many awards, including the 1999 Governor's Take Pride in Idaho award for outstanding contributions toward promoting recreation and tourism in the state; and the Summit Award, the highest honor given by the Idaho Trails Council. He also has been honored by the National Wildlife Federation, Izaak Walton League, Oregon Fish and Wildlife Department and the Idaho Wildlife Federation. He is a past president of the Idaho Trails Council and the Southwest Idaho Mountain Biking Association.

Stephen has written six books about Idaho. He was a contributing author to *Snake: The Plain and Its People* and *Water in the West*.

ACKNOWLEDGMENTS

First, I want to say that I felt honored to work with Morley to write his life story. He's always enthusiastic and upbeat every day. His work toward the conservation of birds of prey is a monument to the way environmental challenges can be won with civility and dignity.

This project wouldn't have been possible without an essential partnership with Bill and Joan Mattox of the Conservation Research Foundation. Through the Foundation's non-profit status, we raised substantial funds to pay for the production of the manuscript, travel expenses and photographs. On behalf of CRF, Morley and myself, thanks to the following contributors:

Patrons – Roy E. Disney, Avian Power Line Interaction Committee, The Idaho Power Company, Lake Hazel Partnership, Velma V. Morrison, Newman's Own, Inc., PacifiCorp. and the Margaret W. Reed Foundation.

Supporters – Archives of American Falconry, The Brookover Family, Bill, Pat and Kurt Burnham, Tom Cade, S. Kent Carnie, Edgar Collins, Allen Derr and Judy M. Peavey-Derr, In Memory of Dennis Dixon, DVM, Marjorie J. Ewing, Paul Fritz, Gil and Peg Gilbertson, Ann and Wes Kluckhohn, Kevin Learned, Ph.D., Jack B. Little, Forbes J. Mack, Joan and Bill Mattox, Walter C. Minnick, The Peregrine Fund, Lynn D. Russell, Gerald H. Schroeder, Dr. Loy T. Swinehart, The Trauber Family.

Contributors – Bruce P. Budge, Vicki and Bill Cutshall, Richard C. Fields, Stephanie Hunt, Robert A. Maynard, Patricia McDaniel, Patricia M. Olsson, Rotary Club of Boise, Virginia B. Wojno.

I also want to thank Norman, Tyler and Susie Nelson for helping me with all aspects of the book project, even when it was painful, emotionally. And I wish to thank Betty Ann McCarthy in particular for consenting to many interviews and answering my phone calls. She proved to be a fountain of knowledge, and I thoroughly enjoyed her sense of humor.

Finally, many thanks to countless other sources for spending time with me to discuss Morley's life and conservation achievements. Everyone, without exception, was eager to help.

Other Books From Caxton Press

J. R. Simplot: A Billion the Hard Way

ISBN 0-87004-399-4 (cloth) $24.95
6x9, 288 pages, rare photos, index

On Sidesaddles to Heaven
The Women of the Rocky Mountain Mission

ISBN 0-87004-384-6 (paper) $19.95
6x9, 268 pages, illustrations, index

Dreamers: On the Trail of the Nez Perce

ISBN 0-87004-393-5 (cloth) $24.95
6x9, 450 pages, photographs, maps, index

Yellow Wolf: His Own Story

ISBN 0-87004-315-3 (paper) $16.95
6x9, 328 pages, illustrations, maps, index

Outlaws of the Pacific Northwest

ISBN 0-87004-396-x (paper) $18.95
6x9, photographs, map, 216 pages, index

Massacre Along the Medicine Road

ISBN 0-87004-387-0 (paper) $22.95
ISBN 0-87004-389-7 (cloth) $32.95
6x9, 500 pages, maps, photos, bibliography, index